Digital Communication Systems: Signals, Channels, and Signaling

Digital Communication Systems: Signals, Channels, and Signaling

Edited by:

Stefano Spezia

ARCLER PRESS

www.arclerpress.com

Digital Communication Systems: Signals, Channels, and Signaling

Stefano Spezia

Arcler Press
224 Shoreacres Road
Burlington, ON L7L 2H2
Canada
www.arclerpress.com
Email: orders@arclereducation.com

ISBN: 978-1-77407-768-9 (Hardcover)

DECLARATION

Some content or chapters in this book are open access copyright free published research work, which is published under Creative Commons License and are indicated with the citation. We are thankful to the publishers and authors of the content and chapters as without them this book wouldn't have been possible.

ABOUT THE EDITOR

Stefano Spezia is Ph.D. holder in Applied Physics at the University of Palermo since April 2012. His major research experience is in noise-induced effects in nonlinear systems, especially in the fields of modeling of complex biological systems and simulation of semiconductor spintronic devices. Associate member of the Italian Physical Society and European Physical Society.

TABLE OF CONTENTS

Section 2: Principles of Information Theory

Section 4: Signaling over Band-Limited Channels

Section 5: Signaling over Fading Channels

LIST OF CONTRIBUTORS

Shree Krishna Sharma
SnT - securityandtrust.lu, University of Luxembourg, Luxembourg

Mohammad Patwary
FCES, Staffordshire University, United Kingdom

Symeon Chatzinotas
SnT - securityandtrust.lu, University of Luxembourg, Luxembourg

Qun Zhang
University of Toronto, The Edward S. Rogers Sr. Department of Electrical and Computer Engineering, Toronto, ON, M5S 3G4, Canada

Terence H. Chan
University of South Australia, Institute for Telecommunications Research, Mawson Lakes, SA, 5095, Australia

AlexGrant
Myriota, Adelaide, SA, 5000, Australia

Yuanhang Wang
School of Information and Communication Engineering, University of Electronic Science and Technology of China Chengdu , People's Republic of China

Tianxian Zhang
School of Information and Communication Engineering, University of Electronic Science and Technology of China Chengdu , People's Republic of China

Lingjiang Kong
School of Information and Communication Engineering, University of Electronic Science and Technology of China Chengdu , People's Republic of China

Xiaobo Yang
School of Information and Communication Engineering, University of Electronic Science and Technology of China Chengdu , People's Republic of China

Olivier Rioul
Télécom ParisTech, Institut Mines-Télécom, CNRS LTCI, 46 Rue Barrault, 75013, Paris, France

José Carlos Magossi
School of Technology (FT), University of Campinas (Unicamp), Rua Paschoal Marmo 1.888, 13484-370 Limeira, São Paulo, Brazil

Guoping Zeng
Elevate, 4150 International Plaza, Fort Worth, TX 76109, USA

David Elkouss
QuTech, Delft University of Technology, Lorentzweg 1, 2628 CJ Delft, The Netherlands

David Pérez-García
Departamento de Análisis Matemático and Instituto de Matemática Interdisciplinar, Universidad Complutense de Madrid, 28040 Madrid, Spain
ICMAT, c/ Nicolás Cabrera, Campus de Cantoblanco, 28049 Madrid, Spain

Ahmed K. Ali
Department of Electric and Electronic Engineering, Gaziantep University, Gaziantep 27310, Turkey

Ergun Erçelebi
Department of Electric and Electronic Engineering, Gaziantep University, Gaziantep 27310, Turkey

Rohitha Ujjinimatad
Department of Electronics and Communication Engineering , Ballari Institute of Technology and Management , Bellary 583 104 , Karnataka , India

Siddarama R. Patil
Electronics and Communication Engineering Department , PDA College of Engineering , Gulbarga , Karnataka , India

Gaoyuan Zhang
School of Electronic and Information Engineering, Henan University of Science and Technology, Luoyang 471023, China
National Key Laboratory of Science and Technology on Communications, University of Electronic Science and Technology of China, Chengdu 611731, China

Hong Wen
National Key Laboratory of Science and Technology on Communications, University of Electronic Science and Technology of China, Chengdu 611731, China

Longye Wang
National Key Laboratory of Science and Technology on Communications, University of Electronic Science and Technology of China, Chengdu 611731, China
School of Engineering and Technology, Tibet University, Lhasa 850000, China

Ping Xie
School of Electronic and Information Engineering, Henan University of Science and Technology, Luoyang 471023, China

Liang Song
School of Electronic and Information Engineering, Henan University of Science and Technology, Luoyang 471023, China

Jie Tang
National Key Laboratory of Science and Technology on Communications, University of Electronic Science and Technology of China, Chengdu 611731, China

Runfa Liao
National Key Laboratory of Science and Technology on Communications, University of Electronic Science and Technology of China, Chengdu 611731, China

Pham Thanh Hiep
Yokohama National University, 79-5 Tokiwadai, Hodogaya, Yokohama 240-8501, Japan
Le Quy Don Technical University, 236 Hoang Quoc Viet, Nghia Do, Cau Giay, Ha Noi 100000, Vietnam

Ryuji Kohno
Yokohama National University, 79-5 Tokiwadai, Hodogaya, Yokohama 240-8501, Japan

Ashraf A. Tahat
School of Electrical Engineering, Princess Sumaya University for Technology, Amman 11941, Jordan

Nikolaos P. Galatsanos
Electrical and Computer Engineering Department, University of Patras, Rio 26500, Greece

Claudio Ferreira Dias
Department of Digital Hardware Design, Instituto Eldorado, Eldorado, Alan Turing - 275, Campinas 13083-898, Brazil

Eduardo Rodrigues de Lima
Department of Digital Hardware Design, Instituto Eldorado, Eldorado, Alan Turing - 275, Campinas 13083-898, Brazil

Gustavo Fraidenraich
Department of Communications, University of Campinas, Albert Einstein - 400, Campinas 13083-852, Brazil

Mohammad Soleymani
Signal and System Theory Group, Universität Paderborn, 33098 Paderborn, Germany

Ignacio Santamaria
Department of Communications Engineering, University of Cantabria, 39005 Santander, Cantabria, Spain

Christian Lameiro
Signal and System Theory Group, Universität Paderborn, 33098 Paderborn, Germany

Peter J. Schreier
Signal and System Theory Group, Universität Paderborn, 33098 Paderborn, Germany

Yuanyuan Ma
College of Information Science and Engineering, Shandong Agricultural University, Shandong 271018, China

Bjørn Olav Hogstad
Department of Mathematical Science, Norwegian University of Science and Technology, Teknologiveien 22, 2815 Gjøvik, Norway

Matthias Pätzold
Faculty of Engineering and Science, University of Agder, 4898 Grimstad, Norway

Pedro M. Crespo
CEIT and Tecnun, University of Navarra, 20018 San Sebastián, Spain

Omar Altrad
Simon Fraser University, Burnaby, BC V5A 1S6, Canada

Sami Muhaidat
Khalifa University, Abu Dhabi 127788, UAE
Centre for Communication Systems Research (CCSR), University of Surrey, Surrey, GU2 7XH, Guildford, UK

Wamberto José Lira de Queiroz
Federal University of Campina Grande and Institute for Advanced Studies in Communications, Campina Grande, Brazil

Danilo Brito Teixeira de Almeida
Federal University of Campina Grande and Institute for Advanced Studies in Communications, Campina Grande, Brazil

Francisco Madeiro
Polytechnic School of Pernambuco, University of Pernambuco, Recife, Brazil

Waslon Terllizzie Araújo Lopes
Department of Electrical Engineering, Federal University of Paraíba, João Pessoa, Brazil

Zhanji Wu
School of Telecommunication Engineering, Beijing University of Posts and Telecommunications, Xitucheng Road 10, Beijing, China

Xiang Gao
School of Telecommunication Engineering, Beijing University of Posts and Telecommunications, Xitucheng Road 10, Beijing, China

Jie Lian
Department of Electrical and Computer Engineering, University of Virginia, Charlottesville, VA 22903, USA

Yan Gao
Information Engineering School, Xi'an University, Xi'an 710065, China

Peng Wu
Xi'an Modern Control Technology Research Institute, Xi'an 710065, China

Dianbin Lian
Electronics and Information School, Northwestern Polytechnical University, Xi'an 710072, China

LIST OF ABBREVIATIONS

AWGN	Additive white Gaussian noise
AMPS	Advanced Mobile Phone Service
AF	Amplify-and-forward
ANNs	Artificial neural networks
ADSL	Asymmetric Digital Subscriber Line
AGC	Automatic gain control
AMC	Automatic modulation classification
AMI	Average mutual information
BS	Base station
BFSK	Binary frequency-shift keying
BER	Bit error rate
BICM	Bit-interleaved coded modulation
CFO	Carrier frequency offset
CFOE	Carrier frequency offset effect
CPO	Carrier phase offset
CO	Central Office
CPU	Central processing unit
CIR	Channel impulse response
CSIT	Channel state information at the transmitter side
CSI	Channel state information
CSS	Chirp spread spectrum
CRAN	Cloud Radio Access Network
CRN	Cognitive radio network
CBF	Coordinated Beamforming
CoMP	Coordinated Multiple Point
CRLB	Cramér-Rao Lower Bound
CDF	Cumulative distribution function
DA	Data-aided

DAB/DVB	
DFT	Discrete Fourier transform
DMT	Discrete Multi-Tone
EM	Expectation-maximization
FFT	Fast Fourier transform
FCC	Federal Communications Commission
FCF	Frequency correlation function
FDMA	Frequency Division Multiple Access
FEQ	Frequency domain equalizer
FMDCSK	Frequency modulated differential chaos shift keying
FSK	Frequency-shift keying
GLRT	Generalized Likelihood Ratio Test
GVD	group velocity dispersion
HNB	Hidden naive Bayes
HOC	High-order cumulants
HOM	High-order moment
IRCI	Inter-range-cell interference
ISI	Intersymbol interference
ILPC	Interval linear phase compression
IDMA	Interweave Division Multiple Access
IFFT	Inverse fast Fourier transform
LDPC	Low-density parity check
LPWAN	Low Power Wide Area Networks
MLE	Maximum likelihood estimation
ML	Maximum Likelihood
MSE	Mean Square Error
MAC	Medium access control
MOM	Method of moments
MSK	Minimum shift keying
NFT	nonlinear Fourier transform
NOMA	Non-orthogonal multiple access
NMSE	Normalized mean square error
OAMM	Orbital Angular Momentum Multiplexing
OAM	Orbital Angular Momentum

OFDMA	Orthogonal Frequency Division Multiple Access
PER	Packet-error rate
PAPR	Peak-to-average power ratio
PSK	Phase-shift keying
PDF	Probability density function
PRG	Proportional Rate Greedy
QRD	QR Decomposition
QAM	Quadrature amplitude modulated
RFI	Radio frequency interference
ROC	Receiver operating characteristic
RVM	Relevance vector machine
SSD	Signal space diversity
SINR	Signal to Interference-plus-Noise Ratio
SDCD	Single differential coherent detection
SIMO	Single input multiple output
SISO	Single-input single-output
TDD	Time Division Duplex
TDMA	Time Division Multiple Access
TIN	Treating interference as noise
UHF	Ultra-high frequency
UCR	Underlay cognitive radio
UFMC	Universal Filtered MultiCarrier
WF	Water-filling
ZF	Zero Forcing

PREFACE

During the last two decades, there has been an explosive growth in the telecommunications industry, characterized by an exponential increase in demand due to the addition of more and more individual computing devices. For such reasons, one of the research goals is to provide broadband communication capabilities, by using several communication channels, among those, wired, mobile terrestrial and satellite communication, and optical.

Section 1 of this book introduces the modern digital communication systems, by describing the multiple access techniques for next-generation wireless, the quasi-spatially periodic signals for optical fiber communications, and the radar–communications signal using interval linear phase compression modulation.

Section 2 treats the principles of information theory, starting from Shannon's formula and Hartley's rule, through the basic concepts in probability theory and of mutual information.

Section 3 focuses on the signaling over channels affected by additive white Gaussian noise. In particular, it examines M-ary Quadrature Amplitude Modulation, Binary Phase-Shift Keying.

Section 4 discusses the signaling over band-limited channels. In particular, it considers, a water-filling method for full-duplex multiple-hop multiple-input multiple-output (MIMO) relay system, and a channel estimation method for discrete multitone communication system based on sparse Bayesian learning relevance vector machine.

Finally, the last Section 5 widely reviews the signaling over fading channels, considering several models of fading, diversity techniques, MIMO systems, orthogonal frequency division multiplexing.

SECTION 1:
INTRODUCTION TO MODERN DIGITAL COMMUNICATION SYSTEMS

1

Multiple Access Techniques for Next Generation Wireless: Recent Advances and Future Perspectives

Shree Krishna Sharma[1] **, Mohammad Patwary**[2] **, Symeon Chatzinotas**[1]

[1]SnT - securityandtrust.lu, University of Luxembourg, Luxembourg
[2]FCES, Staffordshire University, United Kingdom

ABSTRACT

The advances in multiple access techniques has been one of the key drivers in moving from one cellular generation to another. Starting from the first generation, several multiple access techniques have been explored in different generations and various emerging multiplexing/multiple access techniques are being investigated for the next generation of cellular networks. In this context, this paper first provides a detailed review on the existing Space Division Multiple Access (SDMA) related works. Subsequently, it highlights the main features and the drawbacks of various existing and emerging multiplexing/multiple access techniques. Finally, we propose a novel concept of clustered orthogonal signature division multiple access for the next generation of cellular networks. The proposed concept envisions

Citation: Shree Krishna Sharma Mohammad Patwary Symeon Chatzinotas (2016) "Multiple Access Techniques for Next Generation Wireless: Recent Advances and Future Perspectives" WS EAI http://dx.doi.org/10.4108/eai.19-1-2016.151002

to employ joint antenna coding in order to enhance the orthogonality of SDMA beams with the objective of enhancing the spectral efficiency of future cellular networks.

Keywords: 5G, Multiple Access, Multiplexing, SDMA, Cellular Networks, Orthogonal Multiple Access

INTRODUCTION

The next generation of wireless networks is expected to provide better quality of service, lower latency, low energy consumption, and higher throughput [1]. In order to meet these requirements, several techniques such as ultra-high cell densification, bandwidth extension beyond 6 GHz, i.e., mm-wave and the increased spectral efficiency utilizing massive Multiple Input Multiple Output (MIMO) techniques have been considered as the promising solutions in the recent literature.

The advances in multiple access techniques has been one of the main drivers towards moving to a new generation of wireless starting from the first generation (1G) to the current fourth generation (4G). The 1G cellular system, i.e., Advanced Mobile Phone Service (AMPS), was based on Frequency Division Multiple Access (FDMA) technology while the second generation system (2G) was based on Time Division Multiple Access (TDMA). In parallel to the TDMA, Code Division Multiple Access (CDMA) based systems were developed and eventually they were adopted in third generation (3G) cellular systems. Subsequently, Orthogonal Frequency Division Multiplexing Access (OFDMA) came as a candidate technique for 4G due to its several advantages such as intrinsic orthogonality, receiver circuit simplicity, better spatial diversity and multiplexing capabilities. The current 4G technology, also called LTE (Long Term Evolution), and is evolving towards LTE-Advanced (LTE-A) with the inclusion of several advanced features such as coordinated multipoint transmission and carrier aggregation [2].

In order to meet the increasing capacity demands of future wireless networks, there has been a great interest of moving towards higher frequency range such as millimeter-wave (mm-wave) frequencies [3]. Although a significant capacity gain is expected to achieve with mm-wave communication systems over the current 4G wireless networks, several aspects of cellular systems need to be redesigned in order to fully achieve the desired capacity gain. Among these aspects, multiple access scheme is of significant importance and the investigation of suitable multiplexing/ multiple access schemes is crucial in designing future high-frequency wireless networks [4].

The OFDMA, which is the multiuser version of the OFDM scheme, is a widely used technology in today's 4G wireless networks. Despite its significant benefits, the main drawback for making it less attractive for future 5G wireless is that each single subcarrier in an OFDM system is shaped using a rectangular window in the time domain, leading to sinc-shaped subcarriers in the frequency domain [5]. Furthermore, its important characteristic that the spectrum can be divided into multiple parallel orthogonal sub-bands with the highest possible efficiency, is applicable only for the case of frequency synchronization and perfect time alignment within the duration of the cyclic prefix. The LTE technology is able to partially address this issue by employing a closed loop ranging mechanism and demanding oscillator requirements. However, from the energy efficiency point of view, this is not an efficient approach.

In addition to the commonly used time, and frequency multiplexing/ multiple access schemes, the concept of Space Division Multiple Access (SDMA) has received important attention in the cellular literature [6, 8–10]. In this scheme, multiple users can be served simultaneously in the same channel by the superposition of the beam, thus leading to the better utilization of the available spectrum. In this paper, we first provide a review on the existing works in the area of the SDMA technique. We then discuss the emerging multiplexing/multiple access techniques such as Orbital Angular Momentum Multiplexing (OAMM), Polarization Division Multiple Access (PDMA), Interweave Division Multiple Access (IDMA) and Sparse Code Multiple Access (SCMA). Subsequently, we propose a novel clustered orthogonal signature multiple access scheme for the next generation of cellular networks.

In addition to the commonly used time, and frequency multiplexing/ multiple access schemes, the concept of Space Division Multiple Access (SDMA) has received important attention in the cellular literature [6, 8–10]. In this scheme, multiple users can be served simultaneously in the same channel by the superposition of the beam, thus leading to the better utilization of the available spectrum. In this paper, we first provide a review on the existing works in the area of the SDMA technique. We then discuss the emerging multiplexing/multiple access techniques such as Orbital Angular

Momentum Multiplexing (OAMM), Polarization Division Multiple Access (PDMA), Interweave Division Multiple Access (IDMA) and Sparse Code Multiple Access (SCMA). Subsequently, we propose a novel clustered orthogonal signature multiple access scheme for the next generation of cellular networks.

The remainder of this paper is organized as follows: Section 2 provides an overview of the existing works on the applications of SDMA technique in wireless networks. Section 3 highlights the main features of the emerging multiplexing/multiple access techniques while Section 4 proposes a novel clustered orthogonal signature division multiple access. Finally, Section 5 concludes this paper.

SPACE DIVISIONMULTIPLE ACCESS

Introduction

As compared to the traditional mobile communication systems, the radio capacity in an SDMA-based system can be increased by employing antenna arrays at the Base Station (BS) side, and subsequently forming adaptive directed beams in both uplink and downlink directions [6]. This scheme provides the possibility to serve multiple users simultaneously in the same channel by the superposition of the beams and thus allows to enhance the capacity of the system. The importance of SDMA in wireless is not only because of its multiple access capability but also it enhances the spectral efficiency by enabling the frequency reuse within a particular cell [7].

The SDMA technology can be combined with the Orthogonal Frequency Division Multiple Access (OFDMA) technology in order to enhance the spectral spectral efficiency of future wireless systems, resulting in a joint SDMA-OFDMA system. The allocation of time, frequency and space resources to different user terminals in a joint SDMA-OFDMA system is a highly complex resource allocation problem [11]. However, in an SDMA system, the pre-processing of users' signals assuming the channel knowledge at the BS can enable the simultaneous transmission of several data streams to many users while minimizing the multiuser interference. Furthermore, SDMA based MIMO-OFDMA system can support multiple users in both frequency domain and the spatial domain, hence providing finer granularity of the resource allocation than the pure FDMA or pure SDMA based systems [12]. Moreover, the capacity gain which can be achieved due to multiuser diversity can be more significant in SDMAbased systems. However, in practice, SDMA techniques have to deal with the following two main problems [13].

- In the scenarios where two or more users come close to each other or the spatial signatures of the users become almost identical due to the underlying scattering environment (insufficient scattering, keyhole channels [14]), the channel matrices of these users may become highly correlated. Furthermore, channel correlation may also arise due to mutual coupling between the transmit and/or receive antenna elements [15]. In these correlated scenarios, channel correlation may become a source of link failure or outage while employing multiuser detection techniques.
- Since the mobile users are generally located at different distances from the BS, a near-far problem may arise in an uplink SDMA based system. This, in turn, causes the channel matrix observed by the BS to be heavily unbalanced. Such an unbalanced channel matrix may result in the degradation of the total system capacity due to a high eigenvalue spread or the condition number.

With regard to the aforementioned problems, the two main constraints that limit the performance of SDMA based wireless systems are [16]: (i) users sharing the same radio frequency channel, i.e., co-channel users, should be located in different angular locations, and (ii) the difference in their received power levels should not be very large. With respect the first constraint, the angular separation between co-channel users should exceed the angular resolution of the employed directional antenna in order to assign orthogonal SDMA beams.

Considering the first constraint, the contribution in [17] analyzed the capacity of an adaptive SDMA system for a given angular user density distribution which can be either obtained from the measurements or from the scenarios dealing with the user mobility. It has been concluded that the capacity of a wireless system can be enhanced by creating multiple independent beams per traffic channel with the help of an adaptive antenna array at the BS. Furthermore, it has been illustrated that the consideration of the expected user density and an appropriate selection of the BS sites are of significant importance while planning an SDMA-based cellular network in order to enhance its overall spectrum efficiency.

On the other hand, the implication of the second constraint refers to the near-far problem as encountered in CDMA systems. In this regard, the contribution in [16] investigates the grouping of the mobile users into power classes. In CDMA systems, this objective is usually achieved by means of the power control mechanism, since the users' grouping may further deteriorate the system performance. On the other hand, in a joint TDMA-SDMA system, individual TDMA channels can be assigned to different power classes and hence additional power control mechanisms are unnecessary. A power class can be either static or dynamic and the users belonging to the same class can share the same set of channels [16].

Towards the direction of enabling an efficient use of SDMA technique in future wireless networks, several techniques are being investigated in the literature. The key enabling techniques for future SDMA-based wireless networks along with their corresponding references are listed in Table 1. In the following, we review the existing works in the area of SDMA-based wireless systems considering several aspects such as transmission and reception techniques, resource allocation and scheduling, and research challenges.

Transmission and Reception Schemes

Multiple-input multiple-output (MIMO) wireless systems, which employ multiple antennas at both the transmitter and receiver, have received significant attention due to the promising capacity improvement provided by these systems. In MIMO systems, a transmitter can split the information symbols into multiple streams and sends each data stream via a single antenna. Each receive antenna then receives a different linear combination of the signals from the different transmit antennas. This technology provides a significant capacity benefit and the achievable capacity asymptotically increases linearly with the increasing number of transmit and receive antennas [39].

Multiuser MIMO, which is the multiuser version of the MIMO system, relies on the principle that multiplexing streams for different users on different antennas can achieve large gains, even if each user device contains only a few antennas [40]. Multiuser MIMO provides several advantages such as no requirement of the scattering environment, cheap single-antenna terminals, and simple resource allocation mechanism [33]. In order to further enhance the capacity gains, the concept of Massive MIMO has recently emerged [33, 34]. Massive MIMO system uses a very large number of service antennas at the BS which helps to eliminate the multiuser interference with the help of very sharp beams. This is known with several names in the literature such as Large MIMO, Hyper MIMO, and Full-Dimension MIMO. The main advantages of massive MIMO systems are [33]: (i) system throughput improvement, (ii) reduced latency, (iii) higher energy efficiency, (iv) robustness against jamming, and (v) simplification of the medium access layer. Despite its aforementioned advantages, this technology has to address the following main challenges [33, 34]: (i) pilot contamination becomes more problematic in massive MIMO systems than in the conventional MIMO, (ii) the effect of hardware impairments becomes more severe for massive MIMO systems due to the requirement of low-cost terminals, (iii) significant performance degradation occurs in the presence of channel correlation, and (iv) suitable calibration methods are required for the Time Division Duplex (TDD) operation mode.

In the context of LTE-Advanced and IEEE 802.16m networks, Coordinated Multiple Point (CoMP) transmission (also referred to as collaborative MIMO, network MIMO, etc.) has received important attention in the literature [22, 23]. This CoMP transmission scheme is capable of enhancing cell-edge user performance in an interference-limited environment. Besides enhancing the cell edge user performance, the CoMP approach can also improve the system capacity. In this context, the contribution in [22] evaluates the performance of various CoMP methods considering the ray-tracing based path loss calculation. Furthermore, the authors in [23] specify different scenarios of the CoMP based on the following information sharing levels: (a) no Channel State Information (CSI) sharing, (b) partial CSI sharing, (c) full CSI sharing, (d) no data sharing, (e) partial data sharing, and (f) full data sharing.

Based on the aforementioned scenarios, the CoMP techniques can be broadly categorized into the following two types.

- CoMP Joint Processing/Transmission (CoMP-JPT) [24, 25]: In this scheme, multiple BSs collaborate to convert the interfering signal into a desired signal in the downlink and subsequently transmit data to the edge users in a cooperative manner in order to enhance the cell-edge throughput.
- CoMP Coordinated Scheduling/Beamforming (CoMP-CSB) [26]: In this scheme, multiple BSs collaborate in order to mitigate the Inter-Cell Interference (ICI). Although most of the existing research on BS cooperation schemes focus on CoMP-JPT, CoMP-CSB is more practical since only the exchange of partial information such as CSI over the backhaul is required. In this context, the contribution in [26] proposes a CoMP-CSB

scheme with user selection in order to enhance the celledge users' throughput considering the scenario of partial CSI and no data sharing scheme.

Table 1: Key Enabling Techniques for Future SDMA-based Wireless Networks

Techniques	References
Multiuser MIMO	[10, 18–21]
Coordinated Multiple Point (CoMP)/Collaborative MIMO/Network MIMO	[22, 23]
CoMP Joint Processing/Transmission (CoMP-JPT)	[24, 25]
CoMP Coordinated Scheduling/Beamforming (CoMP-CSB)	[26–28]
Opportunistic Scheduling (OS)	[29–32]
Massive MIMO	[33, 34]
Three dimensional (3D) Beamforming	[35–37]

One of the main enablers of the CoMP technology is Cloud Radio Access Network (C-RAN) architecture in which the baseband processing is centralized, i.e., Baseband Units (BBUs) from the multiple BSs are pooled in such a way that they can be shared among these BSs. [38]. As compared to the traditional RAN architecture, C-RAN needs fewer BBUs, thus decreasing the overall cost of the network operation. Furthermore, C-RAN can support non-uniform traffic and makes the efficient use of the network resources such as BSs. Despite its several advantages, it has to address several challenges from the deployment perspective such as the need of suitable BBU cooperation and visualization techniques, requirement of high bandwidth, low delay and low cost transport network [38].

The main problems in designing transmission schemes for SDMA-based systems are: (i) difficulty in acquiring CSI, and (ii) poor synchronization. In this regard, authors in [42] have studied the performance degradation in SDMA networks that may result due to poor synchronization and the imperfect channel knowledge. Subsequently, a signal model for OFDMbased SDMA networks has been presented whose transmissions are impaired by the carrier frequency offset and the sampling frequency offset. It has been concluded that the poor knowledge of the channel state severely limits the maximum number of users that can be served, and is more important than the fine synchronization in terms of serving maximum number of users. In order to address the CSI acquisition difficulty, the contribution in [43] has studied the optimal statistical precoder design for a simple multiuser case in which the transmitter has two antennas serving only two single-antenna users.

Users' data at the receiver of a multiuser SDMA system can be separated on the basis of their unique spatial signatures in the form of Channel Impulse

Responses (CIRs). In this direction, several SDMA Multi-User Detectors (MUDs) have been proposed in the literature [10, 20, 21]. For an MUD to achieve near-single-user performance, the CIRs need to be accurately estimated. In order to achieve a near optimal performance, joint channel estimation and signal detection schemes have recently received significant attention [8, 9]. Among the existing linear MUD techniques, Minimum Mean Square Error (MMSE)- MUD [20] and the Constrained Least Square (CLS)-MUD [21] have been widely investigated in the literature in the context of Mean Square Error (MSE) performance analysis. Since minimizing the MSE doses not necessarily guarantee that the minimum Bit Error Rate (BER) or Symbol Error Rate (SER) of a communication system, the trend is towards exploring techniques which employ the minimum BER constraint such as in [44]. In the above context, the contribution in [10] proposes a differential evolution algorithm-aided iterative channel estimation and turbo MUD scheme for MIMO-aided OFDM/SDMA systems.

Besides, the contribution in [45] studies and compares various MUD schemes such as Zero Forcing (ZF), MMSE, Maximum Likelihood (ML), QR Decomposition (QRD), and Minimum Bit Error Rate (MBER) considering correlated MIMO channel models based on IEEE 802.16n standard. The ML detection provides the optimal performance but its complexity increases exponentially with the constellation size of the employed modulation and the number of users. On the other hand, the QRD-based MUD scheme can be a substitute to the ML detection due its low complexity and near optimal performance. Although the MMSE MUD minimizes the MSE, this may not guarantee the minimum BER of the system. In [45], it has been concluded that the MBER MUD performs better than the classic MMSE MUD in term of the minimum probability of error by directly minimizing the BER cost function.

Recently, the concept of three dimensional (3D) beamforming has received important attention in order to enhance the capacity of future wireless networks [35–37]. In contrast to 2D beamforming, the 3D beamforming controls the radiation pattern in both elevation and azimuth planes, thus providing additional degrees of freedom while planning a cellular network.

Resource Allocation and Scheduling

In multiuser MIMO-OFDMA systems, adaptive resource allocation in different dimensions such as frequency, time, and space becomes challenging due to the inclusion of the space dimension and a large number of resources to be managed. In this context, the authors in [18] have investigated the performance, complexity, and fairness of suboptimal resource allocation strategies with the objective of maximizing the sum rate. Furthermore, the contribution in [19] analyzes the Symbol Error Rate (SER) performance of a capacity-aware adaptive MIMO beamforming scheme, which iteratively finds the beamforming weight vectors that enhance the capacity of OFDMSDMA systems. Moreover, closed-form expressions for the SER performance of OFDM-SDMA systems have been derived with MIMO-Maximum Ratio Combining (MRC) and the proposed capacity-aware MIMO beamforming scheme.

The system fairness can be measured in terms of Jain's Index of Fairness (JIF), given by [46]

$$\mathrm{JIF} = \frac{(\sum_{k=1}^{K} \bar{R}_k)^2}{K \sum_{k=1}^{K} \bar{R}_k^2}, \tag{1}$$

where $\bar{R}_k$ is the mean rate of user k and the value of JIF ranges from 0 to 1. The higher the JIF, the more fair is the throughput distribution among users. Considering this fairness metric, the contribution in [46] studies the fair resource allocation problem of SDMA/ Multiple Input Single Output (MISO)/OFDMA systems. The Proportional Rate Greedy (PRG) algorithm proposed in [46] allocates powers among the selected users for each subcarrier considering user fairness into account.

Coordinated Beamforming (CBF) or coordinated scheduling is regarded as an effective way of mitigating ICI in OFDM-based systems. In order to take full advantage of multiuser diversity, an efficient scheduler should be able to schedule a set of users which experience favorable channel realizations in each time slot. In a cell containing multiple users, only the users having strong channel norms are usually selected as candidates for scheduling. In this context, the contribution in [27] proposes a CBF scheme based on leakage-controlled MMSE precoding. In addition, a regularized factor to maximize the Signal to Interference-plus-Noise Ratio (SINR) of the leakage controlled-MMSE precoding has been derived and the achievable throughput loss has been analyzed.

Furthermore, the contribution in [28] analyzes the sum-rate performance of joint opportunistic scheduling and the receiver design for multiuser MIMO-SDMA downlink systems. In this approach, the BS exploits the limited feedback on the effective SINRs, and schedules simultaneous data transmission on multiple beams to the user terminals which have the largest effective SINRs. Moreover, considering smart antennas at the access points and single antennas at the user terminals, the authors in [47] investigate the use of joint optimal downlink beamforming, power control and access point allocation in a multicell SDMA system. Additionally, the contribution in [48] provides an overview of the scheduling algorithms proposed for multiuser MIMO based 4G wireless networks.

In an opportunistic transmission scheme, channel fluctuations need to be tracked in order to schedule the BS transmissions for the user with the best channel. The main concept behind opportunistic beamforming is to induce rapid channel fluctuations by employing multiple antennas at the BS, which subsequently helps to improve the multiuser diversity gain. In general, the use of spatial diversity is harmful to the multiuser diversity gain. In this regard, the multi-channel multiuser diversity scheduling scheme proposed in [41] simultaneously exploits both diversities. In addition, the contribution in [13] investigates the effect of distributed BS antennas on the reduction of intrauser correlations while analyzing the performance of a MIMO-OFDM system.

Opportunistic Scheduling (OS) can be considered as another promising technique in an SDMA-based system in order to enhance the system throughput by exploiting multiuser diversity with the limited channel feedback [32]. Existing OS schemes can be classified into two categories, namely, Time-Sharing (TS) and SDMA-based OS schemes. In a TS-OS scheme, only the user terminal with the best instantaneous channel conditions is scheduled in one slot being independent of the number of beams employed by the BS. On the other hand, an SDMA-based OS serves multiple terminals simultaneously with multiple orthonormal beams in each time slot. The sum-rate of SDMA-based OS grows linearly with M whereas for the TS-OS, it increases only linearly with min(M, N), M and N being the number of transmit and receive antennas, respectively [29]. In this context, the contribution in [30] proposes the SDMA-based OS for systems with single-antenna mobile terminals while for the mobile terminals having multiple receive antennas, the contribution in [29] proposes to allow each antenna compete for its desired beam as if it was an individual terminal. In the latter case, each beam is assigned to a specific receive antenna of a chosen terminal but the signals captured by the undesired antennas of the mobile terminal are discarded, thus leading to inefficient utilization of multiple antennas.

To address the aforementioned issue, the contribution in [31] proposes various linear combining techniques exploiting signals received by all receive antennas and considers the improved effective SINR as a scheduling metric. In the similar context, the contribution in [32] provides a a systematic approach for deriving asymptotic throughput and scaling laws using SINR based on the extreme value theory. Consequently, with the help of a comparison between the Signal-toInterference Ratio (SIR) and SINR-based analyses, it has been argued that the SIR-based analysis is more computationally efficient for SDMA-based systems, and subsequently more effective in order to capture the high-order behavior of the asymptotic system performance.

A comprehensive overview on SDMA/OFDMA scheduling challenges are highlighted in [11], which further proposes an SDMA-OFDMA Greedy Scheduling Algorithm (sGSA) for WiMAX systems. The proposed solution in [11] considers feasibility constraints in order to allocate resources for multiple mobile terminals on a per packet basis by employing the following two approaches: a) a cluster-based SDMA grouping algorithm, and b) a computationally efficient frame layout scheme. The later approach allocates multiple SDMA groups per frame based on their packet Quality of Service (QoS) utility. In addition, the contribution in [49] considers Opportunistic Beamforming (OB) with finite number of single-antenna users under the constraint that the feedback overhead from the mobile terminals to the BS is constant. The impact of the fading variances of the users and the spatial correlation on the sum rate of TDMA and SDMA based OB has been analyzed. It has been concluded that for a small number of spread out users and moderate to high Signal to Noise Ratio (SNR) values, SDMA-OB scheme performs worse than the TDMA-OB. In addition, authors in [50] investigate a multiuser two-way relay system using SDMA communications and proposes an optimal scheduling method that maximizes the sum rate while ensuring fairness among users. Subsequently, rate and angle-based sub-optimal scheduling methods have been studied in order to reduce the computational load at the relay.

Research Challenges

The performance of an SDMA based system may degrade due to the imperfect channel knowledge and poor synchronization. Furthermore, the difficulty of obtaining instantaneous CSI knowledge at the transmitters may prevent the practical implementation of many multiuser SDMA systems. In the following, we highlight the main research challenges in SDMA-based wireless networks.

- A major challenge, common to all SDMA systems, is the requirement of CSI knowledge at the transmitter to enable the transmission of multiple streams without any harmful interference.
- Under the scenarios that the receivers have their perfect CSIs but the transmitter knows only the statistical CSI, the optimization for downlink multiuser MIMO is still not well understood in the literature. This problem has been recently studied in [43] but only for the case of two users having single antennas. Therefore, the generalization of these results for arbitrary number of antennas with arbitrary number of users is an important future research topic.
- Massive MIMO has been considered as one of the key enablers for future 5G wireless networks. As the number of BS antennas in future SDMA networks increases, the system gets almost entirely limited from the reuse of pilots in neighboring cells. This leads to the pilot contamination and antenna correlation problems, which appear to be fundamental challenges for designing very large MIMO systems.
- In order to take full advantage of the SDMA scheme, it is important to investigate suitable techniques which can simultaneously improve both the spatial and multiuser diversity gains.
- In order to make the best use of the available resources in different dimensions, suitable optimal scheduling algorithms need to be investigated by employing a cross-layer design with the cooperation between the Medium Access (MAC) and physical layers.
- Inter-user interference is the main limiting factor in multiuser SDMA systems. In order to mitigate this, suitable resource allocation (joint carrier and power allocation) strategies need to be investigated.
- Besides the effects of co-channel interference, channel fading and the noise, there may arise distortion in the system performance due to the presence of residual hardware impairments caused due to several reasons such as phase noise, analog to digital converter inaccuracies, oscillator mismatch, etc [51]. In this context, it's important to investigate the residual hardware aware

adaptive beamforming schemes in order to improve the system performance in the presence of practical impairments.

- Opportunistic or cognitive radio communication has been considered as one of the key techniques to enhance the spectral efficiency of future 5G networks [52, 53]. In this context, SDMAbased cognitive wireless networks should employ suitable opportunistic spectrum access/user selection/shceduling algorithms while providing sufficient protection to the already existing licensed systems.
- The investigation of suitable transceiver design strategies such as precoding/beamforming for physical layer multicasting and multi-group multicasting [54] considering a large number of antennas at the BS side is another important research challenge.

EMERGINGMULTIPLEXING/MULTIPLEACCESS SCHEMES

In this section, we briefly discuss the main emerging multiplexing/multiple access schemes.

Orbital Angular Momentum Multiplexing

Recently, Orbital Angular Momentum Multiplexing (OAMM)) has been shown as an important candidate for high capacity millimeter wave communications [4]. In this multiplexing method, one important property of an electromagnetic wave that each beam has a unique helical phase front is utilized in order to obtain multiplex multiple beams. The orthogonality of the beams is defined by a different Orbital Angular Momentum (OAM) state number which is the amount of phase front "twisting". Authors in [4] demonstrated that this scheme can enhance the system capacity as well as the spectral efficiency of mm-wave wireless communication links by transmitting multiple data streams with a single aperture tranmit/receive pair.

The OAMM implementation is completely different from the implementation of the traditional radio frequency spatial multiplexing and therefore it requires a significant architectural change [4]. The traditional spatial multiplexing scheme requires multiple spatially separated transmitter and receiver aperture pairs for the transmission of multiple data streams whereas the multiplexed beams in the OAM scheme are completely coaxial throughout the transmission medium and it uses only single transmitter and receiver aperture.

Polarization Division Multiplexing/Multiple Access

There is an emerging concept of polarization modulation technique for carrying information bearing signals [55]. This approach uses circular polarization of the propagating electromagnetic carrier as a modulation characteristic in contrast to amplitude, frequency and/or phase modulation attributes used in the conventional schemes. The circular modulation techniques have the capability of providing inherent benefits of circular polarization as well as the diversity gain in wireless fading channels. Moreover, the concepts of Polarization Division Multiplexing (PDM) and phase division multiplexing widely used in the optical communications can be regarded as other promising approaches in order to enhance the multiplexing gain of 5G wireless systems on the top of the currently used frequency/time/code multiplexing schemes.

In a PDM scheme, two data streams can be multiplexed with orthogonal polarizations at the transmitter side in order to enhance the channel capacity. The main problem that may arise in wireless fading channels with this approach is that channel depolarization may induce correlation between two corresponding data streams at the receiver which are expected to be received on orthogonal polarizations [56]. This effect can be partially mitigated by utilizing PDM in combination with multi-antenna techniques such as space-time block coding and beamforming [57].

Like PDM, the main concept behind Polarization Division Multiple Access (PDMA) is to transmit two independent data streams at the same time and at the same frequency to two different users by employing orthogonal polarizations [56]. Recently, the contribution in [58] analyzed the capacity of the PDMA scheme and expressed the relation between PDMA channel capacity and Cross Polar Discrimination (XPD) in a mathematical form. Furthermore, authors in [56] have investigated a PDMA scheme for the downlink of a cellular system by employing collaborative transmitreceive polarization and polarization filtering detection for non-line of sight wireless fading channels. It has been concluded that the proposed PDMA scheme has a great potential to be utilized as a new multiple access scheme in the next generation of cellular wireless communication systems.

Interweave Division Multiple Access (IDMA)

Interweave Division Multiple Access (IDMA) is an asynchronous multiple access scheme in which different interleavers are used to distinguish users in contrast to the use of different codes in a conventional CDMA system. In a conventional CDMA scheme, interleavers are placed before the spreaders and they are effective only when used in conjunction with channel coding. Interleavers, which are usually placed between Forward Error Correction (FEC) coding and spreading, are used to combat the fading effect in CDMA, whereas the arrangement of interleaving and spreading is reversed in IDMA, and different interleavers distinguish distinct data streams.

IDMA can be considered as a special case of random waveform CDMA, and the accompanying chip-bychip estimation algorithm is essentially a low-cost iterative soft cancellation technique. Furthermore, the computational cost per user is independent of the number of users, which is significantly lower than that of the MMSE technique. Authors in [59] have shown that IDMA with equal power level can achieve near single user performance in multiuser environments.

IDMA inherits many benefits from CDMA such as path diversity, mitigation of intra-cell interference, and a common spreading sequence. In [60], authors investigated the performance of the MIMO assisted multicarrier IDMA scheme with multiuser detection and showed that this scheme can provide better performance with the aid of VBLAST/ZF/MAP detection technique. In addition, authors in [61] have investigated the following three design aspects for multicarrier IDMA technique: (i) multiplexing versus diversity tradeoff, (ii) coding versus spreading tradeoff, and (iii) complexity versus performance tradeoff. Moreover, authors in [62] have recently studied a quantize and forward strategy for the half-duplex IDMA relay channel considering multiple users, single relay, and single destination.

Grouped IDMA. Grouped-IDMA is a version of the IDMA scheme in which active users are arranged into several groups and each group is characterized by an orthogonal code. This scheme inherits the advantages of IDMA and orthogonal CDMA, utilizing the group specific orthogonal spreading code for group separation. It has been shown in [63] that the groupedIDMA achieves better performance than the simple IDMA when the number of users is relatively large, especially in low/medium SNR region.

In the grouped IDMA scheme, each group of users works in the same way as the IDMA and each group is assigned to a group-specific orthogonal code. Therefore, inter-group interference can be eliminated by carrying out de-spreading at the receiver. Furthermore, interference among different users in the same group can be reduced by a chip-by-chip detection algorithm used in IDMA [63].

For a wireless system with K number of users divided into G groups, the grouped IDMA can be interpreted as

- IDMA when G = 1
- Orthogonal CDMA when G = K
- Grouped IDMA when 1 ≤ G ≤ K

Universal Filtered Multi-Carrier (UFMC)

As stated earlier, OFDM technology is more demanding in terms of synchronization and energy consumption. To alleviate the drawbacks, Filter Bank based MultiCarrier (FBMC) has been recently considered as an alternative to OFDM. In contrast to the OFDM, this technique applies a filtering functionality to each of the subcarriers. Therefore, side-lobes with FBMC become much lower and thus the intercarrier interference issue is far less harmful than in OFDM. Despite several benefits of FBMC technique, practical system configurations renders most of them [5]. In this context, there is an emerging concept of Universal Filtered MultiCarrier (UFMC) which can benefit from the advantages of FBMC while addressing its drawbacks.

UFMC is a filtering operation applied to a group of consecutive subcarriers (e.g., a given allocation of a single user) in order to reduce out-of-band sidelobe levels. This subsequently minimizes the potential inter-channel interference between adjacent users in the case of asynchronous transmissions. In this technique, filtering operation is applied to a group of consecutive subcarriers instead of per subcarrier filtering employed in the FBMC technique [64]. The UFMC can significantly reduce the effect of sidelobe interference and is better suitable for fragmented spectrum operation. It has been shown in [64] that the UFMC outperforms the cyclic prefix-based OFDM for both perfect and imperfect frequency synchronization between user equipments and the BSs.

Sparse Code Multiple Access (SCMA)

Due to the closed loop nature of the multiuser MIMO system, it suffers from different practical limitations such as channel aging and high feedback overhead required to feedback CSI from the serving users to the BS. To address these limitations, open loop multiplexing schemes such as non-orthogonal code domain multiple access can be considered as promising solutions [65]. In the category of non-orthogonal multiple access schemes, Sparse Code Multiple Access (SCMA) has been recently considered as a promising candidate. In this scheme, input information bits are mapped to multi-dimensional complex codewords which are selected from the predefined sets of the codebook. Subsequently, code-domain layers are allocated to different users without requiring the CSI knowledge of the paired user terminals [65, 66].

In comparison to MU-MIMO, SCMA-based multiuser wireless system is more robust in the presence of time varying channel. Furthermore, the CSI feedback problem is removed due to its open loop nature. In addition, higher data rate and the robustness to mobility are two major advantages of multiuser SCMA. Moreover, compared to the spatial domain processing in MU-MIMO schemes, code-domain multiplexing has a significant advantage in terms of the transmit-side computational complexity [65]. On the other hand, the main drawback of SCMA approach is that it requires a non-linear receiver in order to detect the corresponding layer of each user, thus resulting in the decoding complexity. However, the sparsity feature of SCMA codewords allows to utilize low complexity detection algorithms such as message passing algorithm [67].

Table 2: Main Features and drawbacks of Several Existing and Emerging Multiple Access/Multiplexing Schemes

Techniques	Main Features	Drawbacks
Frequency Division Multiple Access (FDMA)	i. All users can transmit in parallel ii. No need of time synchronization	i. Each station only gets a fraction of total bandwidth ii. Need of tunable transmitters and receivers
Time Division Multiple Access (TDMA)	i. Each user can use the total bandwidth ii. No need of tunable receivers	i. Need of time synchronization ii. Transmission over a fraction of the total time
Code Division Multiple Access (CDMA)	i.Each user can transmit over total bandwidth all the time ii. More users per MHz of bandwidth	i. Higher receiver complexity ii. Near-far effect
Orthogonal Frequency Division Multiple Access (OFDMA))	i. Intrinsic orthogonality ii. Receiver circuit simplicity	i. High peak-to-average ratio ii. Sensitive to frequency offset iii. Poor performance in highly asynchronous access scenarios
Space Division Multiple Access (SDMA)	i. Higher spectral efficiency and multiple access capabilities ii. Useful in combination with TDMA, FDMA, CDMA or OFDMA	i. Near-far effect ii. Poor synchronization iii. Loss of orthogonality in the presence of practical imperfections

Orbital Angular Momentum Multiplexing (OAMM)	i. Orthogonality of the beams defined by a different OAM state number ii. Suitable for mmwave communications iii. Higher system capacity	i. Requires a significant architectural change to implement ii. Intermodal crosstalk
Polarization Division Multiple Access (PDMA)	i. Two independent data streams transmitted with orthogonal polarizations ii. Can be combined with any existing multiple access schemes	i. Depolarization effect in fading channel ii. Need of extra receiver circuitry for polarization filtering detection
Interweave Division Multiple Access (IDMA)	i. Chip-interleaving process is used for user separation ii. Reverse arrangement of interleaving and spreading as compared to CDMA iii. Increased diversity against fading iv. Low receiver cost v. Suitable for low rate transmissions	i. Higher receiver complexity for wideband systems ii. Iterative processing required iii. Entire interleaver matrix need to be transmitted to the receiver iv. Need of memory optimization in transmitter and receiver side
Universal Filtered Multi-Carrier (UFMC)	i. Blockwise filtering provides additional flexibility ii. Shorter filter lengths and reduced sidelobe interference iii. Highly suitable for CoMP transmission, and Internet of Things iv. Better suited for fragmented spectrum than OFDM	i. Need of complex receiver circuitry ii. Not significant gains over OFDM for low rate transmissions
Sparse Code Multiple Access (SCMA)	i. Non-orthogonal code domain multiple access ii. Open loop multiplexing and no need of feedback channel iii. More robust in presence of time varying channel and the CSI feedback problem	i. Requires non-linear receiver ii. Complex codebook design since multiple layers are multiplexed with different codebooks

PROPOSED CLUSTERED ORTHOGONALSIGNATURE DIVISIONMULTIPLE ACCESS

In order to fully realize the potentials of recently emerging wireless technologies such as massive MIMO and mm-wave communications, existing multiple access techniques may not be sufficient and we need to investigate new multiple access schemes for future wireless networks. In this context, several multiplexing/multiple access techniques discussed in the previous section are under investigation. Since the beams become very narrower in mm-wave communications and sufficient antenna spacing becomes an issue in massive MIMO systems, there is a high probability of the orthogonality loss between two co-channel users separated in the spatial domain. In this regard, the conventional concept of the SDMA technique should be adapted in future massive MIMO systems which will be possibly implemented in the mm-wave frequency range.

Herein, our proposition is to employ clustered orthogonal signature/power/code division multiple access on the top of the existing multiple access schemes such as SDMA. Unlike the grouped IDMA scheme, the idea here is to group users located within a cluster or a beam and to provide orthogonal codes to these groups. By employing the orthogonality over the conventional SDMA system, we can significantly enhance the number of users which can be supported by a given set of frequency resources in a particular cluster/beam. The proposed clustered orthogonal signature division multiple access scheme envisions to address the aforementioned drawbacks of several approaches highlighted in Table 2. This will further address the problem of loss of orthogonality which may arise in many existing SDMA-based approaches due to the time varying nature of the wireless channel.

One of the enablers for the proposed approach is joint antenna coding approach. By employing the orthogonal coding in combination with the dynamic 3D beamforming approach, we can separate users in different groups depending on the available radio resources. For future dense networks, the conventional SDMA may not be able to guarantee orthogonality between the beams where one beam implies one specific signature. In this context, the proposed idea is to enhance the orthogonality of the SDMA beams with the help of suitable orthogonal coding scheme by employing a joint antenna coding scheme. In order to design the orthogonal codes, there exist several possibilities in the literature such as Hadamard code [68], Gold code [69], and Polyphase orthogonal codes [70].

The proposed approach can be implemented in a two tier manner, meaning that one multiple access scheme in the first tier and another in the second tier. For example, in a heterogeneous network comprising of macro cells and small cells, the backhaul part, i.e., the link from the small cell BSs to the macro cell BS, may employ one multiple access scheme and the access part, i.e., from the end users to the small cell BS, may use another multiple access scheme. In this way, we suggest the following two approaches for implementing the proposed clustered orthogonal signature division multiple access.

- ***Code/signature division multiple access after employing SDMA:*** In this approach, the first tier uses signature (code) division multiple access and the second tier uses SDMA.
- ***SDMA after employing code/signature division multiple access:*** In this approach, the first tier uses SDMA and the second tier uses code/signature division multiple access.

While devising a good multiple access scheme in a time varying wireless environment, the main objective should be to maximize the overall orthogonality. Depending on the deployed environment, if the beam sharpness and the surrounding environment cannot provide sufficient orthogonality, a suitable orthogonal coding can be implemented to enhance the overall orthogonality. The following advantages are foreseen in future wireless networks by employing the proposed multiple access scheme: (i) better flexibility to wireless design engineers, (ii) better performance in the presence of time varying wireless channels, (iii) easier to implement from the practical perspectives, and (iv) allocation of the available resources in an optimized way.

However, from the practical perspectives, the following factors need to be further investigated while realizing the proposed multiple access concept.

- Availability of the required digital signal processing hardware
- Energy efficiency of the system and its implementation complexity
- Flexibility of accommodating users/services into the system: In most of the current wireless systems, we need to tune system parameters according to the requirements.
- Orthogonality of the code: Higher the orthogonality of the code, better becomes the system performance at the cost of the restriction in the number of users/services that can be supported.

CONCLUSIONS

In order to address the issue of spectrum shortage in future wireless networks, investigation of suitable multiple access/multiplexing scheme is of significant importance. In this regard, this paper has reviewed various features of the widely discussed SDMA scheme. In addition, it has highlighted the main features and the drawbacks of other several emerging multiple access/ multiplexing schemes. More importantly, it has proposed a novel concept of clustered orthogonal multiple access scheme as an important candidate for future dense cellular networks.

In our future work, we plan to validate the proposed concept with the help of system level simulations. Furthermore, the comparison of the proposed twotier approaches in terms of the overall orthogonality improvement is the part of our ongoing works.

REFERENCES

1. Andrews, J., Buzzi, S., Choi, W., Hanly, S., Lozano, A., Soong, A. and Zhang, J. (2014) What will 5G be? IEEE J. Sel. Areas in Commun. 32(6): 1065–1082.
2. Viswanathan, H. and Weldon, M. (2014) The past, present, and future of mobile communications. Bell Labs Technical J. 19(8-21).
3. Rangan, S., Rappaport, T. and Erkip, E. (2014) Millimeter-wave cellular wireless networks: Potentials and challenges. Proc. IEEE 102(3): 366–385.
4. Yan, Y., et al (2014) High-capacity millimeter-wave communications with orbital angular momentum multiplexing. Nature Communications 5(4876).
5. Schaich, F. and Wild, T. (2014) Waveform contenders for 5G-OFDM vs. FBMC vs. UFMC. In 6th Int. Symp. on Communications, Control and Signal Process.: 457–460.
6. Rapajic, P. (1998) Information capacity of the space division multiple access mobile communication system. In IEEE 5th Int. Symp. on Spread Spectrum Techniques and Applications, 3: 946–950 vol.3.
7. Zetterberg, P. and Ottersten, B. (1995) The spectrum efficiency of a base station antenna array system for spatially selective transmission. In IEEE Trans. Vehicular Technol., 44(3): 651–660.
8. Jiang, M., Akhtman, J. and Hanzo, L. (2007) Iterative joint channel estimation and multi-user detection for multiple-antenna aided OFDM systems. IEEE Trans. Wireless Commun. 6(8): 2904–2914.
9. Ylioinas, J. and Juntti, M. (2009) Iterative joint detection, decoding, and channel estimation in turbocoded MIMO-OFDM. IEEE Trans. on Vehicular Technol. 58(4): 1784–1796.
10. Zhang, J., Chen, S., Mu, X. and Hanzo, L. (2012) Turbo multi-user detection for OFDM/SDMA systems relying on differential evolution aided iterative channel estimation. IEEE Trans. Commun. 60(6): 1621–1633.
11. Zubow, A., Marotzke, J., Camps-Mur, D. and PerezCosta, X. (2012) Sgsa: An SDMA-OFDMA scheduling solution. In 18th European Wireless Conf.: 1–8.
12. Chan, P. and Cheng, R. (2007) Capacity maximization for zero-forcing MIMO-OFDMA downlink systems with multiuser diversity. IEEE Trans. on Wireless Commun. 6(5): 1880–1889.

13. Dawod, N., Hafez, R. and Marsland, I. (2006) A multiuser zeroforcing system with reduced near-far problem and MIMO channel correlations. In Canadian Conf. on Electrical and Computer Engg.: 936–939.
14. Levin, G., and Loyka, S. (2008) On the Outage Capacity Distribution of Correlated Keyhole MIMO Channels. In IEEE Trans. Info. Theory 54(7): 3232–3245.
15. Sharma, S.K., Chatzinotas, S., and Ottersten, B. (2013) SNR Estimation for Multi-dimensional Cognitive Receiver under Correlated Channel/Noise. In IEEE Trans. Wireless Commun. 12(12): 6392–6405.
16. Tangemann, M. (1995) Near-far effects in adaptive SDMA systems. In IEEE Int. Symp. PIMRC, 3: 1293– 1297.
17. Tangemann, M. (1994) Influence of the user mobility on the spatial multiplex gain of an adaptive SDMA system. In IEEE Int. Symp. PIMRC: 745–749 vol.2.
18. Maciel, T. and Klein, A. (2010) On the performance, complexity, and fairness of suboptimal resource allocation for multiuser MIMO-OFDMA systems. IEEE Trans. on Veh. Technol. 59(1): 406–419.
19. Sulyman, A. and Hefnawi, M. (2010) Performance evaluation of capacity-aware MIMO beamforming schemes in OFDM-SDMA systems. IEEE Trans. on Commun. 58(1): 79–83.
20. Vandenameele, P., Van der Perre, L., Engels, M., Gyselinckx, B. and De Man, H. (2000) A combined OFDM/SDMA approach. IEEE J. Sel. Areas in Commun. 18(11): 2312–2321.
21. Thoen, S., Deneire, L., Van der Perre, L., Engels, M. and De Man, H. (2003) Constrained least squares detector for OFDM/SDMA-based wireless networks. IEEE Trans. on Wireless Commun. 2(1): 129–140.
22. Berger, S., Lu, Z., Irmer, R. and Fettweis, G. (2013) Modelling the impact of downlink CoMP in a realistic scenario. In IEEE Wireless Communications and Networking Conference: 3932–3936.
23. Bell, A.S. and Lucent, A. (2008) Collaborative MIMO for LTE-A downlink. Tech. Rep. R1-082501, 3GPP TSG RAN WG1 Meeting 53b.
24. Zhang, R. (2010) Cooperative multi-cell block diagonalization with per-base-station power constraints. IEEE J. on Sel. Areas in Commun. 28(9): 1435–1445.
25. Chatzinotas, S., Imran, M. A., and Hoshyar, R. (2009) On the multicell processing capacity of the cellular MIMO uplink channel in correlated rayleigh fading environment. IEEE Trans. Wireless Commun. 8 (7):

3704– 3715.

26. Jang, U., Son, H., Park, J. and Lee, S. (2011) CoMP-CSB for ICI nulling with user selection. IEEE Trans. Wireless Commun. 10(9): 2982–2993.
27. Zhou, T., Peng, M., Wang, W. and Chen, H.H. (2013) Low-complexity coordinated beamforming for downlink multicell SDMA/OFDM systems. IEEE Trans. on Veh. Technol. 62(1): 247–255.
28. Pun, M.O., Koivunen, V. and Poor, H. (2011) Performance analysis of joint opportunistic scheduling and receiver design for MIMO-SDMA downlink systems. IEEE Trans. on Commun. 59(1): 268–280.
29. Sharif, M. and Hassibi, B. (2007) A comparison of timesharing, DPC, and beamforming for MIMO broadcast channels with many users. IEEE Trans. on Commun. 55(1): 11–15.
30. Sharif, M. and Hassibi, B. (2005) On the capacity of MIMO broadcast channels with partial side information. IEEE Trans. on Info. Theory 51(2): 506–522.
31. Pun, M.O., Koivunen, V. and Poor, H. (2007) Opportunistic scheduling and beamforming for MIMO-SDMA downlink systems with linear combining. In IEEE Int. Symp. PIMRC: 1–6.
32. Pun, M.O., Koivunen, V. and Poor, H. (2008) SINR analysis of opportunistic MIMO-SDMA downlink systems with linear combining. In IEEE Int. Conf. on Commun.: 3720–3724.
33. Larsson, E., Edfors, O., Tufvesson, F. and Marzetta, T. (2014) Massive MIMO for next generation wireless systems. IEEE Communications Mag. 52(2): 186–195.
34. Lu, L., Li, G., Swindlehurst, A., Ashikhmin, A. and Zhang, R. (2014) An overview of massive MIMO: Benefits and challenges. IEEE J. Sel. Topics in Signal Process. 8(5): 742–758.
35. Mohammad Razavizadeh, S., Ahn, M. and Lee, I. (2014) Three-dimensional beamforming: A new enabling technology for 5G wireless networks. IEEE Signal Process. Mag. 31(6): 94–101.
36. Halbauer, H., Saur, S., Koppenborg, J. and Hoek, C. (2013) 3D beamforming: Performance improvement for cellular networks. Bell Labs Technical J. 18(2): 37–56.
37. Sharma, S.K., Chatzinotas, S. and Ottersten, B. (2015) 3D beamforming for spectral coexistence of satellite and terrestrial networks. In IEEE Vehicular Technol. Conf..

38. Checko, A. and et al (2015) Cloud RAN for Mobile Networks-A Technology Overview. IEEE Communications Surveys & Tutorials, 17(1): 405-426.

39. Lozano, A. and Tulino, A. (2002) Capacity of multipletransmit multiple-receive antenna architectures. IEEE Transactions on Information Theory 48(12): 3117–3128.

40. Choi, R.U., Ivrlac, M., Murch, R. and Utschick, W. (2004) On strategies of multiuser MIMO transmit signal processing. IEEE Transactions on Wireless Communications 3(6): 1936–1941.

41. Aktas, D. and El Gamal, H. (2003) Multiuser scheduling for MIMO wireless systems. In Vehicular Technology Conference, 2003. VTC 2003-Fall. 2003 IEEE 58th, 3: 1743–1747 Vol.3.

42. Oberli, C. and Rios, M. (2007) OFDM-based SDMA networks: Signal model under imperfect synchronization and channel state information. In 2th International OFDM Workshop.

43. Raghavan, V., Hanly, S. and Veeravalli, V. (2013) Statistical beamforming on the grassmann manifold for the two-user broadcast channel. IEEE Trans. on Info. Theory 59(10): 6464–6489.

44. Alias, M., Chen, S. and Hanzo, L. (2005) Multipleantenna-aided OFDM employing genetic-algorithmassisted minimum bit error rate multiuser detection. IEEE Trans. Veh. Technol. 54(5): 1713–1721.

45. Das, S. and Bagadi, K.P. (2011) Comparative analysis of various multiuser detection techniques in SDMAOFDM system over the correlated MIMO channel model for IEEE 802.16n. World Academy of Science, Engg. and Technol. 53(1).

46. Lu, W., Ji, F. and Yu, H. (2011) A general resource allocation algorithm with fairness for SDMA/MISO/OFDMA systems. IEEE Commun. Letters 15(10): 1072–1074.

47. Stridh, R., Bengtsson, M. and Ottersten, B. (2001) System evaluation of optimal downlink beamforming in wireless communication. In IEEE Veh. Technol. Conf., 1: 343–347.

48. Ajib, W. and Haccoun, D. (2005) An overview of scheduling algorithms in MIMO-based fourthgeneration wireless systems. IEEE Network 19(5): 43–48.

49. Jorswieck, E., Svedman, P. and Ottersten, B. (2008) Performance of TDMA and SDMA based opportunistic beamforming. IEEE Trans. Wireless Commun. 7(11): 4058–4063.

50. Joung, J. and Sayed, A. (2010) User selection methods for multiuser two-way relay communications using space division multiple access. IEEE Trans. Wireless Commun. 9(7): 2130–2136.
51. Papazafeiropoulos, A.K., Sharma, S.K. and Chatzinotas, S. (2015) Impact of transceiver impairments on the capacity of dual-hop relay massive MIMO systems. In IEEE GLOBECOM Workshop.
52. Sharma, S., Bogale, T., Chatzinotas, S., Ottersten, B., Le, L.B. and Wang, X. (2015) Cognitive radio techniques under practical imperfections: A survey. IEEE Commun. Surveys Tutorials 17(4): 1858–1884.
53. Sharma, S., Patwary, M., Chatzinotas, S., Ottersten, B. and Abdel-Maguid, M. (2015) Repeater for 5G wireless: A complementary contender for spectrum sensing intelligence. In IEEE Int. Conf. on Commun.: 1416–1421.
54. Christopoulos, D., Chatzinotas, S. and Ottersten, B. (2014) Weighted Fair Multicast Multigroup Beamforming Under Per-antenna Power Constraints. IEEE Trans. Signal Process. 62(19): 5132-5142.
55. ul Abidin, Z., Xiao, P., Amin, M. and Fusco, V. (2012) Circular polarization modulation for digital communication systems. In Int. Symp. on Commun. Systems, Networks Digital Signal Process.: 1–6.
56. Kwon, S.C. and Stuber, G. (2014) Polarization division multiple access on NLOS wide-band wireless fading channels. IEEE Trans. Wireless Commun. 13(7): 3726– 3737.
57. Deng, Y., Burr, A. and White, G. (2005) Performance of MIMO systems with combined polarization multiplexing and transmit diversity. In IEEE Veh. Technol. Conf., 2: 869–873 Vol. 2.
58. Kwon, S.C. (2014) Optimal power and polarization for the capacity of polarization division multiple access channels. In IEEE Globecom: 4221–4225.
59. Ping, L., Liu, L., Wu, K. and Leung, W. (2006) Interleave division multiple-access. IEEE Trans. Wireless Commun. 5(4): 938–947.
60. Nagaradjane, P., Chandrasekaran, S., Vishvaksenan, K. and Ramakrishnan, M. (2010) MIMO multi carrier interleave division multiple access system with multiuser detection. In Int. Conf. on Wireless Commun. and Sensor Comput.: 1–4.
61. Zhang, R. and Hanzo, L. (2008) Three design aspects of multicarrier interleave division multiple access. IEEE Trans. on Veh. Technol. 57(6): 3607–3617.

62. Liu, L., Li, Y., Su, Y. and Sun, Y. (2015) Quantize-andforward strategy for interleave division multiple-access relay channel. IEEE Trans. on Veh. Technol. PP(99): 1–1.

63. Tu, Y., Fan, P. and Zhou, G. (2006) Grouped interleavedivision multiple access. In First Int. Conf. on Commun. and Networking in China: 1–5.

64. Vakilian, V., Wild, T., Schaich, F., ten Brink, S. and Frigon, J.F. (2013) Universal-filtered multi-carrier technique for wireless systems beyond LTE. In IEEE Globecom Workshops: 223–228.

65. Nikopour, H., Yi, E., Bayesteh, A., Au, K., Hawryluck, M., Baligh, H. and Ma, J. (2014) SCMA for downlink multiple access of 5G wireless networks. In IEEE Globecom: 3940–3945.

66. Taherzadeh, M., Nikopour, H., Bayesteh, A. and Baligh, H. (2014) SCMA codebook design. In IEEE Veh. Technol. Conf.: 1–5.

67. Hoshyar, R., Razavi, R. and Al-Imari, M. (2010) LDSOFDM an efficient multiple access technique. In IEEE Veh. Technol. Conf.: 1–5.

68. Del Rio, A. and Rifa, J. (2013) Families of Hadamard BBZ2BBZ4Q8-codes. IEEE Trans. on Info. Theory 59(8): 5140–5151.

69. Das, B., Sarma, M., Sarma, K. and Mastorakis, N. (2015) Design of a few interleaver techniques used with Gold codes in faded wireless channels. In Int. Conf. on Signal Process. and Integrated Networks: 237–241.

70. Liu, Y.C., Chen, C.W. and Su, Y. (2013) New constructions of zero-correlation zone sequences. IEEE Trans. on Info. Theory 59(8): 4994–5007.

2

Quasi-Spatially Periodic Signals for Optical Fibre Communications

Qun Zhang [2], Terence H. Chan[1] & AlexGrant[3]

[1]University of South Australia, Institute for Telecommunications Research, Mawson Lakes, SA, 5095, Australia
[2] University of Toronto, The Edward S. Rogers Sr. Department of Electrical and Computer Engineering, Toronto, ON, M5S 3G4, Canada
[3] Myriota, Adelaide, SA, 5000, Australia

ABSTRACT

We study quasi-spatially periodic signals (QSPSs) as a class of input signals of interest, which maintain their shapes quasi-periodically (with a phase change and a time shift) during propagation in an optical fibre. Instead of the computationally expensive nonlinear Fourier transform (NFT), the

Citation: Zhang, Q., Chan, T.H. & Grant, A. "Quasi-Spatially Periodic Signals for Optical Fibre Communications". Sci Rep 8, 15596 (2018). https://doi.org/10.1038/s41598-018-33589-3

property of quasi-periodic shape invariant could be used as an alternative for decoding at the receiver. In this paper, properties of QSPSs and the signal design problem are studied, including the trade-off between various system parameters.

INTRODUCTION

In the ever increasing demand for data traffic, which grows at a pace of 60% per year[1], fibre optical communication plays a significant role in modern information technological infrastructure. Compared to copper wires, optical fibres are promising media, especially for long distance data transmission, because they provide ultra high speed and low loss transmission. The fibre loss is only 0.2 dB/km, which means that we only loss half of the signal power after a 15 km propagation.

Although the desirable communication capability an optical fibre has, its physical properties are unfortunately complicated. The signal transmission through it is nonlinear, which makes it difficult to analyse. The traditional tools developed for linear time-invariant (LTI) systems, such as signal detection and error control coding, do not work well in optical fibre channels. The reason is that the underlying physics of nonlinear fibre-optic channels are not satisfied by the "flawed" traditional linear approaches. The fibre nonlinearity could be approximated as a small perturbation in a low signal-to-noise ratio (SNR) regime, however, becomes significant when the SNR is high. As a result, fibre nonlinearity is treated as a detrimental effect in existing applications.

To meet the rapidly increasing demand for the data transmission bandwidth, many multiplexing techniques such as optical time division multiplexing (OTDM)[2], wavelength division multiplexing (WDM)[3,4], and orthogonal frequency division multiplexing (OFDM)[5] have been proposed. However, interference between different users is inevitable under those multiplexing techniques. Take the WDM for an example, it was shown[4] that the spectral efficiency (SE) for an optical fibre network increases at the low SNR regime, similar to the trend of the linear capacity bound, however, decreases when the signal power is further increased. The reason for the unnatural SE decrease in the high SNR regime is that the interference between different users increases significantly at the high input power regime, which decreases the capacity. In order to tackle this issue, Yousefi and Kschischang recently proposed the nonlinear frequency division multiplexing (NFDM)[6,7,8] via the use of the nonlinear Fourier transform (NFT)[9] (also known as the direct scattering transform[10]). The fibre nonlinearity and the group velocity dispersion (GVD) effects are fully characterised in integrable nonlinear optical fibre communication systems in the nonlinear spectral domain with the help of the NFT. Under the NFDM, a fibre optical channel is subdivided to multiple multiplicative sub-channels in the nonlinear frequency domain. As a result, interference among users can be minimised. Some progresses of the NFT-based communications can be found in[11,12,13,14,15,16].

Signal propagation along an optical fibre is characterised by the normalised nonlinear Schrödinger equation (NLSE)[17]

$$j\frac{\partial q(t,z)}{\partial z} = \frac{\partial^2 q(t,z)}{\partial t^2} + 2|q(t,z)|^2 q(t,z). \quad (1)$$

Here, $j=\sqrt{-1}$, z and t are spatial and temporal variables respectively, and q(t, z) is the complex envelope of the normalised electrical field after propagating a distance z in the optical fibre. We note that fibre loss is assumed to be perfectly compensated by the ideal distributed Raman amplification in (1). However, the scenario of lumped amplification was also studied previously. For example, the NLSE with periodic perturbation was studied[18], in which the robustness of the solitons suffered from periodic perturbations was characterised by the resonance condition related to the soliton period. The periodic perturbations[18] could be due to the lumped amplifications applied repeatedly every certain distance along the transmission.

For each fixed $z \in \mathbb{R}$, the NFT of q(t, z) consists of continuous and discrete parts:

$\{Q^{(c)}(\lambda, z):\ \lambda \in \mathbb{R})\}$ and $\{Q^{(d)}(\zeta_k, z),\ \zeta_k \in \mathbb{C}^+: k = 1, 2, \ldots, N\}$.

We call $\{\lambda:\ \lambda \in \mathbb{R}\}$ and ζ_k the continuous spectrum and a discrete eigenvalue, while $Q^{(c)}(\lambda, z)$ and $Q^{(d)}(\zeta_k, z)$ their corresponding continuous spectral function and discrete spectral amplitudes. In this paper, we assume that the discrete eigenvalues are simple zeros of the associated scattering function. For detailed treatment of the NFT, please refer to[10,19,20]. Fast algorithms for the NFT and its inverse were studied in[21,22,23]. Recently, an alternative decoding method[24] is proposed, which directly uses scattering data instead of solving the eigenvalues.

In the NFDM, encoding is conducted in the nonlinear frequency domain. Each user is allocated a portion of nonlinear frequency spectrum (discrete and/or continuous), over which data is encoded and modulated. To illustrate, assume that there is only one user (and hence it has exclusive use of all the nonlinear spectrum). The encoding function f is defined over the set $\{1,\ldots,2^R\}$ such that for a given set of eigenvalues $\{\zeta_k: k = 1, 2, \ldots, N\} \subsetneq \mathbb{C}^+ \triangleq \{c \in \mathbb{C}: \mathrm{Im}(c) > 0\}$,

$$f(i) = (x_{k,i}, k = 1, 2, \ldots, N), \quad i \in \{1, \ldots, 2^R\},$$

where $x_{k,i} \in \mathbb{C}$ is a spectral amplitude, and R is the number of bits per transmission in each symbol. The physical interpretation for f is very simple. If the transmitted symbol is i, then the channel input is a function q(t, 0) with spectrum $\{\zeta_k: k = 1, 2, \ldots, N\} \subsetneq \mathbb{C}^+$, and spectral function (amplitudes)

$$Q^{(c)}(\lambda, 0) = 0, \qquad \forall \lambda \in \mathbb{R};$$
$$Q^{(d)}(\zeta_k, 0) = x_{k,i}, \ \zeta_k \in \mathbb{C}^+, \ k = 1, 2, \ldots, N.$$

To transmit, the signals designed in the nonlinear frequency domain are transformed into their time-domain forms by the inverse nonlinear Fourier transform (INFT), and be subsequently transmitted along the fibre. As the continuous spectrum is null, obtaining the INFT is reduced to solving a set of algebraic equations. At receiver, the received time domain signals are first converted into their spectral domain representations by the NFT. Users can then decode their messages by extracting the portion of the signals in their allocated spectra.

We notice that the computationally expensive NFT (and also its inverse) are involved in the signal de/modulation. In this paper, we consider the class of quasi-spatially periodic signals (QSPSs, whose definition will be given shortly), which guarantees that the "shape" of the channel output signal remains largely the same as that of the input. In this paper, our aim is to report this interesting property, which suggests a potential alternative decoding method instead of the NFT.

To illustrate the concept, we consider the following example. Let the length of an optical fibre be $\mathfrak{L}$. (In some application scenarios, the length of a fibre is fixed and known, e.g. fibres in chips). Suppose the possible channel inputs are $q_1(t, 0) = \beta\,\mathrm{sech}(\beta t)$, $q_2(t, 0) = 2\beta\,\mathrm{sech}(\beta t)$, and $q_3(t, 0) = 3\beta\,\mathrm{sech}(\beta t)$ with discrete spectrum respectively $\{j\beta/2\}$, $\{j\beta/2, 3j\beta/2\}$, and $\{j\beta/2, 3j\beta/2, 5j\beta/2\}$ where $\beta = \sqrt{k\pi/4\mathfrak{L}}$ and $k \in \mathbb{Z}^+$. As we shall see, the shapes of the three signals remain unchanged at the receiver. Therefore, to decode (i.e., to determine which one of the three signals is actually transmitted), one can use time-domain signal processing technique to achieve this goal.

Fundamental solitons (or abbreviated as solitons)25 are examples for the QSPSs. Solitons have been studied extensively in the literature in various aspects such as noise modelling8·26·27·28·29·30, capacity analysis26·31·32 and Gordon-Haus effect characterising the arrival time jitter of a soliton33 etc. However, there is only a limited set of soliton solutions. It will therefore be useful, and is also a goal of this paper, to determine a wider class of QSPSs. As we shall see, all QSPSs are N-solitons15·34 (but not vice versa). In particular, we identify conditions for when an N-soliton is quasi-spatially periodic. Signal design is also studied.

The contributions of this paper are as follows. First of all, the QSPS is defined, and its spectral composition is analysed. Using the theorem we derived, it is easy to identify whether a given spectrum corresponds to a QSPS. Furthermore, we propose a trade-off between various system parameters, including minimum period, signal degree-of-freedom, and signal energy constraint. Compared to our earlier conference version35, we broaden the notion of the periodicity by allowing both a time shift and a phase change of the signal in each period of distance, and consider a practical problem of the effect of the time domain pulse truncation for the QSPSs.

In this paper, we consider vanishing inputs, i.e.

$$\lim_{|t|\to 0} q(t, 0) = 0, \quad q(t, 0) \in L^1(\mathbb{R}). \tag{2}$$

We summarise notations as follows. Denote the NFT of a signal q(t, z), $z\in \mathbb{R}$ by $(Q^{(c)}(\lambda, z), Q^{(d)}(\zeta_k, z))$, where $\lambda\in \mathbb{R}$ is its continuous spectrum, and $\zeta_k \in \mathbb{C}^+, k = 1, 2, \ldots, N$ are eigenvalues which form its discrete spectrum. $Q^{(c)}(\lambda, z)$ and $Q^{(d)}(\zeta_k, z)$ are respectively its continuous and discrete spectral amplitudes.

RESULTS

This section is organised as follows. We first present the necessary and sufficient condition for a signal to be quasi-spatially periodic in the first subsection. Then we discuss the problem of signal design, including the trade-off between various parameters.

Quasi-Spatially Periodic Signals

In this subsection, we define quasi-spatially periodic signals (QSPSs), and study their properties. In particular, we obtain the relation between the minimum period of a QSPS and its spectrum.

Concept of Quasi-Spatially Periodic Signals.

Definition 1

(Quasi-Spatially Periodic Signals). A signal q(t, z) propagated in an optical fibre is called a QSPS with period p>0 if for any $z_0 \in \mathbb{R}$, there exist real numbers t_0 and ϕ such that

$$q(t, z_0 + p) = q(t - t_0(z_0), z_0)e^{j\phi(z_0)}, \quad \forall\ t \in \mathbb{R}. \tag{3}$$

Note that, ϕ in (3) is not unique. If ϕ satisfies (3), then $\phi + 2k\pi$ also satisfies (3) for any integers k. Therefore, we may assume without loss of generality that $0 \leq \phi < 2\pi$. Furthermore, while t_0 and ϕ may depend on the initial position z_0 in general, we shall prove later in Theorem 1 that it is indeed independent of z_0.

According to the definition, a QSPS q(t, z) maintains its shape quasi-periodically in z variable (with only possible phase change and time shift). In the following, we will claim that the QSPSs defined in Definition 1 have a minimum period, unless for the special case when the signal is a soliton.

To see this, let $\mathcal{P}(q) \triangleq \{p\colon\ p > 0$ such that q(t, z) satisfies (3)}. We call infP(q) its minimum period. In this paper, we only consider a QSPS which has a positive minimum period[1].Footnote1

Lemma 1.

If infP(q)>0, then the minimum of P(q) exists (i.e., infP(q)∈P(q)).

Proof.

To see this, first notice that if the minimum of P(q) does not exist, then we can construct a strictly decreasing sequence of positive real numbers $p_n \in \{\mathcal{P}(q)\colon\ n = 1, 2, \ldots\}$ such that $p_n \to \inf\mathcal{P}(q)$, as $n \to \infty$. Next, we will show that $p_m - p_{m+1} \in \mathcal{P}(q)$ for all $m \in \mathbb{N}^+$.

First, since q(t, z) is a QSPS with periods p_m and p_{m+1}, for any $z_0 \in \mathbb{R}$, there exist ϕ_m, ϕ_{m+1}, t_m and t_{m+1} such that

$$q(t, z_0) = q(t - t_{m+1}, z_0 - p_{m+1})e^{j\phi_{m+1}} \tag{4}$$

and

$$q(t, z_0 + p_m - p_{m+1}) = q(t - t_m, z_0 - p_{m+1})e^{j\phi_m}. \quad (5)$$

Now, notice that by (4), we have

$$q(t + t_{m+1} - t_m, z_0)e^{-j\phi_{m+1}} = q(t - t_m, z_0 - p_{m+1}). \quad (6)$$

Substitute (6) into (5), we have

$$q(t, z_0 + p_m - p_{m+1}) = q(t + t_{m+1} - t_m, z_0)e^{j(\phi_m - \phi_{m+1})}. \quad (7)$$

Thus, q(t, z) also has a period of $p_m - p_{m+1}$.

As $\lim_{m\to\infty}(p_m - p_{m+1}) = 0$, this implies infP(q)=0 and a contradiction occurs.

Hence, as long as the infimum of P(q) is positive, it must have a minimum. Throughout this paper, q's minimum period refers to infP(q), while its period is any element in P(q).

Lemma 2.

Let q(t, z) be a QSPS with minimum period p. Then any period of q(t, z) must be an integer multiple of p.

Proof.

Let p_1 be a period of q(t, z). Denote $p_1 = np + l$, where $n \in \mathbb{Z}$, and $0 \leq l < p$. Suppose to the contrary that $l \neq 0$. By Definition 1, it is clear that $np \in P(q)$. So $l \in P(q)$ according to (7), which in turn indicates that l is a period that is smaller than the minimum period. A contradiction thus arises and the lemma is proved.

Properties of Quasi-Spatially Periodic Signals

We now study the properties of QSPSs. In particular, we study an equivalent definition of QSPSs. We also show the composition of their spectra, and how do the eigenvalues relate to the minimum period. We first review some properties of the NFT6.

Proposition 1

(NFT properties6). Let ϕ, t_0 be real constants (ϕ is independent of t, but might depend on z). Denote $u(t, z) = e^{j\phi}q(t, z)$ and $v(t, z) = q(t - t_0, z)$. Then

- (Spatial evolution): $Q^{(c)}(\lambda, z) = e^{-4j\lambda^2 z}Q^{(c)}(\lambda, 0), Q^{(d)}(\zeta_k, z) = e^{-4j\zeta_k^2 z}Q^{(d)}(\zeta_k, 0)$. The eigenvalues ζ_k keep invariant, k=1, 2, ..., N;
- (Constant phase change): The discrete eigenvalues of u(t, z) and q(t, z) are the same, and $U^{(c)}(\lambda, z)=e^{-j\phi}Q^{(c)}(\lambda, z)$, $U^{(d)}(\zeta_k, z)=e^{-j\phi}Q^{(d)}(\zeta_k, z)$;
- (Time shift): The discrete eigenvalues of v(t, z) and q(t, z) are the same, and $V^{(c)}(\lambda, z) = e^{-2j\lambda t_0}Q^{(c)}(\lambda, z), V^{(d)}(\zeta_k, z) = e^{-2j\zeta_k t_0}Q^{(d)}(\zeta_k, z)$;
- (Parseval's identity): $\int_{-\infty}^{\infty} |q(t, 0)|^2 dt = \widehat{E} + \widetilde{E}$, where $\widehat{E} = \frac{1}{\pi}\int_{-\infty}^{\infty} \log(1 + |Q^{(c)}(\lambda, 0)|^2)d\lambda$, $\widetilde{E} = 4\sum_{k=1}^{N}\mathrm{Im}(\zeta_k)$.

Theorem 1.

Suppose q(t, z) is a QSPS with a period p given in (3). Then the time shift t_0 and phase change ϕ with respect to a period p are independent of the initial position z_0.

Proof.

If q(t, z) is a zero signal, then its time shift and phase change are both arbitrary, and hence are independent of the initial position z_0 considered.

If q(t, z) is not the trivial zero signal, then using the first argument of Proposition 1, for any $z_0 \in \mathbb{R}$, we have

$$Q^{(d)}(\zeta_i, z_0 + p) = e^{-4j\zeta_i^2 p}Q^{(d)}(\zeta_i, z_0). \tag{8}$$

Following (3), and using the second and third arguments of Proposition 1, for any $z_0 \in \mathbb{R}$, we have

$$Q^{(d)}(\zeta_i, z_0 + p) = e^{-j\phi}e^{-2j\zeta_i t_0(z_0)}Q^{(d)}(\zeta_i, z_0). \tag{9}$$

Then by (8) and (9), for any $z_0 \in \mathbb{R}$, we have $e^{-4j\zeta_i^2 p} = e^{-j\phi - 2j\zeta_i t_0(z_0)}$, i=1,2,...,N. So for every $z_0 \in \mathbb{R}$ and $\zeta_i \in \mathbb{C}^+$, there exists an $l(i, z_0) \in \mathbb{Z}$ such that $-4j\zeta_i^2 p = -j\phi - 2j\zeta_i t_0(z_0) + j2\pi l(i, z_0)$. Let $\zeta_i \triangleq \alpha_i + j\beta_i$, where $\beta_i > 0, i = 1, 2, \ldots, N$. We have

$$4(\alpha_i^2 - \beta_i^2)p + 8j\alpha_i\beta_i p = \phi + 2\alpha_i t_0(z_0) + 2j\beta_i t_0(z_0) - 2\pi l(i, z_0).$$

As a result, we have

$$4(\alpha_i^2 - \beta_i^2)p = \phi + 2\alpha_i t_0(z_0) - 2\pi l(i, z_0). \tag{10}$$

and $8j\alpha_i\beta_i p = 2j\beta_i t_0(z_0)$, which gives us

$$\alpha_i = \frac{t_0(z_0)}{4p}, \quad i = 1, 2, \ldots, N. \tag{11}$$

Since the eigenvalues ζ_i are invariant during spatial evolution, their real parts α_i and imaginary parts β_i are the same for all z_0. As a result, for any given period p, t_0 is also independent of z_0. Similarly, according to equations (10) and (11), we have

$$\beta_i^2 = -\frac{\phi}{4p} - \frac{t_0^2}{16p^2} + \frac{\pi}{2p} l_i, \tag{12}$$

where $l_i \triangleq l(i, z_0) \in \mathbb{Z}$. So the term $-\frac{\phi}{4p} + \frac{\pi}{2p} l_i$ must be independent of z_0. So $\phi/4p$ is invariant of z_0 in the sense of mod $\pi/2p$, and hence ϕ (mod 2π) is z_0-invariant.

According to Theorem 1, we prove that the time shift t_0 and phase change ϕ do not depend on the initial position $z_0 \in \mathbb{R}$. This result has an important role of easing the identification of the total time shift and phase change at the receiver. For example, as to be shown in (14) proved in the following corollary, rather than an unknown function, we know the timing centre is shifted k times of the time shift for one period of distance p, if the total propagation distance is k multiples of p. Using Theorem 1 we have the following equivalent description of a QSPS.

Corollary 1.

A signal q(t, z) propagated in an optical fibre is a QSPS with a period $p>0$ if and only if there exist real numbers t_0 and ϕ, such that

$$q(t, z_0 + p) = q(t - t_0, z_0)e^{j\phi}, \tag{13}$$

and furthermore,

$$q(t, z_0 + kp) = q(t - kt_0, z_0)e^{jk\phi} \tag{14}$$

for any integer k, and real numbers t and z_0.

Proof.

Equation (13) is a direct consequence of Theorem 1 using Definition 1. Equation (14) can be proved by mathematical induction of (3) because neither t_0 nor ϕ depend on the initial position.

Theorem 2.

Let q(t, z) be a signal propagated in an optical fibre. Suppose it has discrete spectrum, and is denoted by $\{\zeta_1,\zeta_2,\ldots,\zeta_N\}$. Then the following three statements are equivalent:

- q(t, z) is a QSPS with period p (not necessary the minimum period) given in (3);
- The continuous spectrum of q(t, z) is null, and for i=1,2,…,N, its discrete eigenvalues satisfy

$$\mathrm{Re}(\zeta_i) = \frac{t_0}{4p}, \qquad (15)$$

and

$$|\zeta_i|^2 \equiv \frac{2\pi - \phi}{4p} \left(\mathrm{mod}\ \frac{\pi}{2p}\right), \quad \phi \in [0, 2\pi). \qquad (16)$$

- Its continuous spectrum is null. For i=1,2,…,N, the real parts of the eigenvalues are identical as given in (15), and for each pair of m≠n,

$$a_{mn} \triangleq \frac{2p\,|\mathrm{Im}(\zeta_m)^2 - \mathrm{Im}(\zeta_n)^2|}{\pi} \qquad (17)$$

are positive integers.

Furthermore, suppose that q(t, z) satisfies 1) and hence also 3). Then p is the minimum period if and only if

$$\gcd(a_{mn}\colon\ m \neq n,\ m, n = 1, 2, \ldots, N) = 1. \qquad (18)$$

Proof.

We first prove that 1) implies 2). To show the continuous spectrum is null, we only need to prove that $Q^{(c)}(\lambda, 0) \equiv 0$, for all $\lambda \in \mathbb{R}$. Similar to the method of obtaining (8) and (9) (following (3) and Proposition 1), we have

$$Q^{(c)}(\lambda, p) = e^{-4j\lambda^2 p} Q^{(c)}(\lambda, 0) \qquad (19)\ \lambda \in \mathbb{R}$$

$$= e^{-j\phi} e^{-2j\lambda t_0} Q^{(c)}(\lambda, 0). \qquad (20)$$

Suppose to the contrary that $Q^{(c)}(\lambda, 0)$ is not zero for all $\lambda \in \mathbb{R}$, i.e., there exists a real number λ0∈Rλ0∈R such that $Q^{(c)}(\lambda_0, 0) \neq 0$. Since $Q^{(c)}$

$(\lambda, 0)$ is a continuous function in $\lambda \in \mathbb{R}$, there exists an interval (x_1, x_2) containing λ_n such that $Q^{(c)}(\lambda, 0) \neq 0$, $\lambda \in (x_1, x_2)$. Then by (19) and (20), $e^{-4j\lambda^2 p} = e^{-j\phi - 2j\lambda t_0}$, $\lambda \in (x_1, x_2)$. On the other hand, for every $\lambda \in (x_1, x_2)$, there exists an $m(\lambda) \in \mathbb{Z}$ such that $-4j\lambda^2 p = -j\phi - 2j\lambda t_0 + j2\pi m(\lambda)$. So, we have

$$\left(\lambda - \frac{t_0}{4p}\right)^2 = \frac{\phi}{4p} - \frac{\pi}{2p} m(\lambda) + \frac{t_0^2}{16p^2},$$

or equivalently,

$$\frac{2p}{\pi}\left(\lambda^2 - \frac{t_0}{2p}\lambda - \frac{\phi}{4p}\right) = -m(\lambda). \tag{21}$$

However, one can easily find a $\lambda \in (x_1, x_2)$ such that the left hand side of (21) is not an integer (could even be an irrational number). In that case, there does not exist any integer $m(\lambda)$ satisfying (21), leading to a contradiction. Consider its discrete spectrum, for i=1,2,…,N, equation (15) is proved in Theorem 1 (as in (11)). According to Theorem 1, using (12) and (15), we have (16).

Next, we prove that 2) implies 3). According to (16), we have (12). So we have

$$\mathrm{Im}(\zeta_m)^2 - \mathrm{Im}(\zeta_n)^2 = \beta_m^2 - \beta_n^2 = \frac{\pi}{2p}(l_m - l_n).$$

Using (17), we have

$$a_{mn} = \frac{2p\left|\mathrm{Im}(\zeta_m)^2 - \mathrm{Im}(\zeta_n)^2\right|}{\pi} = |l_m - l_n| \geq 0.$$

Since $l_m \neq l_n$ if $\mathrm{Im}(\zeta_m) \neq \mathrm{Im}(\zeta_n)$ according to (12), 3) is proved.

Now, we prove that 3) implies 1). If the spectrum of a signal q(t, z) satisfies (15) and (17), we have

$$e^{4j\mathrm{Im}(\zeta_m)^2 p} = e^{4j\mathrm{Im}(\zeta_1)^2 p}, \quad m = 1, 2, \ldots, N. \tag{22}$$

Denote $x(t, z) \triangleq e^{-4j|\zeta_1|^2 p} q(t - t_0, z),\ t \in \mathbb{R},\ z \in \mathbb{R}$. Then according to Proposition 1, x(t, z) and q(t, z) have the same discrete eigenvalues because the former is just a constant phase change and time shift of the latter. In addition, for i=1,2,…,N,

$$X^{(d)}(\zeta_i, z) = e^{4j|\zeta_1|^2 p} e^{-2j\zeta_i t_0} Q^{(d)}(\zeta_i, z)$$
$$= e^{4j[\mathrm{Im}(\zeta_1)^2 - \mathrm{Re}(\zeta_i)^2 - 2j\mathrm{Re}(\zeta_i)\mathrm{Im}(\zeta_i)]p} Q^{(d)}(\zeta_i, z) \quad (23)$$

$$= e^{4j[\mathrm{Im}(\zeta_i)^2 - \mathrm{Re}(\zeta_i)^2 - 2j\mathrm{Re}(\zeta_i)\mathrm{Im}(\zeta_i)]p} Q^{(d)}(\zeta_i, z)$$
$$= Q^{(d)}(\zeta_i, z + p) \quad (24)$$

where (23) is true according to (15), and (24) is obtained by (22). Since its continuous spectrum is null, we have

$$q(t, z + p) = x(t, z) = e^{-4j|\zeta_1|^2 p} q(t - t_0, z)$$

for any $t \in \mathbb{R}$, and $z \in \mathbb{R}$. So q(t, z) is a QSPS with period p according to Definition 1.

Now, it remains to prove that p is minimum if and only if (18) holds. Hence, p is the minimum period if and only if p is the smallest positive real number such that

$$a_{mn} \triangleq \frac{2p|\mathrm{Im}(\zeta_m)^2 - \mathrm{Im}(\zeta_n)^2|}{\pi}$$

are integers for all distinct m, n. Let $x \triangleq \gcd(a_{mn}: m \neq n, m, n = 1, 2, \ldots, N)$. So for any distinct m, n,

$$a'_{mn} \triangleq \frac{a_{mn}}{x} = \frac{2(p/x)\,|\mathrm{Im}(\zeta_m)^2 - \mathrm{Im}(\zeta_n)^2|}{\pi}$$

is still an integer. Consequently, if p is the minimal period, then x=1. On the other hand, if p is not the minimum period, then by Lemma 2, p=cp′ where p′ is the minimum period and c is an integer greater than 1. It can be verified directly that x=c>1. Hence, we prove the theorem.

Theorem 2 analyses the spectral components of a QSPS, which is rather simple. It lets us know that a QSPS is equivalent to a spectral domain signal with null continuous spectrum, and its discrete eigenvalues are located in a special manner. Because of the null continuous spectrum, Darboux transform7 is enough to convert the spectral domain signal to its time domain representation, avoiding the computational more expensive Riemann-Hilbert system6. This theorem leads us explicitly how to design a QSPS, and check whether a spectral domain signal is a QSPS. To be more specific, Theorem 2 implies that all QSPSs are N-solitons. However, its converse is not true.

Remark 1.

According to Theorem 2, the eigenvalues of a QSPS should satisfy

$$\frac{|\mathrm{Im}(\zeta_m)^2 - \mathrm{Im}(\zeta_n)^2|}{|\mathrm{Im}(\zeta_l)^2 - \mathrm{Im}(\zeta_m)^2|} = \frac{a_{mn}}{a_{lm}} \in \mathbb{Q} \tag{25}$$

for any $l \neq m \neq n$, $l,m,n \in \{1,2,\ldots,N\}$ when $N \geq 3$. Equation (25) gives us an insight that the eigenvalues cannot be chosen arbitrarily (even if all their real parts are the same) if quasi-spatially periodic property is required when there are more than two discrete eigenvalues.

Example 1

(A non-quasi-spatially-periodic N-soliton). Let $N=3$, and $\zeta_1 = \sqrt[3]{2}j, \zeta_2 = 2j, \zeta_3 = 3j$. We have

$$\frac{|\mathrm{Im}(\zeta_2)^2 - \mathrm{Im}(\zeta_3)^2|}{|\mathrm{Im}(\zeta_1)^2 - \mathrm{Im}(\zeta_2)^2|} = \frac{5}{4 - \sqrt[3]{4}} \notin \mathbb{Q}.$$

As a result, $\{\sqrt[3]{2}j, 2j, 3j\}$ is not the spectrum of any QSPS.

A Spectral Domain Interpretation of Quasi-Periodicity

So far, we define the quasi-periodicity of a QSPS in the time domain (Definition 1 and Corollary 1) as that the signal maintains its shape quasi-periodically with a phase change and a time shift. In the following, we study the quasi-periodicity in the spectral domain, showing that phase changes of all the spectral amplitudes agree with each other periodically.

Definition 2.

The "spectral period" induced by two distinct eigenvalues of a QSPS is one of the distances at which the phase changes of these two corresponding discrete spectral amplitudes in their spectral domain spatial evolution are identical. More precisely, the spectral periods induced by two distinct eigenvalues (ζ_m and ζ_n, $m \neq n$) are elements of the set

$$\{z > 0:\ e^{-4j\mathrm{Re}(\zeta_m^2)z} = e^{-4j\mathrm{Re}(\zeta_n^2)z}\}. \tag{26}$$

We denote the minimum spectral period of ζ_m and ζ_n by $p(\zeta_m,\zeta_n)$.

Similar to the discussion after Definition 1, we can show that $p(\zeta_m,\zeta_n)$ exists for any $m \neq n$.

Corollary 2.

Phase changes of any two spectral amplitudes of a QSPS during the spatial evolution along an optical fibre coincide periodically. Specifically, the minimum spectral period for ζ_m and ζ_n ($m \neq n$) is

$$p(\zeta_m, \zeta_n) = \frac{\pi}{2|\mathrm{Im}(\zeta_m)^2 - \mathrm{Im}(\zeta_n)^2|} = \frac{p}{a_{mn}}, \tag{27}$$

where p is the minimum period when (18) is satisfied.

Proof.

According to Proposition 1 and (26), for $m \neq n$, the corresponding spectral amplitudes have the same phase change when $e^{-4j[\mathrm{Re}(\zeta_m)^2 - \mathrm{Im}(\zeta_m)^2]z} = e^{-4j[\mathrm{Re}(\zeta_n)^2 - \mathrm{Im}(\zeta_n)^2]z}$, where $z \in \mathbb{R}$, which is equivalent to $4\mathrm{Im}(\zeta_m)^2 z \equiv 4\mathrm{Im}(\zeta_n)^2 z \pmod{2\pi}$ according to Theorem 2, which is further equivalent to

$$z \equiv 0 \left(\mathrm{mod}\ \frac{\pi}{2|\mathrm{Im}(\zeta_m)^2 - \mathrm{Im}(\zeta_n)^2|} \right).$$

According to Theorem 2, we have (27), where p is the minimum spectral period when (18) is satisfied.

As a simple observation from Corollary 2, if a QSPS only has two eigenvalues ζ_1 and ζ_2 with minimum period p, we have $p=p(\zeta_1,\zeta_2)$, because $a_{12}=1$ observed from (18). This also shows that the period of a QSPS (in the spectral domain) is the minimal integer multiple of the minimum spectral period for any pair of eigenvalues.

Trade-off between Energy, Minimum Period and Degree-of-Freedom

So far, we have studied the properties of QSPSs. In particular, we obtain the relationship between the eigenvalues and the minimum period of a QSPS. In the following, we also take energy into consideration, and consider the signal design for QSPSs. Specifically, we derive a trade-off between various parameters including minimum period, signal energy and degree-of-freedom. Note that we refer degree-of-freedom of a QSPS as the number of its discrete eigenvalues, as it is often the number of users allowed for interference free multiuser communications under the NFDM scheme. In this subsection, we call a QSPS with N eigenvalues an N-QSPS.

Theorem 3.

Let q(t, z) be an N-QSPS with minimum period p and energy E. If E is large enough (E satisfies (29)), there exists a positive real number β_1 (which is the smallest among all the imaginary parts of the eigenvalues) and integers as,s=1,2,…,N−1 such that

$$\frac{E}{4} = \beta_1 + \sum_{k=2}^{N} \sqrt{\beta_1^2 + \frac{\pi}{2p}\sum_{s=1}^{k-1} a_s} . \qquad (28)$$

Proof.

Let the discrete spectrum of q(t, z) be $\{\zeta_1,\zeta_2,\ldots,\zeta_N\}$. By Theorem 2, all the discrete eigenvalues have the same real parts. Therefore, we denote that $\zeta_k=\alpha+j\beta_k$ with $\beta_k>0$ for k=1,…,N. In fact, we assume without loss of generality that $\beta_1<\beta_2<\cdots<\beta_N$. Let $a_s = \frac{2p(\beta_{s+1}^2 - \beta_s^2)}{\pi}, s = 1, 2, \ldots, N-1$. By Theorem 2, a_s is an integer. In addition, $\beta_k^2 - \beta_1^2 = (\sum_{s=1}^{k-1} a_s)\frac{\pi}{2p}$ for $k = 2, 3, \ldots, N$. Hence,

$$\beta_k = \sqrt{\beta_1^2 + \frac{\pi}{2p}\sum_{s=1}^{k-1} a_s}, \quad k = 2, 3, \ldots, N.$$

Finally, (28) follows from Proposition 1.

When we design an optical system, N is the degree-of-freedom and can be regarded as the number of subchannels. Roughly speaking, the larger the N is, the more information can be transmitted across to the receiver. On the other hand, since p is the minimum period, the transmitter and the receiver must be separated by a distance of an integer multiple of p. Therefore, it is desirable to minimise p for the sake of flexibility. The natural question thus is: Can we design a system which minimises p and maximises N simultaneously? Unfortunately, the answer is no according to the next corollary.

Corollary 3

(Trade-off). Let q(t, z) be an N-QSPS with minimum period p, energy E and degree-of-freedom N. Then

$$E > 4\sqrt{\frac{\pi}{2p}}\left(\sum_{k=1}^{N-1}\sqrt{k}\right). \tag{29}$$

Consequently, we cannot design a communication system which minimises minimum period, and maximises the number of subchannels at the same time using QSPSs.

Proof.

Let q(t, z) be quasi-spatially periodic with minimum period p, energy E and N eigenvalues. According to Theorem 3 (equation (28)), E can be minimised by minimising β_1 and also a_s. Clearly, it is required that $\beta_1>0$ and $a_1,\ldots,a_{N-1}$ satisfying (18). Let $a_s=1$ for all s=1,2,...,N−1. It can be checked easily that (29) is true. In fact, for any $\varepsilon>0$, one can easily construct a QSPS such that

$$E = 4\sqrt{\frac{\pi}{2p}}\left(\sum_{k=1}^{N-1}\sqrt{k}\right) + \varepsilon.$$

Theorem 3 tells us that constructing a QSPS with minimum period p and energy E is equivalent to solving the equation (28). In fact, we can always find an N-QSPS with minimum period p and energy E as long as (29) holds.

When N=2, the choice of the imaginary parts of the eigenvalues is in fact unique:

$$\beta_1 = j\left(\frac{1}{8}E - \frac{\pi}{pE}\right), \quad \beta_2 = j\left(\frac{1}{8}E + \frac{\pi}{pE}\right)$$

if and only if (29) holds for N=2. However, the uniqueness may not hold in general when N>2.

Theorem 3 provides us a starting point of designing QSPSs. Actually, besides minimum period, energy and degree-of-freedom, there are many other factors affecting signal design. If noise is taken into consideration, we often do not expect the SNR to be low. So the β_1 in (28) should not be chosen very small. In addition, since the signal detection at the receiver relies on the quasi-periodicity of the signal, the position of the detector is subtle. As a result, we expect signals whose shape do not change a lot within a small difference of position. Moreover, we prefer signals to have short duration to increase the efficiency of the communication system, which is also relevant to the choice of signal shape. Theorem 3 provides us an insight that we still have much choices to find optimal pulse shapes for information transmission to obtain a better performance even if the minimum period and energy are fixed when the signals have more than two eigenvalues.

Pulse Truncation

So far, we have studied properties and design of QSPSs. In practice, signals are always time-limited (i.e., with finite duration). Therefore, when time-unlimited QSPSs are transmitted in practice, they are often "truncated". However, truncation causes the discrete eigenvalues to be distorted leading to that nontrivial continuous spectrum would appear. So it is interesting to study how the "quasi-periodicity" of the effect QSPSs will be affected by truncation.

To achieve the goal, we ran a simulation for the noiseless propagation of a "truncated QSPS". In this example, we consider a QSPS which has eigenvalues {j, 2j} and corresponding spectral amplitudes both equal to 1. Also, its period is p=π6≈0.52360p=π6≈0.52360, according to Theorem 2. We first implement the INFT using the Darboux transformation8 to obtain the time domain representation of the signal with the time support [−8, 8] and denote it as q(t, 0), whose magnitude is plotted in Figure 1. We then truncate q(t, 0) such that it is strictly time limited in [−2.5, 2.5]. Denote the truncated pulse by $q_{trun}(t, 0)$. The truncated pulse $q_{trun}(t, 0)$ is then sent to a simulated fibre of normalized distance L=2. Note also that L>3p. Then we compute the NFT of $q_{trun}(t, 0)$, and its eigenvalues are (0.998j, 1.999966j). The result agrees with15 that the smaller eigenvalue is distorted more severely than the bigger one. According to Theorem 2, the period of $q_{trun}(t, 0)$ is p_{trun}=0.52297.

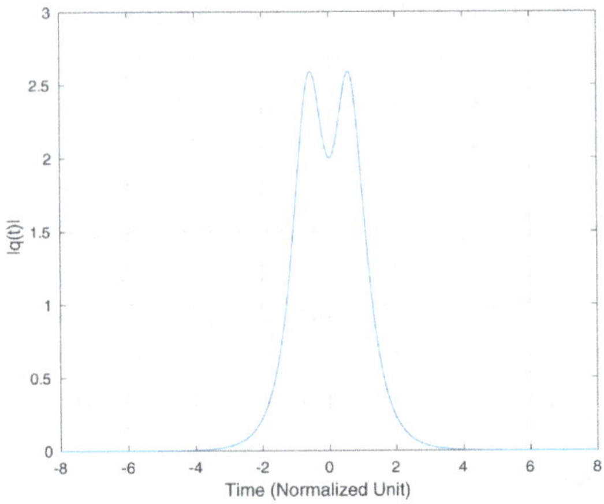

Figure 1: Magnitude of the QSPS q(t, 0) with Eigenvalues (j, 2j).

As before, we let $q_{trun}(t, z)$ and $q(t, z)$ be the signals obtained by respectively propagating $q_{trun}(t, 0)$ and $q(t, 0)$ for a distance of z along the fibre. Then Figure 2 is the heatmap of $|q_{trun}(t, z)|$, showing that the "truncated QSPS" is roughly quasi-periodic. Figure 3 shows the absolute difference $|q_{trun}(t, z)-q(t, z)|$ when $z=kp_{trun}$ for k=1, 2, 3. In most of the times, the difference is within −10 dB. Our simulations also show that further truncation of q(t, 0) will increase the error.

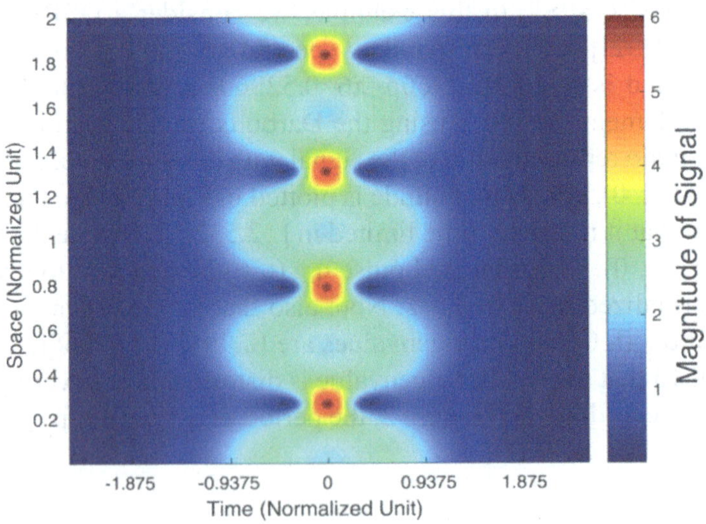

Figure 2: Heatmap of the Magnitude of the Truncated QSPS with Eigenvalues (j, 2j) as Propagated.

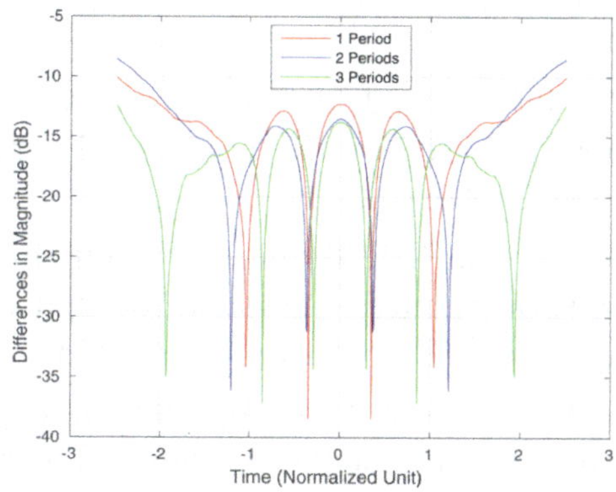

Figure 3: Difference Between the Pulse in Each Period of Distance and the Transmitted Truncated QSPS with Eigenvalues (j, 2j).

DISCUSSION

In this paper, we study a class of signals, which have the quasi-spatially periodic property. We show that all QSPSs have a null continuous spectrum, and the property of quasi-spatially periodic only depends on their eigenvalues, but not spectral amplitudes. The results are of theoretical importance because they help understand the structure of their spectra for further use as input signals in optical fibre communication systems.

In addition, the property of quasi-spatially periodic motivates us a novel idea of decoding that could possibly replace the costly NFT. Specifically, the minimum period of a QSPS could be designed beforehand such that the same shape is received at the output with only a phase change and possibly a time shift. In high SNR regime, it is expected that time domain signal processing technique replace the role of the NFT for decoding with the knowledge of the shapes of signals. Having the property of quasi-spatially periodic, it is beneficial to achieve this goal for optical fibre communications.

METHODS

Mathematical derivation and proofs, as well as numerical simulation are the methods used in this paper. Proofs are given for all results presented in this paper.

ACKNOWLEDGEMENTS

This work was supported in part by ARC Discovery Project DP150103658.

REFERENCES

1. Derevyanko, S., Prilepsky, J. & Turitsyn, S. Capacity estimates for optical transmission based on the nonlinear Fourier transform. Nat. Commun. 7, 12710 (2016).
2. Kinsel, T. S. & Denton, R. T. Terminals for a high-speed optical pulse code modulation communication system: II. Optical multiplexing and demultiplexing. Proc. IEEE 56, 146–154 (1968).
3. Bergano, N. S. & Davidson, C. R. Wavelength division multiplexing in long-haul transmission systems. J. Lightw. Technol. 14, 1299–1308 (1996).
4. Essiambre, R. J., Kramer, G., Winzer, P. J., Foschini, G. J. & Goebel, B. Capacity limits of optical fiber networks. J. Lightw. Technol. 28, 662–701 (2010).
5. Armstrong, J. OFDM for optical communications. J. Lightw. Technol. 27, 189–204 (2009).
6. Yousefi, M. I. & Kschischang, F. R. Information transmission using the nonlinear Fourier transform, Part I: Mathematical tools. IEEE Trans. Inf. Theory 60, 4312–4328 (2014).
7. Yousefi, M. I. & Kschischang, F. R. Information transmission using the nonlinear Fourier transform, Part II: Numerical methods. IEEE Trans. Inf. Theory 60, 4329–4345 (2014).
8. Yousefi, M. I. & Kschischang, F. R. Information transmission using the nonlinear Fourier transform, Part III: Spectrum modulation. IEEE Trans. Inf. Theory 60, 4346–4369 (2014).
9. Tao, T. & Thiele, C. Nonlinear Fourier analysis. CoRR, http://arxiv.org/abs/1201.5129 (2012).
10. Zakharov, V. E. & Shabat, A. B. Exact theory of two-dimensional self-focusing and one-dimensional self-modulation of waves in nonlinear media. Soviet Physics-JETP 34, 62–69 (1972).
11. Wahls, S. & Poor, H. V. Inverse nonlinear Fourier transforms via interpolation: The Ablowitz-Ladik case. In 21st International Symposium on Mathematical Theory of Networks and Systems (MTNS 2014) (Groningen, The Netherlands, 2014).
12. Prilepsky, J. E., Derevyanko, S. A., Blow, K. J., Gabitov, I. & Turitsyn, S. K. Nonlinear inverse synthesis and eigenvalue division multiplexing in optical fiber channels. Phys. Rev. Lett. 113, 013901 (2014).

13. Le, S. T., Prilepsky, J. E. & Turitsyn, S. K. Nonlinear inverse synthesis for high spectral efficiency transmission in optical fibers. Opt. Express 22, 26720–26741 (2014).
14. Le, S. T. et al. Demonstration of nonlinear inverse synthesis transmission over transoceanic distances. J. Lightwave Technol. 34, 2459–2466 (2016).
15. Hari, S., Yousefi, M. I. & Kschischang, F. R. Multieigenvalue communication. J. Lightw. Technol. 34, 3110–3117 (2016).
16. He, G., Wang, L., Li, C., Liu, S. & Hu, W. Spectral function modulation based on nonlinear frequency division multiplexing. Sci. Rep. 7, 6058 (2017).
17. Agrawal, G. P. Nonlinear Fiber Optics, Fifth Edition (Springer, 2013).
18. Hasegawa, A. & Kodama, Y. Guilding-center soliton. Phys. Rev. Lett. 66, 161–164 (1991).
19. Ablowitz, M. J. & Segur, H. Solitons and the Inverse Scattering Transform (Society for Industrial and Applied Mathematics (SIAM), 2000).
20. Ablowitz, M. J., Prinari, B. & Trubatch, A. D. Discrete and Continuous Nonlinear Schrödinger Systems (Cambridge University Press, 2003).
21. Wahls, S. & Poor, H. V. Introducing the fast nonlinear Fourier transform. In 2013 IEEE International Conference on Acoustics, Speech and Signal Processing (ICASSP 2013) (Vancouver, BC, Canada, 2013).
22. Wahls, S. & Poor, H. V. Fast numerical nonlinear Fourier transforms. IEEE Trans. Inf. Theory 61, 6957–6974 (2015).
23. Wahls, S. & Poor, H. V. Fast inverse nonlinear Fourier transform for generating multi-solitons in optical fiber. In 2015 IEEE International Symposium on Information Theory (ISIT 2015) (Hong Kong, China, 2015).
24. Gui, T., Chan, T. H., Lu, C., Lau, A. P. T. & Wai, P.-K. A. Alternative decoding methods for optical communications based on nonlinear fourier transform. J. Lightwave Technol. 35, 1542–1550 (2017).
25. Kivshar, Y. & Agrawal, G. P. Optical Solitons: From Fibers to Photonic Crystals (Academic Press, 2003).
26. Meron, E., Feder, M. & Shtaif, M. On the achievable communication rates of generalized soliton transmission systems. CoRR, http://arxiv.org/abs/1207.0297v2 (2012).

27. Derevyanko, S. A., Turitsyn, S. K. & Yakushev, D. A. Non-Gaussian statistics of an optical soliton in the presence of amplified spontaneous emission. Opt. Lett. 28, 2097–2099 (2003).
28. Derevyanko, S. A., Turitsyn, S. K. & Yakushev, D. A. Fokker-Planck equation approach to the description of soliton statistics in optical fiber transmission systems. J. Opt. Soc. Am. B 22, 743–752 (2005).
29. Zhang, Q. & Chan, T. H. A spectral domain noise model for optical fibre channels. In 2015 IEEE International Symposium on Information Theory (ISIT 2015) (Hong Kong, China, 2015).
30. Zhang, Q. & Chan, T. H. A Gaussian noise model of spectral amplitudes in soliton communication systems. In 16th IEEE International Workshop on Signal Processing Advances in Wireless Communications, (SPAWC 2015) (Stockholm, Sweden, 2015).
31. Shevchenko, N. A. et al. A lower bound on the per soliton capacity of the nonlinear optical fibre channel. In 2015 IEEE Information Theory Workshop (ITW 2015) (Jeju, South Korea, 2015).
32. Zhang, Q. & Chan, T. H. Achievable rates of soliton communication systems. In 2016 IEEE International Symposium on Information Theory (ISIT 2016) (Barcelona, Spain, 2016).
33. Gordon, J. P. & Haus, H. A. Random walk of coherently amplified solitons in optical fiber transmission. Opt. Lett. 11, 665–667 (1986).
34. Hari, S. & Kschischang, F. R. Bi-directional algorithm for computing discrete spectral amplitudes in the NFT. J. Lightw. Technol. 34, 3529–3537 (2016).
35. Zhang, Q., Chan, T. H. & Grant, A. Spatially periodic signals for fiber channels. In 2014 IEEE International Symposium on Information Theory (ISIT 2014) (Honolulu, USA, 2014).

3

CP-based OFDM Radar–communications Signal Using Interval Linear Phase Compression Modulation

Yuanhang Wang; Tianxian Zhang; Lingjiang Kong; Xiaobo Yang [1]

School of Information and Communication Engineering, University of Electronic Science and Technology of China Chengdu , People's Republic of China

ABSTRACT

In this study, a signal design algorithm for radar–communications system is investigated. An arbitrary pulse length orthogonal frequency-division multiplexing with sufficient cyclic prefix (CP-based OFDM) signal is designed as the radar–communications signal, the communication symbols are embedded by proposing an interval linear phase compression (ILPC) modulation. Firstly, the authors construct the model of CP-based OFDM signal, and analyse the impact of the communication symbols embedding. Then, to decrease the impact caused by the communication symbols embedding, the ILPC modulation is proposed. To guarantee the performances of the CP-based OFDM radar–communications signal, a time/frequency

Citation: Wang, Yuanhang; Zhang, Tianxian; Kong, Lingjiang; Yang, Xiaobo: "CP-based OFDM radar–communications signal using interval linear phase compression modulation", The Journal of Engineering, 2019, 2019, (21), p. 7902-7906, https://doi.org/10.1049/joe.2019.0768

domain iterative filtering and clipping method is used. This algorithm can be used to design a CP-based OFDM signal that has a balanced performance in communications and radar. Finally, numerical results are provided to verify the validity of the proposed algorithm.

INTRODUCTION

Recently, the coexistence of radar and communications has been the focus of research due to its advantages of reducing equipment size and improving resource utilisation [1 – 3]. However, the implementation of radar–communications system is subject to the design of radar–communications signal. The radar–communications signal can solve the competition over a frequency spectrum between radar and communications and achieve the sharing of the same resources from a single platform [4 – 6]. Thus, it is necessary to design a well-behaved radar–communications signal. In these years, some researchers have focused on the design of radar–communications signal based on orthogonal frequency-division multiplexing (OFDM), which have been recently introduced in a number of papers [5 – 8]. However, the impact of the communication symbols embedding is still not fully considered and the feature of OFDM signals with sufficient cyclic prefix (CP) has not been considered [8] proposed an arbitrary pulse length CP-based OFDM pulse where the CP part is all zeros, the inter-range-cell interference (IRCI) free and ideally zero range sidelobes for range reconstruction can be obtained by using it. However, it is designed for radar without considering the communication function. Thus, the potential of CP-based OFDM signals is still not fully exploited as a radar–communications signal.

As we know, for communication applications, multipath effects will lead to inter-symbol interference (ISI). For radar applications, IRCI will decrease the performances of range reconstruction. These problems can be solved by designing a CP-based OFDM radar–communications signal. In [8], the CP-based OFDM pulse has no restrictions on the phase of the complex weights transmitted over the subcarriers, which provides an opportunity for communication symbols embedding. Thus, in theory, it is feasible to design a CP-based OFDM radar–communications signal. However, the traditional communication symbols embedding method by phase-shift keying (PSK) limits the degree of freedom of the CP-based OFDM pulse strictly, so that the requirements for peak-to-average power ratio (PAPR) and the constant module of the complex weights transmitted over the subcarriers are hard to satisfy at the same time, which will affect the performance of CP-based

OFDM signal seriously. Therefore, the design of CP-based OFDM signal for radar–communications system is urgently needed. However, to the best of the author's knowledge, the design algorithm for the CP-based OFDM radar–communications pulses is not yet available in the open literature. This is indeed the main topic of this paper.

Unlike the work [8] which design the CP-based OFDM pulses for radar without considering the communication function, in this paper, for the radar–communications system, the design algorithm for the CP-based OFDM radar–communications pulses is investigated. Firstly, we construct the model of CP-based OFDM signal for radar–communications system, and evaluate the impact of the communication symbols embedding. Then, to balance the impact of the communication symbols embedding, the interval linear phase compression (ILPC) modulation is proposed. To guarantee the performances of the CP-based OFDM radar–communications pulses, a time/frequency domain iterative filtering and clipping method is used. Finally, numerical results are provided to verify the validity of the proposed algorithm.

The remainder of this paper is organised as follows. In Section 2, we construct the model of CP-based OFDM signal for radar–communications system and describe the problem of interest. Then, we propose the communication symbols embedding method by ILPC, and the corresponding CP-based OFDM radar–communications signal design flow based on ILPC and time/frequency domain iterative filtering and clipping method is given. In Section 3, we show some simulation results. In Section 4, we conclude this paper.

METHODOLOGY

CP-based OFDM Radar–communications Signal Model

Consider an OFDM signal with N subcarriers and bandwidth of B Hz. Let $\boldsymbol{S} = [S_0, S_1, \ldots, S_{N-1}]^{\mathrm{T}}$ represent the complex weights transmitted over the subcarriers, $(\cdot)^{\mathrm{T}}$ denotes the transpose, and $\sum_{k=0}^{N-1} |S_k|^2 = 1$. Inverse fast Fourier transform (IFFT) for the vector $\boldsymbol{S}$, after the A/D converter we can get the corresponding time-domain OFDM signal as follows:

$$\tilde{s}(t) = \frac{1}{\sqrt{N}} \sum_{k=0}^{N-1} S_k \exp\{j2\pi k \Delta f t\}, \quad t \in [0, T + T_{CP}], \tag{1}$$

where $\Delta f = B/N = 1/T$ is the subcarrier spacing, T is the time duration of OFDM signal without CP, and T_{CP} is the time duration of CP. From [8] we know that to avoid energy waste caused by adding a CP and achieve an arbitrary length of the transmitted pulse, $\tilde{s}(t)$ should satisfy the zero CP as follows:

$$\tilde{s}(t) = 0, t \in [0, T_{CP}] \cup [T, T + T_{CP}]. \tag{2}$$

A time filtering is used to guarantee zero CP.

Notice that the design signal in [8] is just suitable for radar, but not appropriate for communications. Consider a radar–communications system using the $\tilde{s}(t)$ in (1) as shown in Fig. 1 .

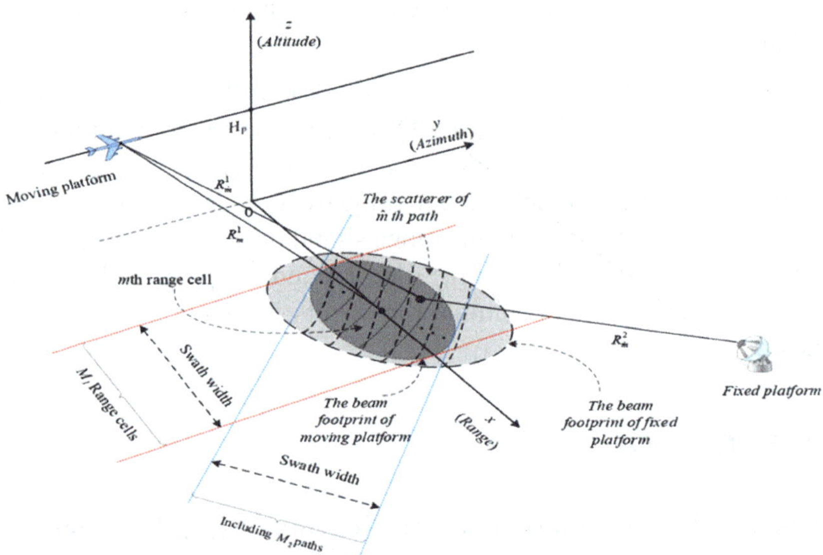

Figure *1*: Simple diagram of radar–communications system.

This system consists of a moving platform and a fixed platform, the moving platform is responsible for the transmission of radar–communications signal and the reception of radar echo signal, the fixed platform is responsible for the reception of a communication signal. To achieve the radar–communication functions, the communication symbols are needed to be embedded by a modulation function. After modulation, the transmitted signal $s(t)$ can be expressed as follows:

$$s(t) = \frac{1}{\sqrt{N}}\sum_{k=0}^{N-1} F(S_k)\exp\{j2\pi k\Delta f\}, \quad t \in [0, T + T_{CP}], \tag{3}$$

where s(t) still satisfy the zero CP constraint as (2), F(·) represents the modulation function, $\sum_{k=0}^{N-1}|F(S_k)|^2 = 1$. From (2), we notice that, despising the part of the signal with s(t)=0 , the actual time duration of $s(t)$ is $T-T_{CP}$.

Assume that the swath width for the moving platform can be divided into M1 range cells. For the receiver of the moving platform, the received signal of radar function from M1 range cells in the swath can be expressed as

$$\begin{aligned} r(t) &= \sum_{m=0}^{M_1-1} r_m(t) \\ &= \frac{1}{\sqrt{N}}\sum_{m=0}^{M_1-1} g_m \exp\left\{-j4\pi f_c \frac{R_m^1}{c}\right\} \\ &\quad \times \sum_{k=0}^{N-1} F(S_k)\exp\left\{\frac{j2\pi k}{T}\left[t - \frac{2R_m^1}{c}\right]\right\} \\ &\quad + \bar{w}(t), \quad t \in \left[\frac{2R_0^1}{c}, \frac{2R_{M_1-1}^1}{c} + T - T_{CP}\right], \end{aligned} \tag{4}$$

where f_c is the carrier frequency, g_m is the radar cross-section (RCS) coefficient caused from the scatterers in the mth range cell within the radar beam footprint, R_m^1 is the instantaneous slant range between the moving platform and the mth range cell, c is the speed of light, and $\bar{w}(t)$ is the noise. According to [8], M_1 range cells correspond to M_1 paths.

Assume that there are M_2 paths in the communication channel between the moving platform and the fixed platform. Similar to (4), for the communication receiver of the fixed platform, the received signal of communication from M_2 paths can be expressed as

$$\tilde{r}(t) = \sum_{\tilde{m}=0}^{M_2-1} r_{\tilde{m}}(t)$$

$$= \frac{1}{\sqrt{N}} \sum_{\tilde{m}=0}^{M_2-1} h_{\tilde{m}} \exp\left\{-j2\pi f_c \frac{(R_{\tilde{m}}^1 + R_{\tilde{m}}^2)}{c}\right\}$$

$$\times \sum_{k=0}^{N-1} F(S_k) \exp\left\{\frac{j2\pi k}{T}\left[t - \frac{(R_{\tilde{m}}^1 + R_{\tilde{m}}^2)}{c}\right]\right\}$$

$$+w(t), \quad t \in \left[\frac{R_0^1 + R_0^2}{c}, \frac{R_{M_2-1}^1 + R_{M_2-1}^2}{c} + T - T_{\mathrm{CP}}\right], \tag{5}$$

where $h_{\tilde{m}}$ is the channel coefficient caused from the scatterers in the $\tilde{m}$th path, $R_{\tilde{m}}^1$ is the instantaneous slant range between the moving platform and the scatterer cell of the $\tilde{m}$th path, $R_{\tilde{m}}^2$ is the instantaneous slant range between fixed platform and the scatterer cell of the $\tilde{m}$th path.

We know that M_1 range cells in radar correspond to M_1 paths. To avoid the IRCI between different range cells in radar, the time duration of CP should satisfy $T_{\mathrm{CP}} \geq (M_1 - 1)T_s$ [8], where $T_s = 1/B$ is sampling interval length for A/D converter. Similarly, to avoid the ISI between different paths in communications, the time duration of CP should satisfy $T_{\mathrm{CP}} \geq (M_2 - 1)T_s$. Thus, the time duration of CP for radar–communications signal s(t)st should be at least equal to

$$T_{\mathrm{CP}} \geq (M - 1)T_s \tag{6}$$

where $M=\max\{M_1,M_2\}$. At the receiver of a fixed platform, the received communication signal is sampled by the A/D converter, after fast Fourier transform (FFT), the received communication signal can be expressed as

$$\tilde{\boldsymbol{Y}} = \boldsymbol{\Lambda}\tilde{\boldsymbol{S}} + \boldsymbol{W} = \begin{pmatrix} \alpha_0 F(S_0) + W_0 \\ \alpha_1 F(S_1) + W_1 \\ \vdots \\ \alpha_{N-1} F(S_{N-1}) + W_{N-1} \end{pmatrix}, \tag{7}$$

where Λ is the diagonal matrix of eigenvalues of the channel matrix $\tilde{\boldsymbol{H}}$ [9], $\alpha_0, \alpha_1, \ldots, \alpha_{N-1}$ are eigenvalues of the channel matrix $\tilde{\boldsymbol{H}}$, $\tilde{\boldsymbol{S}} = [F(S_0), F(S_1), \ldots, F(S_{N-1})]^{\mathrm{T}}$, $\boldsymbol{W} = [W_0, W_1, \ldots, W_{N-1}]^{\mathrm{T}}$ is the noise vector. This result shows that ISI caused by multipath effects can be solved by adding sufficient CP. Assume that the channel is estimated accurately. In practice, training sequences can be periodically transmitted to update the channel estimate and adjust phase synchronisation between the transmitter and the communication receiver. Thus, S~S~ can be obtained easily, after demodulation, the communication information can be obtained.

At the receiver of moving platform, similar to (7), after the A/D converter and FFT, the received radar echo signal can be expressed as

$$\boldsymbol{Y} = \tilde{\boldsymbol{\Lambda}}\tilde{\boldsymbol{S}} + \tilde{\boldsymbol{W}} = \begin{pmatrix} \beta_0 F(S_0) + \tilde{W}_0 \\ \beta_1 F(S_1) + \tilde{W}_1 \\ \vdots \\ \beta_{N-1} F(S_{N-1}) + \tilde{W}_{N-1} \end{pmatrix}, \tag{8}$$

where $\tilde{\boldsymbol{\Lambda}}$ is a diagonal matrix of eigenvalues of the RCS matrix HH [8], $\beta_0, \beta_1, \ldots, \beta_{N-1}$ are eigenvalues of the RCS matrix H . At the radar receiver of the moving platform, $\tilde{S}$ is known, so the diagonal matrix $\tilde{\boldsymbol{\Lambda}}$ can be easily obtained, and this result shows that the IRCI free can be achieved by using the CP-based OFDM signal. However, the method of communication symbols embedding is a difficult point for the design of radar–communications signal. We know that the PAPR is an important indicator for OFDM signal, the PAPR should be as low as possible. From (8) we know that, to calculate $\tilde{\boldsymbol{\Lambda}}$, noise vector $\tilde{W} = [\tilde{W}_0, \tilde{W}_1, \ldots, \tilde{W}_{N-1}]^T$ is magnified by $\tilde{S} = [F(S_0), F(S_1), \ldots, F(S_{N-1})]^T$, the module of $F(S_k)$ should be as constant module as possible for all *k* for achieving the better signal-to-noise ratio (SNR) at the radar receiver [8]. A time/frequency domain iterative filtering and clipping method is used to optimise PAPR and the module of $F(S_k)$. Thus, the information embedding scheme should meet the following two requirements:

(i) To optimise PAPR and the module of $F(S_k)$ effectively, the restriction on $\tilde{S}$ caused by the information embedding should be as small as possible.

(ii) Communication performance, such as data rate and the bit error rate (BER) must be guaranteed.

The traditional PSK modulation cannot meet these requirements; a well-behaved scheme for communication symbols embedding is urgently needed. This is the goal of the remainder of this paper.

Design of CP-based OFDM Radar–communications Signal

In this section, our main goal is to design a CP-based OFDM pulse s(t) to satisfy the requirements of radar–communications system. s(t) is desired to have a low PAPR, and the module of S_k should be as constant module as possible for all *k* for achieving the better SNR at the radar receiver, and the BER and date rate must be considered. The details are given in the following subsections.

Impact of Communication Symbols Embedding

In this subsection, to clearly describe the design algorithm, the impact of communication symbols embedding is discussed. We know the embedding of communication information has an impact on PAPR of s(t) and the SNR at the radar receiver, the PAPR of the transmitted OFDM pulse can be defined as

$$\text{PAPR} = \frac{\max_{n=0,1,\ldots,LN-1} |\tilde{s}_n|^2}{(1/LN)\sum_{n=0}^{LN-1} |\tilde{s}_n|^2}, \tag{9}$$

where $\tilde{s}_n$ is oversampled discrete-time sequence of s(t) can be implemented by the *LN*-point IFFT of the sequence $[F(S_0), F(S_1), \ldots, 0, 0, \ldots, 0]^{\mathrm{T}}$ of length *LN*.

According to [8], to evaluate the change of SNR after the range compression caused by the non-constant module of $\tilde{S}$, we define the SNR degradation factor as

$$\lambda = \frac{\text{SNR}_m}{\text{SNR}_{\max,m}} = \frac{N^2}{\sum_{k=0}^{N-1} |F(S_k)|^{-2}}, \tag{10}$$

where SNR_m is the actual SNR of the signal returned from the mth range cell, $\text{SNR}_{\max,m}$ is the optimal SNR of the signal returned from the mth range cell.

Unlike the traditional CP-based OFDM radar signal design method in [8] which only needs to consider PAPR and λ two constraints. For radar–communication signal design, the embedding of communication information must be considered. Information is embedded by $F(\cdot)$, $F(\cdot)$ is the phase modulation for the elements in S, actually. So, for CP-based OFDM radar–communication signal design, the embedding of communication information will cause an additional constraint, which increases the difficulty for the design of radar–communications signal. Thus, for CP-based OFDM radar–communication signal design, there are three restrictions: the PAPR constraint, the constraint and the constraint from the embedding of communication information. The above three constraints interact with each other, to optimise PAPR and λ effectively, the constraint caused by the information embedding should be as relaxed as possible, but the performance of communication must be considered.

If the phases of more elements in ***S*** are modulated, which helps to increase the data rate, but the constraint is more strict, the performances of PAPR and λ will be affected more seriously. Assume that $\tilde{N}$ communication symbols are embedded during a pulse period. The achievable data rate in bits per second (bps) can be expressed as

$$v = I \times \tilde{N} \times \text{PRF} \tag{11}$$

where I is the size of the constellation bits, PRF is the pulse repetition frequency of the CP-based OFDM signal. From (11), we can know the achievable data rate v mainly depends on PRF, in radar applications with a high PRF, a data rate in the range of Mbps can still be easily achieved. So $\tilde{N}$ and I can be properly reduced, such as that not every element in ***S*** must be modulated. When I is small, according to the constellation diagram of PSK, the signal has a strong anti-noise ability, so the phase can no longer be modulated to a fixed value, but a fixed space with a certain degree of freedom, by that, the constraint is further relaxed and the BER will not be greatly affected. The above fact provides the possibility for ILPC. The details are given in the next subsection.

Information Embedding Scheme

In this subsection, the information embedding scheme by ILPC modulation is proposed, the constraint is relaxed from two ways:

(i) Not every S_k must be modulated, S_k to be modulated can be selected at a fixed interval γ as follows:

$$\begin{cases} \hat{S}_k = F(S_k), & k = \gamma\mu \\ S_k, & \text{else} \end{cases}, \tag{12}$$

where μ=0,1,…,⌊(N−1)/γ⌋.

(ii) The phase of $\hat{S}_k$ can be relaxed, unlike the traditional PSK, in the ILPC, the phase of $\hat{S}_k$ is compressed linearly into a fixed area as follows:

If symbol ξ_j is sent

$$\begin{cases} \hat{p}_k = \hat{p}_k, & \hat{p}_k \in [e_j, e_j + \sigma] \\ \hat{p}_k = \dfrac{\sigma[\hat{p}_k - (e_j + \sigma)]}{2\pi - \sigma} + e_j, & \hat{p}_k \notin [e_j, e_j + \sigma] \end{cases} \tag{13}$$

where $\hat{p}_k = \hat{S}_k / |\hat{S}_k|$ is the phase of $\hat{S}_k$, $\{\xi_1, \xi_2, \ldots, \xi_J\}$ is the symbol set of the communication with J symbols, $J = 2^I$, $[e_j, e_j + \sigma]$ is the corresponding area of ξ_j and satisfy $e_{j+1} - e_j = e_j - e_{j-1} = 2\pi / J, j = 2, 3, \ldots, J$. σ is the size of the area. From (13), we can know the phase of $\hat{S}_k$ is modulated linearly into a fixed area rather than a fixed value, $\hat{p}_k$ is given a certain degree of freedom, the constraint is further relaxed. The constraint can be adjusted by adjusting the size of γ and σ.

At the demodulation end, the received signal is sampled by the A/D converter, after FFT, $\breve{S}$ that corresponds to ξ_j can be obtained. Then, demodulate as follows:

Symbol ξ_j is received, if

$$\breve{p}_k \in \left[\frac{e_j + e_{j-1} + \sigma}{2}, \frac{e_j + e_{j+1} + \sigma}{2}\right], \tag{14}$$

where $\breve{p}_k = \breve{S}_k / |\breve{S}_k|$ is the phase of $\breve{S}$, $\breve{p}_k$ can be expressed as

$$\breve{p}_k = \hat{p}_k + w_k \tag{15}$$

where w_k is the noise. It is easy to know that the performance of BER is closely related to the size of the space between adjacent fixed areas, that can be considered as the fault tolerance space for $\breve{p}_k$ according to (13), the size of fault tolerance space can be expressed as

$$\begin{aligned}\eta &= e_{j+1} - (e_j + \sigma)\\ &= (e_{j+1} - e_j) - \sigma\\ &= \frac{2\pi}{J} - \sigma\end{aligned} \tag{16}$$

According to (16), the performance of BER is inversely proportional to J and σ . But larger σ is conducive to the relaxation of constraint. When σ is relatively large, to guarantee the BER, J should be relatively small, according to (11) the data rate still be able to guarantee. In addition, it is noteworthy that the transmit power of the radar–communications system is much larger than that of a traditional communication system, which is good for the performance of BER. The larger modulation interval γ is conducive to the relaxation of constraint, but according to (11), the date rate is inversely proportional to γ , γ can be properly enlarged, but on the basis of ensuring the performances of PAPR and λ , the date rate should be as large as possible. Thus, these parameters including J, σ and γ should be determined according to the actual situation of the system.

OFDM Sequence Design

In this subsection, an algorithm for OFDM sequence design is proposed. In this algorithm, the method of time/frequency domain iterative filtering and clipping is used to optimise PAPR and λ, the communication information is embedded by using ILPC. To choose a better OFDM sequence, the standard Monte Carlo technique is used. Algorithm 1 (see Fig. 2) shows the design flow.

```
i = 0;
while i < Mon do
    generate the initial sequence of ith trial:
    in each trial, the kth element of initial sequence
    S^(0) is set as S_k^(0) = e^{j2πφ_k}, k = 0,1,···,N−1, where
    φ_k is uniformly distributed over the interval
    [0, 2π];
    for q = 0; q < Q; q++ do
        pad (L−1)N zeros → S̈^(q);.
        LN points IFFT → s̃^(q);
        filtering using (17) to satisfy the zero CP → s̈^(q);
        time domain clipping using (31) in [8] → s̄^(q);
        LN points FFT → S̈^(q);
        filtering using (18) → S⃛^(q);
        frequency domain clipping using (33) in [8]
        → S'^(q);
        Embedded communication symbols using ILPC
        modulation (12) and (13) → S^(q);
    end
    N points IFFT → s̃;
    filtering using (19) to satisfy the zero CP in (2)
    → s̈;
    energy normalization → s̄;
    N points FFT → S;
    calculate PAPR and λ using (9) and (10);
    if PAPR ≤ PAPR_0 and λ ≤ λ_0 then
        Output S;
        Break;
    else
        i = i + 1;
    end
end
end
```

Figure 2: Algorithm 1: The CP-OFDM sequence design algorithm.

An arbitrary-length OFDM radar–communications pulse with a low PAPR and a high λ can be designed by this algorithm. Where Mon is the upper limit of the number of standard Monte Carlo trials, L is the oversampling ratio, $PAPR_d$ and G_f are the scales of time/frequency domain clipping, respectively, [8], $PAPR_0$ and λ_0 is the threshold, and BER and date rate are considered. For simplicity, our proposed iterative algorithm starts with an initial random constant modular sequence $S^{(0)}$. The time domain filter is

$$h(n) = \begin{cases} 0, & 0 \le n \le L(M-1)-1 \\ 1, & L(M-1) \le n \le LN-1 \end{cases}. \tag{17}$$

Similarly, the frequency domain filter is

$$H(k) = \begin{cases} 1, & 0 \le k \le N-1 \\ 0, & N \le k \le LN-1 \end{cases}. \tag{18}$$

To satisfy the zero head condition, the last time domain filter is

$$\tilde{h}(n) = \begin{cases} 0, & 0 \le n \le M-2 \\ 1, & M-1 \le n \le N-1 \end{cases}. \tag{19}$$

RESULTS

In the simulations, we set N=128 , M=96 , the oversampling ratio L=4 . For simplicity, we set the number of iterations Q=100 , the scales of time/frequency domain clipping $PAPR_d$=1dB , G_f=0.05 . We evaluate the PAPR and the SNR degradation factor λ by using the standard Monte Carlo technique with 5×10^5 independent trials, in Fig. 3 we plot the cumulative distribution functions (CDFs) of the PAPR, the curves in Fig. 3 denote that, with the increase of γ and σ , the PAPR decreases, this is consistent with our previous analysis.

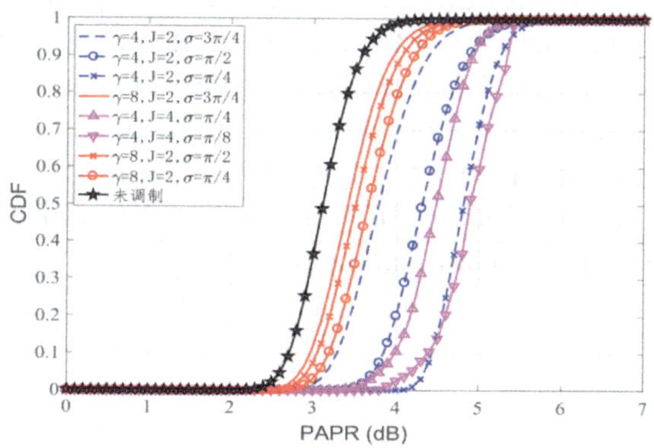

Figure 3: *CDF of PAPR for different* γ , σ *and J with* Q=100Q=100 , $PAPR_d$=1dB , G_f=0.05.

When γ=8 , PAPA increase of <0.5 dB compared to unmodulated. When γ=4 , σ≥π/2 , PAPA increase of <1.2 dB compared to unmodulated. When σ is fixed, with the increase of *J*, the PAPR decreases. In Fig. 4 , we plot the CDF of the SNR degradation factor λ , the curves in Fig. 4 denote that, with the increase of γ and σ , the λ increase, this is consistent with our previous analysis.

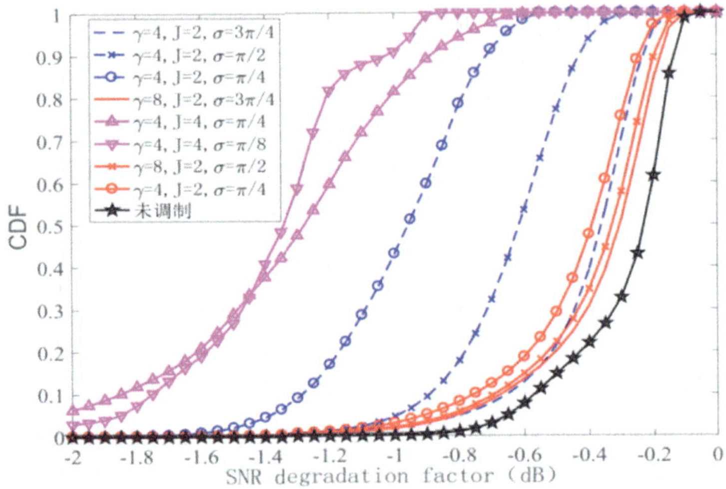

Figure *4*: The CDF of λ for different γ , σ and J with Q=100 , PAPRd=1dB , Gf=0.05.

When γ=8 , λ decrease of <0.2 dB compared to unmodulated. When γ=4 , σ≥π/2 , λ decrease of <0.4 dB compared to unmodulated. λ change is more sensitive than the PAPR for different *J*. According to [8], we can get better PAPR and λ with the increase of the maximum iteration number *Q*. Assume that the channel is estimated accurately, we investigate the BER performance of the proposed information embedding technique using different *J* and σ . To test the BER performance, a total number of 2.048×10^{6} random symbols are used. In Fig. 5 , we plot the BER versus signal-to-noise ratio (SNR) for different *J* and σ , and the Gaussian noise is used.

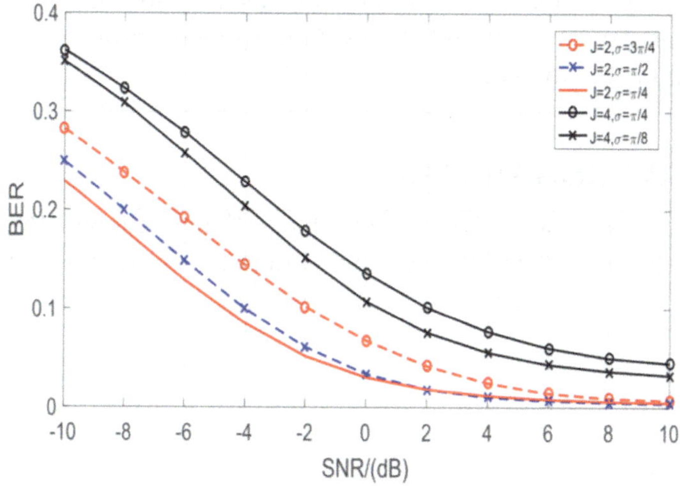

Figure 5: BER versus SNR for different σσ and J.

The curves in Fig. 5 denote that, the smaller the area size σ is, the better the BER performance will be, and the smaller the number of areas *J* is, the better the BER performance will be, this is consistent with (16). It is worth noting that we did not use any error code correction technology, so these results are acceptable. From Fig. 5 , we can see the BER performance is more sensitive to *J* and σ , so when the noise of environment is bad, we had better choose a smaller *J*. The data rate is mainly determined by PRF, so a smaller *J* does not have a big impact on the data rate.

CONCLUSION

In this paper, a signal design algorithm for the radar–communications system has been investigated. The goal of the signal design algorithm is to design a CP-based OFDM radar–communications signal. We have proposed a signal design algorithm based on the ILPC modulation and the time/frequency domain iterative filtering and clipping method. The ILPC modulation was used for communication information embedding, the time/frequency domain iterative filtering and clipping method could be used to guarantee the performances of PAPR and the SNR degradation factor λ . The performance of BER can be accepted, and the achievable data rate is mainly determined by PRF, a data rate in the range of Mbps can be easily achieved with a high PRF. Finally, the simulation results show that the proposed algorithm can design CP-based OFDM pulses that meet the requirements of radar–communications system.

ACKNOWLEDGMENTS

The research was supported by the Chang Jiang Scholars Program, the National Natural Science Foundation of China under grant no. 61501083.

REFERENCES

1. YousifE.KhanF.RatnarajahT.: ‘ On the spectral coexistence of colocated MIMO radars and wireless communications systems’. 2016 IEEE 17th Int. Workshop on Signal Processing Advances in Wireless Communications (SPAWC), Edinburgh, UK, 2016, pp. 1–5.
2. LiB.KumarH.PetropuluA.P.: ‘ A joint design approach for spectrum sharing between radar and communication systems’. 2016 IEEE Int. Conf. on Acoustics, Speech and Signal Processing (ICASSP), Shanghai, China, 2016, pp. 3306–3310.
3. GriffithsH.BluntS.ChenL.: ‘ Challenge problems in spectrum engineering and waveform diversity’. Proc. IEEE Radar Conf. (RadarCon 2013), Ottawa, Canada, April–May 2013, pp. 1–5.
4. HassanienA.HimedB.RiglingB.D.: ‘ A dual-function MIMO radar-communications system using frequency-hopping waveforms’. 2017 IEEE Radar Conf. (RadarConf), Seattle, WA, USA, 2017, pp. 1721–1725.
5. ChiriyathA.R.BlissD.W.: ‘ Effect of clutter on joint radar-communications system performance inner bounds’. 2015 49th Asilomar Conf. on Signals, Systems and Computers, Pacific Grove, CA,USA, 2015, pp. 1379–1383.
6. DengH.HimedB.: ‘ Interference mitigation processing for spectrum-sharing between radar and wireless communications systems’, *IEEE Trans. Aerosp. Electron. Syst.*, 2013, 49, (3), pp. 1911–1919 .
7. HassanienA.AminM.G.ZhangY.D.: ‘ A dual-function MIMO radar-communications system using PSK modulation’. 2016 24th European Signal Processing Conf. (EUSIPCO), Budapest, Hungary, 2016, pp. 1613–1617.
8. ZhangT.XiaX.-G.: ‘ OFDM synthetic aperture radar imaging with sufficient cyclic prefix’, *IEEE Trans. Geosci. Remote Sens.*, 2015, 53, (1), pp. 394–404 .
9. GoldsmithA.: ‘ *Wireless communications*’ (Cambridge University Press, Cambridge, UK, 2005), p. 388 .

SECTION 2:
PRINCIPLES OF INFORMATION THEORY

4

On Shannon's Formula and Hartley's Rule: Beyond the Mathematical Coincidence

Olivier Rioul [1] **and José Carlos Magossi** [2]

[1]Télécom ParisTech, Institut Mines-Télécom, CNRS LTCI, 46 Rue Barrault, 75013, Paris, France
[2]School of Technology (FT), University of Campinas (Unicamp), Rua Paschoal Marmo 1.888, 13484-370 Limeira, São Paulo, Brazil

ABSTRACT

In the information theory community, the following "historical" statements are generally well accepted: (1) Hartley did put forth his rule twenty years before Shannon; (2) Shannon's formula as a fundamental tradeoff between transmission rate, bandwidth, and signal-to-noise ratio came out unexpected in 1948; (3) Hartley's rule is inexact while Shannon's formula is characteristic of the additive white Gaussian noise channel; (4) Hartley's rule is an imprecise relation that is not an appropriate formula for the capacity of a communication channel. We show that all these four statements are somewhat wrong. In fact, a careful calculation shows that "Hartley's rule"

Citation: Rioul, O.; Magossi, J.C. "On Shannon's Formula and Hartley's Rule: Beyond the Mathematical Coincidence". Entropy 2014, 16, 4892-4910. https://doi.org/10.3390/e16094892.

in fact coincides with Shannon's formula. We explain this mathematical coincidence by deriving the necessary and sufficient conditions on an additive noise channel such that its capacity is given by Shannon's formula and construct a sequence of such channels that makes the link between the uniform (Hartley) and Gaussian (Shannon) channels.

Keywords: Shannon's formula; Hartley's rule; additive noise channel; differential entropy; channel capacity; signal-to-noise ratio; pulse-amplitude modulation (PAM); additive white Gaussian noise (AWGN) channel; uniform noise channel; characteristic function; uniform B-spline function; uniform sum distribution; central limit theorem

INTRODUCTION

As researchers in information theory, we all know that the milestone event that founded our field is Shannon's publication of his seminal 1948 paper **[1]** that created a completely new branch of applied mathematics and called it to immediate worldwide attention. What has rapidly become the emblematic classical expression of the theory is *Shannon's formula* **[1,2]**

$$C = \tfrac{1}{2}\log_2\left(1 + \frac{P}{N}\right) \qquad (1)$$

for the information capacity of a communication channel with signal-to-noise ratio *P/N*.

Hereafter we shall always express information capacity in binary units (bits) per *sample*. Shannon's well-known original formulation was in bits per second:

$$C = W\log_2\left(1 + \frac{P}{N}\right) \quad \text{bits/s.}$$

The difference between this formula and **(1)** is essentially the content of the *sampling theorem*, often referred to as Shannon's theorem, that the number of independent samples that can be put through a channel of bandwidth W hertz is $2W$ samples per second. We shall not discuss here whether the sampling theorem should be attributed to Shannon or to other authors that predate him in this discovery; see e.g., **[3]** for a recent account and extensive study on this subject.

The classical derivation of **(1)** was done in **[1]** as an application of Shannon's coding theorem for a memoryless channel, which states that the best coding procedure for reliable transmission achieves a maximal rate of $C = \max_X I(X;Y)$ bits per sample, where X is the channel input with average power $P = \mathbb{E}(X^2)$ and $Y = X + Z$ is the channel output. Here Z denotes the additive Gaussian random variable (independent of X) that models the communication noise with power $N = \mathbb{E}(Z^2)$. By expanding mutual information $I(X;Y) = h(Y) - h(Y|X)$ as a difference of differential entropies, noting that $h(Y|X) = h(Z) = \log_2 \sqrt{2\pi eN}$ is constant, and choosing X Gaussian so as to maximize $h(Y)$, Shannon arrived at his formula $C = \max_X h(Y) - h(Z) = \log_2 \sqrt{2\pi e(P+N)}$ $-\log_2 \sqrt{2\pi eN} = \frac{1}{2}\log_2(1+P/N)$.

Formula (**1**) is also known as the *Shannon–Hartley formula*, and the channel coding theorem stating that (**1**) is the maximum rate at which information can be transmitted reliably over a noisy communication channel is often referred to as the *Shannon–Hartley theorem* (see, e.g., [**4**]). The reason for which Hartley's name is associated to the theorem is commonly justified by the so-called *Hartley's law*, which is described as follows:

During 1928, Hartley formulated a way to quantify information and its line rate (also known as data signalling rate R bits per second) [5]. This method, later known as Hartley's law, became an important precursor for Shannon's more sophisticated notion of channel capacity. (...)

Hartley argued that the maximum number of distinguishable pulse levels that can be transmitted and received reliably over a communications channel is limited by the dynamic range of the signal amplitude and the precision with which the receiver can distinguish amplitude levels. Specifically, if the amplitude of the transmitted signal is restricted to the range of $[-A,+A]$ *volts, and the precision of the receiver is* $\pm\Delta$ *volts, then the maximum number of distinct pulses M is given by* $M = 1 + \frac{A}{\Delta}$*. By taking information per pulse in bit/pulse to be the base-2-logarithm of the number of distinct messages M that could be sent, Hartley [5] constructed a measure of the line rate R as* $R = \log_2(M)$ *[bits per symbol].*

—Wikipedia [**4**]

In other words, within a noise amplitude limited by Δ, by taking regularly spaced input symbol values in the range $[-A,A]$ with step 2Δ:

$-A, -A+2\Delta, \dots, A-2\Delta, A,$

one can achieve a maximum total number of $M = A/\Delta + 1$ possible distinguishable values. This holds in the most favorable case where A/Δ is an integer, where the "+1" is due to the sample values at the boundaries—

otherwise, M would be the integer part of $A/\Delta + 1$. Therefore, error-free communication is achieved with at most

$$C' = \log_2\left(1 + \frac{A}{\Delta}\right) \qquad (2)$$

bits per sample. This equation strikingly resembles (**1**). Of course, the "signal-to-noise ratio" A/Δ is a ratio of amplitudes, not of powers, hence should not be confused with the usual definition P/N; accordingly, the factor 1/2 in Formula (**1**) is missing in (**2**). Also, (**2**) is only considered as an approximation of (**1**):

Hartley's rate result can be viewed as the capacity of an errorless M-ary channel (...). But such an errorless channel is an idealization, and if M is chosen small enough to make the noisy channel nearly errorless, the result is necessarily less than the Shannon capacity of the noisy channel (...), which is the Hartley–Shannon result that followed later [in 1948].

—Wikipedia [**4**]

In the information theory community, the following "historical" statements are generally well accepted:

(1) *Hartley did put forth his rule* (2) *twenty years before Shannon.*

(2) *The fundamental tradeoff* (1) *between transmission rate, bandwidth, and signal-to-noise ratio came out unexpected in 1948: the time was not even ripe for this breakthrough.*

(3) *Hartley's rule is inexact while Shannon's formula is characteristic of the additive white Gaussian noise (AWGN) channel ($C' \neq C$).*

(4) *Hartley's rule is an imprecise relation between signal magnitude, receiver accuracy and transmission rate that is not an appropriate formula for the capacity of a communication channel.*

In this article, we show that all these four statements are somewhat wrong. The organisation is as follows. Sections 2–5 will each defend the opposite view of statements (**1**)–(**4**) correspondingly. Section 6 concludes through a detailed mathematical analysis.

HARTLEY'S RULE IS NOT HARTLEY'S

Hartley [**5**] was the first researcher to try to formulate a theory of the transmission of information. Apart from stating explicitly that the amount of transmitted information is proportional to the transmission bandwidth, he showed that the number$M$ of possible alternatives from a message source over given a time interval grows exponentially with the duration, suggesting a definition of information as the logarithm log$M$. However, as Shannon recalled in 1984:

I started with information theory, inspired by Hartley's paper, which was a good paper, but it did not take account of things like noise and best encoding and probabilistic aspects.

—Claude Elwood Shannon [**6**]

Indeed, no mention of signal vs. noise, or of amplitude limitation A or Δ was ever made in Hartley's paper [**5**]. One may then wonder how (**2**) was coined as Hartley's law.

The oldest reference that attributes (**2**) to Hartley—and incidentally cited in the Wikipedia page [**4**]—seems to be the classical 1965 textbook of Wozencraft and Jacobs, most notably its introduction chapter:

(...) in 1928, Hartley [5] reasoned that Nyquist's result, when coupled with a limitation on the accuracy of signal reception, implied a restriction on the amount of data that can be communicated reliably over a physical channel. Hartley's argument may be summarized as follows. If we assume that (1) the amplitude of a transmitted pulse is confined to the voltage range $[-A,A]$ *and (2) the receiver can estimate a transmitted amplitude reliably only to an accuracy of* $\pm\Delta$ *volts, then, as illustrated in [the] Figure (...), the maximum number of pulse amplitudes distinguishable at the receiver is* $(1 + A/\Delta)$. *(...) [in* the Figure's legend:*] Hartley considered received pulse amplitudes to be distinguishable only if they lie in different zones of width* 2Δ *(...)*
Hartley's formulation exhibits a simple but somewhat inexact interrelation among (...) the maximum signal magnitude A, the receiver accuracy Δ, *and the allowable number of message alternatives. Communication theory is intimately concerned with the determination of more precise interrelations of this sort.*

—John M. Wozencraft; Irwin Mark Jacobs [7]

The textbook was highly regarded and still widely used today. Its introductive text has become famous to many researchers in the field of communication theory and has had a tremendous impact. This would explain why (**2**) is now widely known as Hartley's capacity law.

One may then wonder whether Wozencraft and Jacobs have found such a result themselves while attributing it to Hartley or whether it was inspired from other researchers. We found that the answer is probably in very first tutorial article in information theory that was ever published by E. Colin Cherry in 1951:

Although not explicitly stated in this form in his paper, Hartley [5] has implied that the quantity of information which can be transmitted in a frequency band of width B and time T is proportional to the product: $2BT \log M$, *where M is the number of "distinguishable amplitude levels." [...] He approximates the waveform by a series of steps, each one representing a selection of an amplitude level. [...] For example, consider a waveform to be traced out on a rectangular grid [...], the horizontal mesh-width representing units of time (equal to* $1/2B$ *in order to give the necessary* $2BT$ *data in a time T), and the vertical the "smallest distinguishable" amplitude change; in practice this smallest step may be taken to equal the noise level n. Then the quantity of information transmitted may be shown to be proportional to* $BT \log(1 + a/n)$ *where a is the maximum signal amplitude, an expression given by Tuller [8], being based upon Hartley's definition of information.*

—E. Colin Cherry [**9**]

Cherry attributes (**2**) to an *implicit* derivation of Hartley but cites the explicit derivation of Tuller [**8**]. The next section investigates the contribution of Tuller and others.

INDEPENDENT 1948 DERIVATIONS OF SHANNON'S FORMULA

In the introduction to his classic textbook, Robert McEliece wrote:

With many profound scientific discoveries (for example Einstein's discovery in 1905 of the special theory of relativity) it is possible with the aid of hindsight to see that the times were ripe for a breakthrough. Not so with information theory. While of course Shannon was not working in the vacuum in the 1940's, his results were so breathtakingly original that even the communication specialists of the day were at a loss to understand their significance.

—Robert McEliece [**10**]

One can hardly disagree with this statement when one sees the power and generality of Shannon's results. Just to mention a few examples: the introduction of the formal architecture of communication systems (Shannon's paradigm) with explicit distinction between source, channel and destination; the emphasis on digital representation to make the chance of error as small as desired; the consideration of codes in high dimensions; and the use of probabilistic models for the signal as well as for the noise, *via* information theoretic tools like entropy and mutual information. Shannon's ideas were revolutionary, in keeping with J.R. Pierce's famous quote:

In the end, [1] and the book based on it came as a bomb, and something of a delayed-action bomb.

—John R. Pierce [**11**]

Indeed, [**1**] being so deep and profound, did not have an immediate impact. As Robert Gallager recalls:

(...) two important papers (...) were almost concurrent to [1].
The first subsequent paper was [12], whose coauthors were B. R. Oliver and J. R. Pierce. This is a very simple paper compared to [1], but it had a tremendous impact by clarifying a major advantage of digital communication. (...) It is probable that this paper had a greater impact on actual communication practice at the time than [1].
The second major paper written at about the same time as [1] is [2]. This is a more tutorial amplification of the AWGN channel results of [1]. (...) This was the paper that introduced many communication researchers to the ideas of information theory.
—Robert Gallager [**13**]

In [**12**], Shannon's Formula (**1**) was used without explicit reference to the Gaussian nature of the added white noise, as the capacity of an "ideal system". On the other hand, [**2**] was devoted to a geometric proof of Formula (**1**).

It appears, therefore, that Shannon's Formula (**1**) was *the* emblematic result that impacted communication specialists at the time, as expressing the correct tradeoff between transmission rate, bandwidth, and signal-to-noise ratio. It is one Shannon's result that is the best known and understood among communications engineers. As Verdú has noticed in [**14**], only a few months after the publication of [**2**], M. Golay [**15**] referred to (**1**) as "the now classical expression for the information reception capacity of a channel." In the following years, finding "codes to reach the promised land (**1**)" [**16**] became the "holy grail of information theory" [**14**].

As far as (**1**) is concerned, Shannon, after the completion of [**1**], acknowledged other works:

Formulas similar to (1) *for the white noise case have been developed independently by several other writers, although with somewhat different interpretations. We may mention the work of N. Wiener [17], W. G. Tuller [8], and H. Sullivan in this connection.*

—Claude Elwood Shannon [**1**]

Unfortunately, Shannon gave no specific reference to H. Sullivan. S. Verdú cited many more contributions during the same year of 1948:

*By 1948 the need for a theory of communication encompassing the fundamental tradeoffs of transmission rate, reliability, bandwidth, and signal-to-noise ratio was recognized by various researchers. Several theories and principles were put forth in the space of a few months by A. Clavier [**18**], C. Earp [**19**], S. Goldman [**20**], J. Laplume [**21**], C. Shannon [**1**], W. Tuller [**8**], and N. Wiener [**17**]. One of those theories would prove to be everlasting.*

—Sergio Verdú [**14**]

Lundheim reviewed some of these independent discoveries and concludes:

(...) this result [Shannon's formula] was discovered independently by several researchers, and serves as an illustration of a scientific concept whose time had come.

—Lars Lundheim [**22**]

This can be contrasted to the above citation of R. McEliece.

Wiener's independent derivation [**17**] of Shannon's formula is certainly the one that is closest to Shannon's. He also used probabilistic arguments, logarithmic measures (in base 2) and differential entropy, the latter choice being done "mak[ing] use of a personal communication of J. von Neumann". Wiener considers "the information gained by fixing one or more variables in a problem", e.g., fixing $Y = X + Z$ where X and Z are independent Gaussian. By computing the difference $h(X) - h(X|Y)$, he concludes that "the excess of information concerning X when we know Y is" (**1**). Unlike Shannon, however, his definition of information is not based on any precise communication problem. There is also no relation to Hartley's argument leading to (**2**).

Concerning the idea of information theory, Wiener wrote in his book Cybernetics:

This idea occurred at about the same time to several writers, among them the statistician R. A. Fisher, Dr. Shannon of the Bell Telephone Laboratories, and the author. Fisher's motive in studying this subject is to be found in classical statistical theory; that of Shannon in the problem of coding information; and that of the author in the problem of noise and message in electrical filters. Let it be remarked parenthetically that some of my speculations in this direction attach themselves to the earlier work of Kolmogoroff in Russia, although a considerable part of my work was done before my attention was called to the work of the Russian school.

—Norbert Wiener[**17**]

It is likely that it is the importance of Shannon's formula for which he has made an independent derivation that lead him to declare:

Information theory has been identified in the public mind to denote the theory of information by bits, as developed by C. E. Shannon and myself.
—Norbert Wiener[**23**]

J.R. Pierce comments:

Wiener's head was full of his own work and an independent derivation of (1) *(...) Competent people have told me that Wiener, under the misapprehension that he already knew what Shannon had done, never actually found out.*
—John R. Pierce [**11**]

All other independent discoveries in the year of 1948 were in fact essentially what is now referred to Hartley's rule leading to (**2**). Among these, the first published work in April 1948 was by the French engineer Jacques Laplume [**21**] from Thompson-Houston. He essentially gives the usual derivation that gives (**2**) for a signal amplitude range [0,A]. C. Earp's publication [**19**] in June 1948 also makes a similar derivation of (**2**) where the signal-to-noise amplitude ratio is expressed as a "root-mean-square ratio" for the "step modulation", which is essentially pulse-code modulation. In a footnote, Earp claims that his paper "was written in original form in October, 1946". In an another footnote at the first page, he mentions that

A symposium on "Recent Advances in the Theory of Communication" was presented at the November 12, 1947, meeting of the New York section of the Institute of Radio Engineers. Four papers were presented by A. G. Clavier (...); B.D. Loughlin (...); and J. R. Pierce and C. E. Shannon, both of Bell Telephone Laboratories.

—C.W. Earp [**19**]

André Clavier is another French engineer from LMT laboratories (subsidiary of ITT Corporation), who published "Evaluation of transmission efficiency according to Hartley's expression of information content" [**18**] in December 1948. He again makes a similar derivation of (**2**) as Earp's, expressed with root-mean-square values. As Lundheim notes [**22**],"it is, perhaps, strange that neither Shannon nor Clavier have mutual references in their works, since both [**2**] and [**18**] were orally presented at the same meeting (...) and printed more than a year afterwards."

In May 1948, Stanford Goldman again re-derived (**2**), acknowledging that the equation "has been derived independently by many people, among them W. G. Tuller, from whom the writer first learned about it" [**20**]. William

G. Tuller's thesis was defended in June 1948 and printed as an article in May 1949 [**8**]. His derivation uses again root-mean-square (rms) ratios.

Let S be the rms amplitude of the maximum signal that may delivered by the communication system. Let us assume, a fact very close to the truth, that a signal amplitude change less than noise amplitude cannot be recognized, but a signal amplitude change equal to noise is instantly recognizable. Then, if N is the rms amplitude of the noise mixed with the signal, there are $1 + S/N$ *significant values of signal that may be determined. (...) the quantity of information available at the output of the system [is* $= \log(1 + S/N)$*].*

—William G. Tuller [**8**]

In the 1949 article [**8**] he explains that

The existence of [Shannon's] work was learned by the author in the spring of 1946, when the basic work underlying this paper had just been completed. Details were not known by the author until the summer of 1948, at which time the work reported here had been complete for about eight months.

—William G. Tuller [**8**]

In view of this note it is perhaps not completely fair so say, following J.R. Pierce [**11**] (Shannon's co-author of [**12**]), that

(...) much of the early reaction to Shannon's work was either uninformed or a diversion from his aim and accomplishment. (...) In 1949, William G. Tuller published a paper giving his justification of (1) *[8].*

—John R. Pierce [**11**]

Considering that Tuller's work is—apart from Wiener's—the only work referenced by Shannon in [**1**], and that the oldest reference known (1946) is Tuller's, it should be certainly appropriate to refer to (**2**) as *Tuller's formula* or to (**1**) as the *Tuller–Shannon formula*.

There is perhaps no better conclusion for this section than to cite Shannon's 1949 article [**2**] where he explicitly mentioned (and criticized) Hartley's Law as the property that the maximum amount of information per second is proportional to the bandwidth (without reference to noise limitation), and where he proposed his own interpretation of (**2**) making the link with his formula (**1**):

How many different signals can be distinguished at the receiving point in spite of the perturbations due to noise? A crude estimate can be obtained as follows. If the signal has a power P, then the perturbed signal will have a

power $P + N$. The number of amplitudes that can be reasonably well distinguished is $K\sqrt{\frac{P+N}{N}}$ where K is a small constant in the neighborhood of unity depending on how the phrase "reasonably well" is interpreted. (...) The number of bits that can be sent in this time is $\log_2 M\ [= \frac{1}{2}\log_2 K^2\left(1 + \frac{P}{N}\right)]$.

—Claude Elwood Shannon [2]

It may be puzzling to notice, as Hodges did in his historical book on A. Turing [24], that Shannon's article [2] mentioned a manuscript with a received date of 23 July, 1940! But this was later corrected by Shannon himself in 1984 (cited in [6], Reference 10):

(...) Hodges cites a Shannon manuscript date 1940, which is, in fact, a typographical error.
(...) First submission of this work for formal publication occurred soon after World War II.

—Claude Elwood Shannon [6]

This would mean in particular that Shannon's work leading to his formula was completed in 1946, at about the same time as Tuller's.

HARTLEY'S RULE YIELDS SHANNON'S FORMULA: $C' = C$

Let us consider again the argument leading to (2). The channel input X is taking $M = 1 + A/\Delta$ values in the set $\{-A, -A + 2\Delta, \ldots, A - 2\Delta, A\}$, which is the set of values $(M - 1 - 2k)\Delta$ for $k = 0, \ldots, M - 1$. A maximum amount of information will be conveyed through the channel if the input values are equiprobable. Then, using the well-known formula for the sum of squares of consecutive integers, one finds:

$$P = \mathbb{E}(X^2) = \frac{1}{M}\sum_{k=0}^{n}(M - 1 - 2k)^2 = \Delta^2\frac{M^2 - 1}{3}$$

Interestingly, this is the classical formula for the average power of a M-state pulse-code modulation or pulse-amplitude modulation signal, as was derived by Oliver, Pierce and Shannon in [12].

The input is mixed with additive noise Z with accuracy $\pm\Delta$. The least favorable case would be that Z follows a uniform distribution in $[-\Delta, \Delta]$. Then its average power is

$$N = \mathbb{E}(Z^2) = \frac{1}{2\Delta}\int_{-\Delta}^{\Delta} z^2\,\mathrm{d}z = \frac{\Delta^2}{3}$$

It follows that (**2**) takes the form of a striking identity!

$$C' = \log_2 M = \tfrac{1}{2}\log_2(1 + M^2 - 1) = \tfrac{1}{2}\log_2\Big(1 + \frac{3P}{\Delta^2}\Big) = \tfrac{1}{2}\log_2\Big(1 + \frac{P}{N}\Big) = C.$$

A mathematical coincidence?

One may perhaps argue that if Tuller or others knew about such a coincidence, they would probably have followed Wiener's attitude in claiming paternity of information theory. In any case, such an identification of (**1**) and (**2**) calls for verification that Hartley's rule would in fact be "mathematically correct" as a capacity formula.

HARTLEY'S RULE AS A CAPACITY FORMULA

Consider the *uniform channel*, a memoryless channel with additive white noise Z with uniform density in the interval $[-\Delta, \Delta]$. If X is the channel input, the output will be $Y = X + Z$, where X and Z are independent. We assume that the input has the amplitude constraint $|X| \leq A$. The following calculation was proposed as a homework exercise in the excellent textbook by Cover and Thomas [**25**].

Theorem 1

Assuming A/Δ is integral, the uniform channel has capacity C' given by (2).

(If A/Δ is not integral, then the proof of the theorem shows that $C' \leq \log_2(1 + A/\Delta)$, yet C' cannot be obtained in closed form.)

Proof

From Shannon's coding theorem, the channel's capacity is $C = \max_X I(X; Y)$ bits per sample, where the maximum is taken over all distributions of X such that $|X| \leq A$, *i.e.*, with support $[-A, A]$. By expanding mutual information $I(X; Y) = h(Y) - h(Y|X)$ as a difference of differential entropies, and noting that $h(Y|X) = h(Z) = \log_2(2\Delta)$ is constant, the required capacity C' is obtained by maximizing $h(Y)$.

Now since $|X| \leq A$, by the triangular inequality, the output amplitude is limited by $|Y| \leq |X|+|Z| \leq A+\Delta$. Choosing $X = X^*$ to be discrete uniform taking $M = 1+A/\Delta$ equiprobable values in the set $\{-A, -A+2\Delta, \ldots, A-2\Delta, A\}$, it is immediate to see that $Y = X^* + Z$ will have the uniform density over the interval $[-A-\Delta, A+\Delta]$, which is known to maximize $h(Y)$ under the constraint $|Y| \leq A+\Delta$. Therefore such an X^* achieve the capacity and we have $C' = \max_X h(Y) - h(Z) = \log_2(2(A+\Delta)) - \log_2(2\Delta) = \log_2(1 + A/\Delta)$.

Thus there is a sense in which the "Tuller–Shannon Formula" **(2)** is indeed *correct* as the capacity of a communication channel, except that the communication noise is *not* Gaussian, but uniform, and that the signal limitation is *not* on the power, but on the amplitude (as a side remark, it is interesting to mention that C' is in fact a zero-error capacity and that no coding is actually necessary to achieve it).

The analogy between the Gaussian and uniform channels can be pushed further. Both channels are memoryless and additive, with $Y = X + Z$ where X and Z are independent. Both have "additive" constraints on their inputs of the form $\Phi(X) \leq c$, where additivity means that $\Phi(X) \leq c$ and $\Phi(Z) \leq c'$ imply $\Phi(X + Z) \leq c + c'$. Specifically, in the Gaussian case, $\Phi(X) = \mathbb{E}(X^2)$ and additivity results from the fact that X and Z are uncorrelated; and in the uniform case, $\Phi(X) = |X|$ and additivity is simply a consequence of the inequality $|X + Z| \leq |X| + |Z|$. Also in both cases, the noise $Z = Z^*$ maximizes the differential entropy $h(Z)$ under the constraint $\Phi(Z) \leq c'$, and the input $X = X^*$ that maximizes mutual information $I(X; Y) = I(X;X + Z^*)$ is such that the corresponding output $Y^* = X^* + Z^*$ also maximizes the differential entropy $h(Y)$ under the constraint $\Phi(Y) \leq c + c'$. When $\Phi(X) = \mathbb{E}(X^2)$ (power limitation), both Y^* and Z^* are Gaussian while for $\Phi(X) = |X|$ (amplitude limitation), both Y^* and Z^* have a uniform distribution.

Shannon used these properties for $\Phi(X) = \mathbb{E}(X^2)$ to show that under limited *power*, Gaussian noise is the *worst* possible noise that one can inflict in the channel (in terms of its capacity). To show this, he considered an arbitrary additive noise Z and defined $\tilde{Z}$ as a random variable of the same distribution type as Z^* but with the same differential entropy as Z. Thus for $\Phi(X) = \mathbb{E}(X^2)$, $\tilde{Z}$ is a zero-mean Gaussian variable of average power $\tilde{N} = 2^{2h(Z)}/2\pi e$, which is referred to as the *entropy power* **[1]** of Z. He then established that the capacity associated with the noise Z satisfies **[1]**

$$\tfrac{1}{2}\log_2\left(1 + \alpha\frac{P}{N}\right) \leqslant C \leqslant \tfrac{1}{2}\log_2\left(1 + \frac{P}{N}\right) + \tfrac{1}{2}\log_2\alpha, \qquad (3)$$

where we have noted $\alpha = N/\tilde{N}$. The first inequality was in fact derived by Shannon as a consequence of the *entropy power inequality* (see, e.g., **[26]** for more details on this inequality). Since $h(\tilde{Z}) = h(Z) \leq h(Z^*)$, one has $\tilde{N} \leq N$ so that $\alpha \geq 1$ (with equality $\alpha = 1$ only in the case of Gaussian noise). It follows from the above inequality that the capacity has the lowest value for Gaussian noise.

The uniform channel enjoys a similar property: under

limited *amplitude*, *uniform* noise is the worst possible noise that one can inflict in the channel. To show this, consider the following

Definition 2 (Entropic Amplitude).

Given an arbitrary additive noise Z, let $\tilde{Z}$ *be a random variable of the same distribution type as* Z^* *but with the same differential entropy as Z. Thus for* $\Phi(X) = |X|$, $\tilde{Z}$ *is a zero-mean uniformly distributed variable with amplitude* $\tilde{\Delta}$. *The entropic amplitude of Z is*

$\Delta^{\sim}=2^{h(Z)-1}$.

The squared entropic amplitude is related to the entropy power by the relation $\tilde{\Delta}^2 = \tilde{N}\pi e/2$.

Theorem 3

When $\Phi(X) = |X|$ *(amplitude limitation) under the same conditions as Theorem 1, the capacity C' associated with an arbitrary additive noise Z satisfies*

$$\log_2\left(1+\frac{A}{\Delta}\right) \leqslant C' \leqslant \log_2\left(1+\frac{A}{\Delta}\right) + \log_2 \alpha, \tag{4}$$

where $\alpha = \Delta/\tilde{\Delta} \geq 1$ *(with equality* $\alpha = 1$ *only for uniform noise).*

It follows as announced that the capacity has the lowest value for uniform noise.

Proof

One has $I(X;X+Z) = h(X+Z) - h(Z)$ where $h(Z) = \log_2(2\tilde{\Delta})$; since $|Y| \leq A + \Delta$, $h(Y) \leq \log_2(2(A+\Delta))$. Therefore, $I(X;X+Z) \leq \log_2(2(A+\Delta)) - \log_2(2\tilde{\Delta}) = \log_2\left(1+\frac{A}{\Delta}\right) + \log_2\alpha$. Maximizing $I(X;X+Z)$ over the distribution de X in this inequality gives the second inequality in **(4)**.

To prove the first inequality, notice that $C = \max_X I(X;X+Z) \geq I(X^*;X^*+Z) = h(X^*+Z) - h(Z)$ where, as above, X^* is discrete uniform in the M-ary set $X = \{-A, -A+2\Delta, \ldots, A-2\Delta, A\}$ with $M = 1 + A/\Delta$. Now $Y = X^* + Z$ follows the density $p_Y(y) = \frac{1}{M}\sum_{x\in\mathcal{X}} p_Z(y-x)$ where $p_Z(z)$ is the density of Z. Since $|Z| \leq \Delta$ all terms in this sum have disjoint supports. Therefore,

$$h(X^*+Z) = -\sum_{x\in\mathcal{X}}\int_{-\Delta}^{\Delta}\left(\frac{1}{M}p_Z(y-x)\right)\log_2\left(\frac{1}{M}p_Z(y-x)\right)\mathrm{d}y = \log_2 M - \int p_Z(z)\log_2 p_Z(z)\,\mathrm{d}z$$

which reduces to the simple formula $h(X^*+Z) = \log_2 M + h(Z)$. Therefore, $C \geq h(X*+Z) - h(Z) = \log_2 M = \log_2\left(1+\frac{A}{\Delta}\right)$, which proves the first inequality in **(4)**.

A MATHEMATICAL ANALYSIS

Conditions for Shannon's Formula to Hold

In this section, we consider a memoryless additive noise channel with zero-mean input X and output $Y = X + Z$. Such a channel is defined by:

- the probability density function (pdf) p_Z of the zero-mean noise Z, which is assumed independent of X;
- a constraint set C on the possible distributions of X. The channel capacity is computed under this constraint as

$$C = \max_{X \in \mathcal{C}} I(X;Y) = \max_{X \in \mathcal{C}} h(Y) - h(Z) = \left(\max_{X \in \mathcal{C}} h(X+Z)\right) - h(Z).$$

We let X^* be the input that attains this maximum and let $Y^* = X^* + Z$ be the corresponding output. Thus $C = h(Y^*) - h(Z) = h(X^* + Z) - h(Z)$. We also let $P = \mathbb{E}(X^{*2})$ and $N = \mathbb{E}(Z^2)$ so that P/N denotes the signal-to-noise ratio at the optimum.

Lemma 4

If there exists a number $\alpha > 1$ such that αZ and Y^* share the same distribution, then the channel capacity C is given by Shannon's Formula (1).

Proof

One has $C = h(Y^*) - h(Z) = h(\alpha Z) - h(Z) = \log_2 |\alpha| = \frac{1}{2}\log_2 \alpha^2$. However, $P + N = \mathbb{E}(X^2) + \mathbb{E}(Z^2) = \mathbb{E}(Y^2) = \alpha^2\, \mathbb{E}(Z^2) = \alpha^2 N$ and so $\alpha^2 = 1 + P/N$. This gives (**1**).

Example 1 (Gaussian channel).

Here both Z and $Y^* = X^* + Z$ are zero-mean Gaussian so that the condition of the lemma is satisfied. We recover (1) *as the classical expression for the channel capacity.*

Example 2 (uniform channel).

Here both Z and $Y^* = X^* + Z$ are uniformly distributed over a centered interval so the condition of the lemma is also satisfied. This explains anew the coincidence found in the calculation of Section 4.

In the following we note $\varphi_X(\omega) = \mathbb{E}(e^{i\omega X})$, the characteristic function of any random variable X.

Lemma 5

The condition of Lemma 4 is satisfied if and only if there exists $\alpha > 1$ *such that*

$$\frac{\phi_Z(\alpha\omega)}{\phi_Z(\omega)} = \phi_{X^*}(\omega)$$

Proof

αZ and $Y^* = X^* + Z$ have the same distribution if and only if they share the same characteristic function, which is equal to $\varphi_{\alpha Z}(\omega) = \varphi_Z(\alpha\omega)$ and to $\varphi_{Y^*}(\omega) = \varphi_{X^*}(\omega)\varphi_Z(\omega)$.

In particular the above quotient must be a characteristic function of some random variable. This shows that the distribution of Z should be *divisible*.

Example 3 (Gaussian channel (continued)).

Here $\alpha^2 = \frac{P+N}{N}$ *and*

$$\frac{\phi_Z(\alpha\omega)}{\phi_Z(\omega)} = \frac{e^{-\alpha^2\omega^2 N/2}}{e^{-\omega^2 N/2}} = e^{-\omega^2 P/2}$$

which the characteristic function of $X^* \sim \mathcal{N}(0, P)$.

Example 4 (uniform channel (continued)).

Here $\alpha = \frac{A+\Delta}{\Delta} = M$ *is assumed integral and*

$$\frac{\phi_Z(M\omega)}{\phi_Z(\omega)} = \frac{\mathrm{sinc}(M\Delta\cdot\omega)}{\mathrm{sinc}(\Delta\cdot\omega)} = \frac{\sin(M\Delta\cdot\omega)}{M\sin(\Delta\cdot\omega)} = \frac{1}{M}(e^{-i(M-1)\omega\Delta} + e^{-i(M-3)\omega\Delta} + \cdots + e^{i(M-1)\omega\Delta})$$

Where $\mathrm{sinc}\, x = \frac{\sin x}{x}$ *is the sine cardinal function and where the last equality is the well-known Dirichlet kernel expression. The result is the characteristic function of* X^* *which take* M *equiprobable values in the set* $\{-(M-1)\Delta, -(M-3)\Delta, \ldots, (M-3)\Delta, (M-1)\Delta\}$.

Example 5 (Cauchian channel).

Let Z *be Cauchy distributed with* $p_Z(z) = \frac{1}{\pi}\frac{a}{a^2+z^2}$, *where* $a > 0$. *Then for any* $\alpha > 0$,

$$\frac{\phi_Z(\alpha\omega)}{\phi_Z(\omega)} = \frac{e^{-a\alpha|\omega|}}{e^{-a|\omega|}} = e^{-a(\alpha-1)|\omega|}$$

is the characteristic function of X^*, *which is Cauchy distributed with parameter* $(\alpha - 1)a$. *However, in this particular case,* $P = \mathbb{E}(X^{*2}) = +\infty$ *and* $N = \mathbb{E}(Z^2) = +\infty$ *so that the signal-to-noise ratio is not defined.*

Lemma 6

Let p_Z *and* p_{Y^*} *be the pdf's of* Z *and* Y^*, *respectively. Then* X^* *attains capacity subject to an average cost per channel use of the form* $\mathbb{E}(b(X)) \leq C$, *where*

$$b(x) = \mathbb{E}\Big(\log_2 \frac{p_Z(Z)}{p_{Y^*}(x+Z)}\Big). \tag{5}$$

Thus given the pdf of Y^*, **(5)** defines an adequate constraint set C so that $C = h(Y^*) - h(Z)$.

Proof

Let p_Y be the pdf of $Y = X + Z$. By the information inequality $D(p_Y \| p_Y{}^*) \geq 0$, we obtain

$$h(Y) \leqslant \mathbb{E}\log_2 \frac{1}{p_{Y^*}(Y)} = \mathbb{E}_X\Big(\mathbb{E}_Z \log \frac{1}{p_{Y^*}(X+Z)}\Big).$$

Therefore,

$$I(X;Y) = h(Y) - h(Z) \leqslant \mathbb{E}_X\Big(\mathbb{E}_Z \log \frac{p_Z(Z)}{p_{Y^*}(X+Z)}\Big) = \mathbb{E}(b(X))$$

Equality holds if and only if $p_Y = p_Y{}^*$, that is, when the channel capacity is attained. In this case max $I(X;Y) = \mathbb{E}(b(X))$ should be equal to the capacity C. The assertion follows.

Example 6 (Gaussian channel (continued)).

Here $Z \sim \mathcal{N}(0,N)$ and $Y^* \sim \mathcal{N}(0, P+N)$. Therefore,

$$b(x) = \log_2\sqrt{\frac{P+N}{N}} + \mathbb{E}\log_2 \exp\Big(\frac{(x+Z)^2}{2(P+N)} - \frac{Z^2}{2N}\Big) = C + \frac{\log_2 e}{2}\Big(\frac{x^2+N}{P+N} - 1\Big).$$

The constraint $\mathbb{E}(b(X)) \leq C$ is now equivalent to $\mathbb{E}(X^2) \leq P$ as expected.

Example 7 (uniform channel (continued)).

Here Z is uniformly distributed on the interval $[-\Delta, \Delta]$ and Y^* is uniformly distributed on $[-A-\Delta, A+\Delta]$ where $A = (\alpha - 1)\Delta > 0$. Therefore,

$$b(x) = \log_2 \frac{A+\Delta}{\Delta} + \mathbb{E}\log \frac{1}{\mathbf{1}_{|x+Z| \leqslant A+\Delta}}$$

where **1** *denotes the indicator function. The first term in the r.h.s. is equal to C. If $|x| \leq A$ then $|x + Z| \leq A+\Delta$ a.e. so that the second term equals* log 1 = 0. *Otherwise,* $\mathbf{1}_{|x+z| \leq A+\Delta}$ *vanishes for z in some subinterval of $[-\Delta, \Delta]$ of positive length and the second term is infinite. Hence*

$$b(x) = \begin{cases} C & \text{if } |x| \leqslant A \\ +\infty & \text{otherwise.} \end{cases}$$

The constraint $\mathbb{E}(b(X)) \leq C$ is equivalent to $|X| \leq A$ a.e. as expected.

Theorem 7

Assume that there exists $\alpha > 1$ *such that* $\frac{\phi_Z(\alpha\omega)}{\phi_Z(\omega)}$ *is a characteristic distribution and let C be defined by the condition* $\mathbb{E}(b(X)) \leq C$ *where*

$$b(x) = \mathbb{E}\log_2 \frac{\alpha p_Z(Z)}{p_Z((x+Z)/\alpha)}. \tag{6}$$

Then the channel capacity $C = \log_2 \alpha$ *of the corresponding additive noise channel is given by Shannon's Formula* (1).

Proof

Apply the preceding lemmas, noting that $p_{Y^*}(y) = \frac{1}{\alpha} p_Z(\frac{y}{\alpha})$.

B-Spline Channels of Any Degree

Equipped with Theorem 7 we can construct many additive noise channels whose capacities are given by Shannon's Formula (**1**).

Definition 8 (B-spline Channel).

Let U_Δ *be uniformly distributed over the interval* $[-\Delta, \Delta]$ *and let* $d \in \mathbb{N}$. *Define*

$$Z_d = U_{\Delta,0} + U_{\Delta,1} + \ldots + U_{\Delta,d}$$

where the $U_{\Delta,i}$ *are independent copies of* U_Δ. *The (uniform)* B-spline channel *of degree d is the associated additive noise channel* $Y = X + Z_d$ *with capacity* C_d.

For $d = 0$ one recovers the uniform channel. It is easily seen and well-known that the pdf of Z_d is the uniform *B*-spline function:

$$p_{Z_d}(z) = \frac{1}{2\Delta} \cdot \beta_d\left(\frac{z}{2\Delta}\right)$$

where β_d is the standard central B-spline [**27**] of order *d*, the $(d + 1)$th convolution power of the indicator function of the interval $[-1/2, 1/2]$.

Theorem 9

For all $d \in \mathbb{N}$ *and any choice of a positive integer M, the capacity* C_d *of the B-spline channel of degree d under the input constraint* $\mathbb{E}(b_d(X)) \leq C_d$ *where*

$$b_d(x) = \mathbb{E}\log_2 \frac{M\beta_d\left(\frac{Z}{2\Delta}\right)}{\beta_d\left(\frac{x+Z}{2M\Delta}\right)}. \tag{7}$$

is given by Shannon's Formula (1).

Proof

Since $p_{Z_d}(z) = \frac{1}{2\Delta} \cdot \beta_d(\frac{z}{2\Delta})$ is the $(d+1)$th convolution power of the rectangle function of the interval $[-\Delta, \Delta]$, the corresponding characteristic function is a $(d+1)$th power of a cardinal sine:

$$\phi_{Z_d}(\omega) = \text{sinc}^{d+1}(\Delta \cdot \omega).$$

Let $M > 0$ be an integer. From Example 4, we have

$$\frac{\phi_{Z_d}(M\omega)}{\phi_{Z_d}(\omega)} = \frac{\text{sinc}^{d+1}(M\Delta \cdot \omega)}{\text{sinc}^{d+1}(\Delta \cdot \omega)} = \left(\frac{\sin(M\Delta \cdot \omega)}{M \sin(\Delta \cdot \omega)}\right)^{d+1}$$
$$= \left(\frac{1}{M}\left(e^{-i(M-1)\omega\Delta} + e^{-i(M-3)\omega\Delta} + \cdots + e^{i(M-1)\omega\Delta}\right)\right)^{d+1}.$$

This is the characteristic function of the random variable

$$X_d = X_{M,0} + \cdots + X_{M,d},$$

where the $X_{M,i}$ are i.i.d. and take M equiprobable values in the set $\{-(M-1)\Delta, -(M-3)\Delta, \ldots, (M-3)\Delta, (M-1)\Delta\}$. Hence, Theorem 7 applies with $\alpha = M$ and cost function (7).

Again for $d = 0$ one recovers the case of the uniform channel with input $X_0 = X_{M,0}$ taking M equiprobable values in the set $\{-(M-1)\Delta, -(M-3)\Delta, \ldots, (M-3)\Delta, (M-1)\Delta\}$ (**Figure 1a**). In general, the probability distribution of X_d is the $(d+1)$th discrete convolution power of the uniform distribution. For $d = 1$, the pdf of the noise has a triangular shape and the distribution of X_d is also triangular (**Figure 1b**). As d increases, it becomes closer to a Gaussian shape (**Figure 1c,d**).

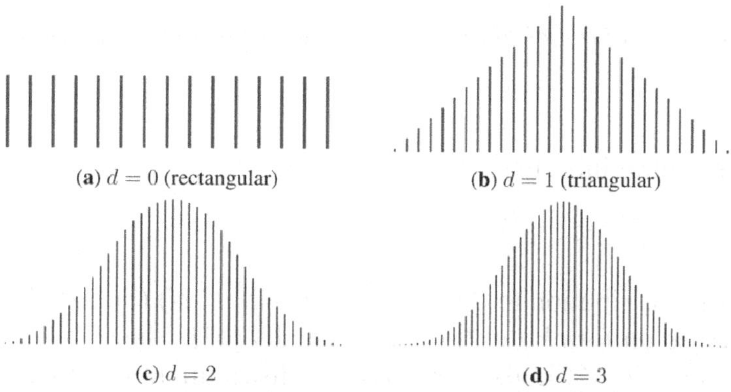

Figure 1: Discrete plots of input probability distributions (of X_d) that attain capacity for $M = 15$ and different values of d.

Convergence as d $\rightarrow +\infty$

To determine the limit behavior as $d \rightarrow +\infty$, we need to apply some normalization on the probability distributions. Since the pdf of Z_d is obtained by successive convolutions of rectangles of length 2Δ, its support $[-(d+1)\Delta, (d+1)\Delta]$ as well as its average power (or variance) $N = (d+1)\Delta^2/3$ increase linearly with $(d+1)$. Similarly, the support and average power P of X^* also increase linearly with $(d+1)$. Although this does not affect the ratio P/N, in order for average powers P and N to converge as $d \rightarrow +\infty$ we need to divide Z_d and X^*, hence their sum Y, by $\sqrt{d+1}$. The capacity will remain unaltered because

$$
\begin{aligned}
I\left(\frac{X}{\sqrt{d+1}}; \frac{Y}{\sqrt{d+1}}\right) &= h\left(\frac{Y}{\sqrt{d+1}}\right) - h\left(\frac{Z}{\sqrt{d+1}}\right) \\
&= h(Y) - \tfrac{1}{2}\log(d+1) - h(Z) + \tfrac{1}{2}\log(d+1) \\
&= h(Y) - h(Z) \\
&= I(X;Y).
\end{aligned}
$$

Therefore, in what follows, we assume that all random variables X, Y,Z have been normalized by the factor $\sqrt{d+1}$. We then say that the additive channel with input X_d, output Y_d, noise Z_d, and cost function $b_d(x)$ converges as $d \rightarrow +\infty$ to the additive channel with input X, output Y, noise Z, and cost function $b(x)$ if $X_d \rightarrow X$, $Y_d \rightarrow Y$, $Z_d \rightarrow Z$ in distribution, and $b_d(x) \rightarrow b(x)$.

Theorem 10

The B-spline channel of degree d converges to the Gaussian channel as $d \rightarrow +\infty$.

Proof

By the central limit theorem,

$$
\frac{Z_d}{\sqrt{d+1}} = \frac{U_{\Delta,0} + U_{\Delta,1} + \ldots + U_{\Delta,d}}{\sqrt{d+1}}
$$

converges in distribution to the Gaussian $Z \sim \mathcal{N}(0,N)$ (in fact, the B-spline pdf converges uniformly to the Gaussian pdf) [27]. Since Y_d has the same distribution as $M \cdot Z_d$, it also converges in distribution to the Gaussian $Y \sim \mathcal{N}(0, P+N)$. Again by the central limit theorem,

$$
\frac{X^*}{\sqrt{d+1}} = \frac{X_0^* + \cdots + X_d^*}{\sqrt{d+1}}
$$

converges in distribution to the Gaussian $\mathcal{N}(0, P)$. Finally, we can write

$$
b_d(x) = \mathbb{E}\left(\log_2 \frac{M p_{Z_d}(Z_d)}{p_{Z_d}(\frac{x+Z_d}{M})}\right) = \mathbb{E}\left(\log_2 \frac{p_{Z_d}(Z_d)}{p_Z(Z_d)}\right) - \mathbb{E}\left(\log_2 \frac{p_{Z_d}(\frac{x+Z_d}{M})}{p_Z(\frac{x+Z_d}{M})}\right) + \mathbb{E}\left(\log_2 \frac{M p_Z(Z_d)}{p_Z(\frac{x+Z_d}{M})}\right)
$$

The first term in the r.h.s. tends to zero by the strengthened central limit theorem of Barron **[28]** in relative entropy. The second term also tends to zero by a similar argument and change of variable. By a calculation identical to that of Example 6, the third term is equal to

$$\log_2 M + \mathbb{E}\log_2 \exp\Bigl(\frac{(x+Z_d)^2}{2(P+N)} - \frac{Z_d^2}{2N}\Bigr) = C + \frac{\log_2 e}{2}\Bigl(\frac{x^2+N}{P+N} - 1\Bigr) = b(x)$$

which shows that $b_d(x) \to b(x)$ as $d \to +\infty$.

Figure 2 shows the graphs of the cost functions $b_d(x)$ for different values of degree d. As the degree increases, the curves converge to the parabola that represents the quadratic cost function $b(x)$ for the Gaussian channel.

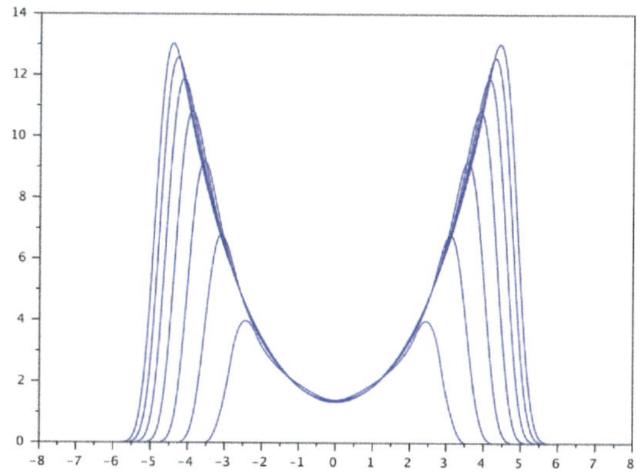

Figure 2: Cost functions $b_d(x)$ for $d = 1$ to 7 (with $M = 4$ and $\Delta = 1$). Convergence holds to the quadratic cost function $b(x)$.

Thus we have built a sequence of additive noise "B-spline" channels indexed by $d \in$ N that makes the transition from the uniform ($d = 0$) to the Gaussian channel ($d \to \infty$). Shannon's Formula (**1**) holds for all these channels.

ACKNOWLEDGMENTS

The authors wish to thank Max H. M. Costa for valuable discussions and suggestions. This work was partially supported by São Paulo Research Foundation (FAPESP) Grant # 2014/13835-6, under the FAPESP thematic project *Segurança e Confiabilidade da Informação: Teoria e Prática*, Grant # 2013/25977-7.

AUTHOR CONTRIBUTIONS

Both authors performed the historical research. Olivier Rioul wrote the paper and carried out the mathematical analysis. Both authors have read and approved the final manuscript.

REFERENCES

1. Shannon, C.E. A mathematical theory of communication. *Bell Syst. Tech. J* 1948, *27*.
2. Shannon, C.E. Communication in the presence of noise. *Proc. Inst. Radio Eng* 1949, *37*, 10–21.
3. Butzer, P.; Dodson, M.; Ferreira, P.; Higgins, J.; Lange, O.; Seidler, P.; Stens, R. Multiplex signal transmission and the development of sampling techniques: The work of Herbert Raabe in contrast to that of Claude Shannon. *Appl. Anal* 2011, *90*, 643–688.
4. Wikipedia. *Shannon–Hartley theorem*. Available online: http://en.wikipedia.org/wiki/Shannon-Hartley_theorem accessed on 28 August 2014.
5. Hartley, R.V.L. Transmission of information. *Bell Syst. Tech. J* 1928, *7*, 535–563.
6. Ellersick, F.W. A conversation with Claude Shannon. *IEEE Commun. Mag* 1984, *22*, 123–126.
7. Wozencraft, J.M.; Jacobs, I.M. *Principles of Communication Engineering*; John Wiley & Sons: New York, NY, USA, 1965; pp. 2–5.
8. Tuller, W.G. Theoretical limitations on the rate of transmission of information. Ph.D. Thesis, Massachusetts Institute of Technology, Cambridge, MA, USA, 1948.
9. Cherry, E.C. A history of information theory. *Proc. Inst. Elect. Eng* 1951, *98*, 383–393.
10. McEliece, R.J. *The Theory of Information and Coding*, 2nd ed; Cambridge University Press: Cambridge, UK, 2002.
11. Pierce, J.R. The early days of information theory. *IEEE Trans. Inf. Theory* 1973, *19*, 3–8.
12. Oliver, B.; Pierce, J.; Shannon, C.E. The Philosophy of PCM. *Proc. Inst. Radio Eng* 1948, *36*, 1324–1331.
13. Gallager, R.; Claude, E. Shannon: A retrospective on his life, work, and impact. *IEEE Trans. Inf. Theory* 2001, *47*, 2681–2695.
14. Verdú, S. Fifty years of Shannon theory. *IEEE Trans. Inf. Theory* 1998, *44*, 2057–2078.
15. Golay, M.J.E. Note on the theoretical efficiency of information reception with PPM. *Proc. Inst. Radio Eng* 1949, *37*, 1031.

16. Slepian, D. Information theory in the fifties. *IEEE Trans. Inf. Theory* 1973, *19*, 145–148.
17. Wiener, N. Time series, Information and Communication. In *Cybernetics*; John Wiley & Sons: New York, NY, USA, 1948; Volume Chapter III, pp. 10–11.
18. Clavier, A.G. Evaluation of transmission efficiency according to Hartley's expression of information content. *Electron. Commun. ITT Tech. J* 1948, *25*, 414–420.
19. Earp, C.W. Relationship between rate of transmission of information, frequency bandwidth, and signal-to-noise ratio. *Electron. Commun. ITT Tech. J* 1948, *25*, 178–195.
20. Goldman, S. Some fundamental considerations concerning noise reduction and range in radar and communication. *Proc. Inst. Radio Eng* 1948, *36*, 584–594.
21. Laplume, J. Sur le nombre de signaux discernables en présence du bruit erratique dans un système de transmission à bande passante limitée. *Comptes rendus de l'Académie des Sciences de Paris* 1948, *226*, 1348–1349. (In French)
22. Lundheim, L. On Shannon and "Shannon's Formula". *Telektronikk* 2002, *98*, 20–29.
23. Wiener, N. What is information theory? *IRE Trans. Inf. Theory* 1956, *2*, 48.
24. Hodges, A. *Alan Turing: The Enigma*; Simon and Schuster: New York, NY, USA, 1983; p. 552.
25. Cover, T.M.; Thomas, J.A. *Elements of Information Theory*, 2nd ed; John Wiley & Sons: New York, NY, USA, 2006.
26. Rioul, O. Information theoretic proofs of entropy power inequalities. *IEEE Trans. Inf. Theory* 2011, *57*, 33–55.
27. Unser, M.; Aldroubi, A.; Eden, M. On the asymptotic convergence of B-spline wavelets to Gabor functions. *IEEE Trans. Inf. Theory* 1992, *38*, 864–872.
28. Barron, A.R. Entropy and the central limit theorem. *Ann. Probab* 1986, *14*, 336–342.

5

A Unified Definition of Mutual Information with Applications in Machine Learning

Guoping Zeng

Elevate, 4150 International Plaza, Fort Worth, TX 76109, USA

ABSTRACT

There are various definitions of mutual information. Essentially, these definitions can be divided into two classes: (1) definitions with random variables and (2) definitions with ensembles. However, there are some mathematical flaws in these definitions. For instance, Class 1 definitions either neglect the probability spaces or assume the two random variables have the same probability space. Class 2 definitions redefine marginal probabilities from the joint probabilities. In fact, the marginal probabilities are given from the ensembles and should not be redefined from the joint probabilities. Both Class 1 and Class 2 definitions assume a joint distribution exists. Yet, they all ignore an important fact that the joint or the joint probability measure is not unique. In this paper, we first present a new unified definition of mutual

Citation: Guoping Zeng. "A Unified Definition of Mutual Information with Applications in Machine Learning" Hindawi Publishing Corporation Mathematical Problems in Engineering Volume 2015, Article ID 201874, 12 pages http://dx.doi.org/10.1155/2015/201874.

information to cover all the various definitions and to fix their mathematical flaws. Our idea is to define the joint distribution of two random variables by taking the marginal probabilities into consideration. Next, we establish some properties of the newly defined mutual information. We then propose a method to calculate mutual information in machine learning. Finally, we apply our newly defined mutual information to credit scoring.

INTRODUCTION

Mutual information has emerged in recent years as an important measure of statistical dependence. It has been used as a criterion for feature selection in engineering especially in machine learning (see [1–3] and references therein).

Mutual information is a concept rooted in information theory. Its predecessor, called the rate of transmission, was first introduced by Shannon in 1948 in a classical paper [4] for the communication system. Shannon first introduced a concept called entropy for a single discrete chance variable. He then defined the joint entropy and conditional entropy for two discrete chance variables using the joint distribution. Finally, he defined the rate of transmission as the difference between the entropy and conditional entropy. While Shannon did not define a chance variable in his paper, it is understood to be a synonym of a random variable.

Since Shannon's pioneering work [4], there have been various definitions for mutual information. Essentially, these definitions can be divided into two classes: (1) definitions with random variables and (2) definitions with ensembles, that is, probability spaces in the mathematical literature.

Class 1 definitions of mutual information depend on the joint distribution of two random variables. More specifically, Kullback ([5], 1959) defined entropy, conditional entropy, and joint entropy using compact mathematical formulas. Pinsker ([6], 1960 and 1964) treated the fundamental concepts of Shannon in a more advanced manner by employing probability theory. His definition of mutual information was more general in that he implicitly assumed the two random variables had different probability spaces. Ash ([7], 1965) explicitly assumed the two random variables had the same probability space and followed Shannon's way to define mutual information. Cover and Thomas ([8], 2006) defined mutual information in a simple way by avoiding mentioning probability spaces.

Class 2 definitions depend on the joint probability measure of the joint sample space of two ensembles. Among such definitions, Fano ([9], 1961), Abramson ([10], 1963), and Gallager ([11], 1968) developed their definitions in a similar way. They first defined the entropy of an ensemble, condition entropy, and the joint entropy of two ensembles. Next, they defined the mutual information of a joint event. Noting that the mutual information of a joint event is a random variable, they calculated the mean value of this random variable and called the result the mean information of two ensembles.

However, there are some mathematical flaws in these various definitions of mutual information. Class 2 definitions redefine marginal probabilities from the joint probabilities. As a matter of fact, the marginal probabilities are given from the ensembles and hence should not be redefined from the joint probabilities. Moreover, except Pinsker's definition, Class 2 definitions either neglect the probability spaces or assume the two random variables have the same probability space. Both Class 1 definitions and Class 2 definitions assume a joint distribution or a joint probability measure exists. Yet, they all ignore an important fact that the joint distribution or the joint probability measure is not unique.

In this paper, we first present a unified definition for mutual information using random variables with different probability spaces. Our idea is to define the joint distribution of two random variables by taking the marginal probabilities into consideration. With our new definition of mutual information, different joint distributions will result in different mutual information. Next, we establish some properties of the newly defined mutual information. We then propose a method to calculate mutual information in machine learning. Finally, we apply our newly defined mutual information to credit scoring.

The rest of the paper is organized as follows. In Section 2, we briefly review the basic concepts in probability theory. In Section 3, we examine various definitions of mutual information. In Section 4, we first propose a new unified definition for mutual information and then establish some properties of the newly defined mutual information. In Section 5, we first propose a method to calculate mutual information in machine learning. We then apply the newly defined mutual information to credit scoring. The paper is concluded in Section 6.

Throughout the paper, we will restrict our focus on mutual information for finite discrete random variables.

BASIC CONCEPTS IN PROBABILITY THEORY

Let us review some basic concepts of probability theory. They can be found in many books in probability theory such as [12].

Definition 1. A probability space is a triple (Ω, F, P),, where

(1) Ω is a set, called a sample space, and elements of Ω are denoted by ω and are called outcomes,

(2) F is an σ-field consisting of all subsets of Ω, and elements of F are called events,

(3) P is called a probability measure, and it is a mapping from F to [0, 1] with (Ω) = 1 such that if A_1, A_2,... are pairwise disjoint,

$$P\left(\bigcup_i A_i\right) = \sum_i P(A_i). \tag{1}$$

Definition 2. A discrete probability space is a probability space (Ω, F, P) such that Ω is finite or countable: $\Omega = \{\omega_1, \omega_2, ...,\}$. In this case, F is chosen to be all the subsets of Ω, and the probability measure P can be defined in terms of a series of nonnegative numbers p_1, p_2,... whose sum is 1. If A is any subset of Ω, then

$$P(A) = \sum_{\omega_i \in A} p_i. \tag{2}$$

In particular,

$$P(\{\omega_i\}) = p_i. \tag{3}$$

For simplicity, we will write $(\{\omega\})$ as $P(\omega)$. From Definition 2, we see that for a discrete probability space the probability measure is characterized by the pointwise mapping $p: \{\omega_1, \omega_2, \ldots,\} \rightarrow [0, 1]$ in (2). The probability of an event A is computed simply by adding the probabilities of the individual points of A.

Definition 3. A randomvariable X on probability space (Ω, F, P) is a Borel measurable function from Ω to $(-\infty,\infty)$ such that, for every Borel set $B, X^{-1}(B) = \{X \in B\} \in$ F. Here, we use notation $\{X \in B\} = \{\omega \in \Omega : X(\omega) \in B\}$.

Definition 4. If X is a random variable, then for every Borel subset B of R, we define a function by $\mu(B) = P(X \in B) = P(X^{-1}(B))$. Then μ_X is a probability measure on (R, B(R), $_X$) and is called the probability distribution of X.

Definition 5. A random variable X is discrete if its range is finite or countable. In particular, any random variable on a discrete probability space is discrete, since Ω is countable.

Definition 6. A (discrete) random variable X on a discrete probability space (Ω, F, P) is a Borel measurable function from Ω to R, where $\Omega = \{\omega_1, \omega_2, \ldots,\}$ and R is the set of real numbers. If the range of X is $\{x_1, x_2, \ldots\}$, then function $f_X : \{x_1, x_2, \ldots\} \rightarrow [0, 1]$ defined by

$$f_X(x_i) = P(X = x_i) \tag{4}$$

is called the probability mass function of X, whereas probabilities $P(X = x_1)$, $P(X = x_2)$, . . . are called the probability distribution of X.

Note that, in Definition 2,

$$\sum_i P(X = x_i) = 1. \tag{5}$$

Thus, a discrete random variable may be characterized by its probability mass function.

VARIOUS DEFINITIONS OF MUTUAL INFORMATION

Since Shannon's pioneering work [4], there have been various definitions for mutual information. Essentially, these definitions can be divided into two classes: (1) definitions with random variables and (2) definitions with ensembles, that is, probability spaces in the mathematical literature.

Shannon's Original Definition

Definition 7. Let x be a chance variable with probabilities $p_1, p_2, \ldots, _n,$ whose sum is 1. Then

$$H(x) = -\sum_{i=1}^{n} p_i \log p_i \tag{6}$$

is called the entropy of x.

Suppose two chance variables, x and y, have m and n possibilities, respectively. Let indices i and j range over all the m possibilities and all the n possibilities, respectively. Let (i) be the probability of i and $p(i, j)$ the probability of the joint occurrence of i and j. Denote the conditional probability of i given j by $(i \mid j)$ and conditional probability of j given i by $p(j \mid i)$.

Definition 8. The joint entropy of x and y is defined as

$$H(x, y) = -\sum_{i,j} p(i, j) \log p(i, j). \tag{7}$$

Definition 9. The conditional entropy of y, $H(y)$, is defined as

$$H_x(y) = -\sum_{i,j} p(i,j)\log p(j \mid i)$$

$$= -\sum_{i,j} p(i,j)\log \frac{p(i,j)}{\sum_j p(i,j)}. \qquad (8)$$

The conditional entropy of x, $H(x)$, can be defined similarly.

Then the following relation holds:

$$H(x,y) = H(x) + H_x(y) = H(y) + H_y(x),$$

$$H(x) - H_y(x) = H(y) - H_x(y)$$

$$= H(x) + H(y) - H(x,y). \qquad (9)$$

Definition 10. The rate of transmission of information R is defined as the difference between (x) and $H_y(x)$. Then R can be written in two other forms:

$$R = H(x) - H_y(x) = H(y) - H_x(y)$$

$$= H(x) + H(y) - H(x,y). \qquad (10)$$

Remark 11. Shannon did not derive the explicit formula for :

$$R = \sum_{i,j} p(i,j)\log \frac{p(i,j)}{p(i)\,p(j)}. \qquad (11)$$

However, he did imply it in Appendix 7 [4].

Class 1 Definitions

Kullback's Definition

Kullback [5] redefined entropy more mathematically in a standalone homework question as follows. Consider two discrete random variables x, y, where

$$p_{ij} = \text{Prob}\left(x = x_i, y = y_j\right) > 0,$$

$$i = 1, 2, \ldots, m, \quad j = 1, 2, \ldots, n,$$

$$p_{i\cdot} = \sum_{j=1}^{n} p_{ij},$$

$$p_{\cdot j} = \sum_{i=1}^{m} p_{ij},$$

$$\sum_{i=1}^{m}\sum_{j=1}^{n} p_{ij} = \sum_{i=1}^{m} p_{i\cdot} = \sum_{j=1}^{n} p_{\cdot j} = 1. \tag{12}$$

Define the joint entropy, entropy, and conditional entropy as follows:

$$H(x, y) = -\sum_i \sum_j p_{ij} \log p_{ij},$$

$$H(x) = -\sum_i p_{i\cdot} \log p_{i\cdot},$$

$$H(y) = -\sum_j p_{\cdot j} \log p_{\cdot j},$$

$$H(y \mid x_i) = -\sum_j \frac{p_{ij}}{p_{i\cdot}} \log \frac{p_{ij}}{p_{i\cdot}},$$

$$H(y \mid x) = \sum_i p_i H(y \mid x_i) = -\sum_i \sum_j p_{ij} \log \frac{p_{ij}}{p_{i\cdot}}. \tag{13}$$

Then $(x, y) = H(x) + H(y \mid x) \leq H(x) + H(y)$ and $H(y) \geq H(y \mid x)$.

Information Conveyed

Ash [7] began with two random variables X and Y and assumed X and Y had the same probability space. He systematically defined the entropy, conditional entropy, and joint entropy following Shannon's path in [4]. At the end, he denoted $(X) - H(X \mid Y)$ by $I(X \mid Y)$ and called it the information conveyed about X by Y.

Information of One Variable with respect to the Other

Information of One Variable with respect to the Other. Pinsker [6] treated the fundamental concepts of Shannon in a more advanced manner by employing probability theory. Suppose ξ is a random variable defined on a probability space $(\Omega, S_\omega, P_\xi)$ and is taking values in a measurable space (X, S_x) and η is a random variable defined on a probability space (Ψ, S_ψ, P_η) and is taking values in a measurable space (Y, S_y). Then, the pair ξ, η of random variables may be regarded as a single random variable (ξ, η) with values in the product space $X \times Y$ of all pairs (x, y) with $x \in X$, $y \in Y$. The distribution $_{(\xi,\eta)}(\cdot) = P_{\xi\eta}(\cdot)$ of (ξ, η) is called the joint distribution of

random variables ξ and η. By the product of the distributions $P(\cdot)$ and $P_\eta(\cdot)$, denoted by $P_{\xi\times\eta}(\cdot)$, we mean the distribution defined on $S_x \times S_y$

$$P_{\xi\times\eta}(E \times F) = P_\xi(E) \times P_\eta(F) \tag{14}$$

for $E \in S_x$ and $F \in S_y$. If the joint distribution $P_\xi(\cdot)$ coincides with the product distribution $P_{\xi\times\eta}(\cdot)$, the random variables ξ and η are said to be independent. If ξ and η are discrete random variables, say, X and Y contain countably many points, $x_1, x_2, \ldots$ and $y_1, y_2, \ldots$, then

$$I(\xi, \eta) = \sum_{i,j} P_{\xi\times\eta}(x_i, y_j) \log \frac{P_{\xi\eta}(x_i, y_j)}{P_\xi(x_i) P_\eta(y_j)}. \tag{15}$$

I is called the information of ξ and η with respect to the other.

A Modern Definition in Information Theory

Of the various definitions of mutual information, the most widely accepted, of recent years, is the one by Cover and Thomas [8].

Let X be a discrete variable with alphabet χ and probability mass function $(x) = \Pr\{X = x\}$, $x \in \chi$. Let Y be a discrete variable with alphabet Y and probability mass function $(y) = \Pr\{Y = y\}$, $y \in$ Y. Suppose X and Y have a joint mass function (joint distribution) $p(x, y)$. Then the mutual information (X, Y) can be defined as

$$I(X, Y) = \sum_{x\in\chi} \sum_{y\in Y} p(x, y) \log \frac{p(x, y)}{p(x)\, p(y)}. \tag{16}$$

Class 2 Definitions

In Class 2 definitions, random variables are replaced by ensembles and mutual information is so-called average mutual information. Gallager [11] adopted a more general and more rigorous approach to introduce the concept of mutual information in communication theory. Indeed, he combined and compiled the results from Fano [9] and Abramson [10].

Suppose that discrete ensemble X has a sample space $\{a_1, a_2, \ldots,_K\}$ and discrete ensemble Y has a sample space $\{b_1, b_2, \ldots, b_L\}$. Consider the joint sample space $\{(a_k, b_j): 1 \le k \le K,\ 1 \le j \le J\}$. A probability measure on the joint sample space is given by the join probability $P_X(a_k, b_j)$, defined for

$1 \leq k \leq K$, $1 \leq j \leq J$. The combination of a joint sample space and probability measure for outcomes x and y is called a joint XY ensemble. Then the marginal probabilities can be found as

$$P_X(a_k) = \sum_{j=1}^{J} P_{XY}(a_k, b_j), \quad k = 1, 2, \ldots, K. \tag{17}$$

In more abbreviated notation, this is written as

$$P(x) = \sum_{y} P(x, y). \tag{18}$$

Likewise,

$$P_Y(b_j) = \sum_{k=1}^{K} P_{XY}(a_k, b_j), \quad j = 1, 2, \ldots, J. \tag{19}$$

In more abbreviated notation, this is written as

$$P(y) = \sum_{x} P(x, y). \tag{20}$$

If $P(a_k)>0$, the conditional probability that outcome y is b_j, given that outcome of x is a_k, is defined as

$$P_{Y|X}(b_j \mid a_k) = \frac{P_{XY}(a_k, b_j)}{P_X(a_k)}. \tag{21}$$

The mutual information between events $x=a_k$ and $y=b_j$ is defined as

$$I_{X:Y}(a_k, b_j) = \log \frac{P_{X|Y}(a_k \mid b_j)}{P_X(a_k)} = \log \frac{P_{XY}(a_k, b_j)}{P_X(a_k) P_Y(b_j)}$$
$$= \log \frac{P_{X|Y}(a_k \mid b_j)}{P_X(a_k)} = I_{Y:X}(b_j, a_k). \tag{22}$$

Since the mutual information defined above is a random variable on the XY joint ensemble, the mean value, which is called the average mutual information denoted by (X, Y), is given by

$$I(X, Y) = \sum_{k=1}^{K} \sum_{j=1}^{L} P_{XY}(a_k, b_j) \log \frac{P_{XY}(a_k, b_j)}{P_X(a_k) P_Y(b_j)}. \tag{23}$$

Remark 12. By means of an information channel consisting of a transmitter of alphabet A with elements a_i and total elements t and a receiver of alphabet B with elements b_j and total elements r, Abramson [10] denoted $H(A) - H(A \mid B) = \sum_{A,B} P(a, b) \log(P(a, b)/P(a)P(b))$ by $I(A, B)$ and called it mutual information of A and B.

The mutual information (X, Y) between 2 continuous random variables X and Y [8] (also called rate of transmission in [1]) is defined as

$$I(X,Y) = \iint P(x,y) \log \frac{P(x,y)}{P(x)P(y)} dx\, dy, \qquad (24)$$

where $P(x, y)$ is the joint probability density function of X and Y and $P(x)$ and $P(y)$ are the marginal density functions associated with X and Y, respectively. The mutual information between 2 continuous random variables is also called the differential mutual information.

However, the differential mutual information is much less popular than its discrete counterpart. On the one hand, the joint density function involved is unknown in most cases and hence must be estimated [13, 14]. On the other hand, data in engineering and machine learning are mostly finite, and so mutual information between discrete random variables is used.

A NEW UNIFIED DEFINITION OF MUTUAL INFORMATION

In Section 3, we reviewed various definitions of mutual information. Shannon's original definition laid the foundation of information theory. Kullback's definition used random variables for the first time and was more mathematical and more compact. Although Ash's definition followed Shannon's path, it was more systematic. Pinsker's definition was most mathematical in that it employed probability theory. Gallager's definition was more general and more rigorous in communication theory. Cover and Thomas's definition is so succinct that it is now a standard definition in information theory.

However, there are some mathematical flaws in these various definitions of mutual information. Class 2 definitions redefine marginal probabilities from the joint probabilities. As a matter of fact, the marginal probabilities are given from the ensembles and hence should not be redefined from the joint probabilities. Except Pinsker's definition, Class 2 definitions either neglect the probability spaces or assume the two random variables have the same probability space. Both Class 1 definitions and Class 2 definitions assume a joint distribution or a joint probability measure exists. Yet, they all ignore an important fact that the joint distribution or the joint probability measure is not unique.

Unified Definition of Mutual Information

Let X be a finite discrete random variable on discrete probability space (Ω_1, F_1, P_1) with $\Omega_1 = \{\omega_1, \omega_2,...,_n\}$ and range $\{x_1, x_2, ...,x_K\}$ with $K \leq n$. Let Y be a discrete random variable on probability space (Ω_2, F_2, P_2) with $\Omega_2 = \{\rho_1, \rho_2,...,_m\}$ and range $\{y_1, y_2,...,y_L\}$ with $L \leq m$.

If X and Y have the same probability space (Ω, F, P), then thc joint distribution is simply

$$P_{XY}(X = x, Y = y) = P(\omega \in \Omega : X(\omega) = x,\ Y(\omega) = y). \tag{25}$$

However, when X and Y have different probability spaces and so different probability measures, the joint distribution is more complicated.

Definition 13. The joint sample space of random variables X and Y is defined as the product $\Omega_1 \times \Omega_2$ of all pairs (ω_i, ρ_j), $i = 1, 2, \ldots, n$ and $j = 1, 2, \ldots, m$. The joint σ-field $F_1 \times F_2$ is defined as the product of all pairs (A_1, A_2), where A_1 and A_2 are elements of F_1 and F_2, respectively. A joint probability measure P_{XY} of P_1 and P_2 is a probability measure on $F_1 \times F_2 : P_{XY}(A \times B)$ such that for any $A \subseteq \Omega_1$ and $B \subseteq \Omega_2$

$$P_1(A) = P_{XY}(A \times \Omega_2) = \sum_{j=1}^{m} P_{XY}(A \times \{\rho_j\}),$$

$$P_2(B) = P_{XY}(\Omega_1 \times B) = \sum_{i=1}^{m} P_{XY}(\{\omega_i\} \times B), \tag{26}$$

($\Omega_1 \times \Omega_2$, $F_1 \times F_2$, P_{XY}) is called the joint probability space of X and Y, and $PX(\{X = x_i\} \times \{Y = y_j\})$ for $i = 1, 2, \ldots, n$ and $j = 1, 2, \ldots, m$ the joint distribution of X and Y.

Combining Definitions 2 and 13, we immediately obtain the following results.

Proposition 14. A sequence of nonnegative numbers $\{p_i : 1 \leq i \leq K;\ 1 \leq j \leq L\}$ whose sum is 1 can serve as a probability measure on $F_1 \times F_2 : P_{XY}(\omega_i, \rho_j) = p_{ij}$. The probability of any event $A \times B \subseteq \Omega_1 \times \Omega_2$ is computed simply by adding the probabilities of the individual points of $(\omega, \rho) \in A \times B$. If, in addition, for $i = 1, 2, \ldots, K$ and $j = 1, 2, \ldots, L$, the following hold:

$$\sum_{j=1}^{L} p_{ij} = P_X(\omega_i),$$

$$\sum_{i=1}^{K} p_{ij} = P_Y(\rho_j), \tag{27}$$

then $P_{XY}(\omega_i, \rho_j)=p_{ij}$ is a joint distribution of X and Y.

For convenience, from now on we will shorten $P_X(\{X = x_i\} \times \{Y = y_j\})$ as $P_{XY}(x_i, y_j)$.

This two-dimensional measure should not be confused with one-dimensional joint distribution when X and Y have the same probability space.

Remark 15. If $(\Omega_1, F_1, P_1) = (\Omega_2, F_2, P_2)$, instead of using the two-dimensional measure $P_X(\{X = x_i\} \times \{Y = y_j\})$, we may use the one-dimensional measure $P_1(\{X = x_i\}$ and $\{Y = y_j\})$. Then, (26) always hold. In this sense, our new definition of joint distribution reduces to the definition of joint distribution with the same probability space.

Definition 16. The conditional probability $Y=y_j$, given $X = x_i$, is defined as

$$P_{Y|X}\left(Y = y_j \mid X = x_i\right) = \frac{P_{XY}\left(x_i, y_j\right)}{P_1\left(X = x_i\right)}. \tag{28}$$

Theorem 17. For any two discrete random variables, there is at least one joint probability measure called the product probability measure or simply product distribution.

Proof. Let random variables X and Y be defined as before. Define a function from $\Omega_1 \times \Omega_2$ to [0, 1] as follows:

$$P_{XY}\left(\omega_i, \rho_j\right) = P_1\left(\omega_i\right) P_2\left(\rho_j\right). \tag{29}$$

Then

$$\sum_{i=1}^{n}\sum_{j=1}^{m} P_{XY}\left(\omega_i, \rho_j\right) = \sum_{i=1}^{n}\sum_{j=1}^{m} P_1\left(\omega_i\right) P_2\left(\rho_j\right)$$

$$= \sum_{i=1}^{n} P_1\left(\omega_i\right) \sum_{j=1}^{m} P_2\left(\rho_j\right) = 1. \tag{30}$$

Hence, P_{XY} can serve as a probability measure on $F_1 \times F_2$ by Definition 2. The probability of any event $A\times B \subseteq \Omega_1 \times \Omega_2$ is computed simply by adding the probabilities of the individual points of $(\omega, \rho) \in A \times B$. Moreover, for any $A = \{\omega_{i_1}, \omega_{i_2}, \ldots, \omega_{i_s}\} \in \Omega_1$ of s elements,

$$P_{XY}(A\times\Omega_2)=\sum_{j=1}^{m}P_{XY}(A\times\{\rho_j\})$$
$$=\sum_{j=1}^{m}\sum_{u=1}^{s}P_{XY}(\{\omega_{i_u}\}\times\{\rho_j\})$$
$$=\sum_{j=1}^{m}\sum_{u=1}^{s}P_1(\omega_{i_u})P_2(\rho_j)$$
$$=\sum_{j=1}^{m}P_2(\rho_j)\sum_{u=1}^{s}P_1(\omega_{i_u})$$
$$=\sum_{u=1}^{s}P_1(\omega_{i_u})=P_1(A). \tag{31}$$

Similarly, $P_X(\Omega_1\times B)=P_2(B)$ for any $B\in\Omega_2$. Hence, $P_X(\{X=x_i\}\times\{Y=y_j\}) = P_1(X=x_i)P_2(Y=y_j)$ is a joint probability measure of X and Y by Definition 13.

Definition 18. Random variables X and Y are said to be independent under a joint distribution $P_X(\cdot)$ if $P_{XY}(\cdot)$ coincides with the product distribution $P_{X\times Y}(\cdot)$.

Definition 19. The joint entropy (X, Y) is defined as

$$H(X,Y)=-\sum_{i=1}^{K}\sum_{j=1}^{L}P_{XY}(x_i,y_j)\log P_{XY}(x_i,y_j). \tag{32}$$

Definition 20. The conditional entropy is as follows:

$$H(Y\mid X)=-\sum_{i=1}^{K}\sum_{j=1}^{L}P_{XY}(x_i,y_j)\log P_{Y|X}(Y=y_j\mid X=x_i)$$
$$=-\sum_{i=1}^{K}\sum_{j=1}^{L}P_{XY}(x_i,y_j)\log\frac{P_{XY}(x_i,y_j)}{P_1(X=x_i)}. \tag{33}$$

Definition 21. The mutual information (X, Y) between X and Y is defined as

$$I(X,Y)=\sum_{i=1}^{K}\sum_{j=1}^{L}P_{XY}(x_i,y_j)\log\frac{P_{XY}(x_i,y_j)}{P_1(X=x_i)P_2(Y=y_j)}. \tag{34}$$

As other measures in information theory, the base of logarithm in (34) is left unspecified. Indeed, (X, Y) under one base is proportional to that under another base by the change-of-base formula. Moreover, we take 0 log 0 to be 0. This corresponds to the limit of $x \log x$ as x goes to 0.

It is obvious that our new definition covers Class 2 definitions. It also covers Class 1 definitions by the following arguments. Let $\Omega_1 = \{a_1, a_2,\ldots,_K\}$ and $\Omega_2 = \{b_1, b_2,\ldots,b_L\}$. Define random variables $X:\Omega_1 \rightarrow$ R and $Y:\Omega_2 \rightarrow$ R as one-to-one mappings as

$$X(a_i) = x_i, \quad i = 1, 2, \ldots, K,$$

$$Y(b_j) = y_j, \quad j = 1, 2, \ldots, L. \tag{35}$$

Then we have

$$P_{XY}(x_i, y_j) = P_{XY}(a_i, b_j). \tag{36}$$

It is worth noting that our new definition of mutual information has some advantages over various existing definitions. For instance, it can be easily used to do feature selection as seen later. In addition, our new definition leads different values for different joint distribution as demonstrated in the following example.

Example 22. Assumerandom variables X and Y have the following probability distributions:

$$P_1(Y = 0) = \frac{1}{3}, \qquad P_1(Y = 1) = \frac{2}{3},$$

$$P_2(X = 1) = \frac{1}{3}, \quad P_2(X = 2) = \frac{1}{3}, \quad P_2(X = 3) = \frac{1}{3}. \tag{37}$$

We can generate four different joint probability distributions to lead 4 different values of mutual information. However, under all the existing definitions, a joint distribution must be given in order to find mutual information:

(1) $(1, 0) = 0$, $P(1, 1) = 1/3$, $P(2, 0) = 1/3$, $P(2, 1) = 0$, $P(3, 0) = 0$, $P(3, 1) = 1/3$;

(2) $(1, 0) = 0$, $P(1, 1) = 1/3$, $P(2, 0) = 0$, $P(2, 1) = 1/3$, $P(3, 0) = 1/3$, $P(3, 1) = 0$;

(3) $(1, 0) = 1/3$, $P(1, 1) = 0$, $P(2, 0) = 0$, $P(2, 1) = 1/3$, $P(3, 0) = 0$, $P(3, 1) = 1/3$;

(4) $(1, 0) = 1/9$, $P(1, 1) = 2/9$, $P(2, 0) = 1/9$, $P(2, 1) = 2/9$, $P(3, 0) = 1/9$, $P(3, 1) = 2/9$.

Properties of Newly Defined Mutual Information

Before we discuss some properties of mutual information, we first introduce Kullback-Leibler distance [8].

Definition 23. The relative entropy or Kullback-Leibler distance between two discrete probability distributions $P = \{p_1, p_2, \ldots,_n\}$ and $Q = \{q_1, q_2, \ldots, q_n\}$ is defined as

$$D(P \parallel Q) = \sum_i p_i \log \frac{p_i}{q_i}. \tag{38}$$

Lemma 24 (see [8]). Let P and Q be two discrete probability distributions. Then $(P \,\square\, Q) \geq 0$ with equality if and only if $p_i = q_i$ for all i.

Remark 25. The Kullback-Leibler distance is not a true distance between distributions since it is not symmetric and does not satisfy the triangle inequality either. Nevertheless, it is often useful to think of relative entropy as a "distance" between distributions.

The following property shows that mutual information under a joint probability measure is the Kullback-Leibler distance between the joint distribution P_{XY} and the product distribution $P_X P_Y$.

Property 1. Mutual information of random variables X and Y is the Kullback-Leibler distance between the joint distribution P_{XY} and the product distribution $P_1 P_2$.

Proof. Using a mapping from 2-dimensional indices to one-dimensional index,

$$(i, j) \longrightarrow (i-1) * L + j \triangleq n$$
$$\text{for } i = 1, \ldots, K, \quad j = 1, 2, \ldots, L, \tag{39}$$

and using another mapping from one-dimensional index back to two-dimensional indices,

$$i = \left\lceil \frac{n}{L} \right\rceil, \qquad j = n - (i-1) * L$$
$$\text{for } n = 1, 2, \ldots, L, L+1, \ldots, 2L, \ldots, (K-1)L + \cdots + KL, \tag{40}$$

we rewrite $I(X, Y)$ as

$$\begin{aligned} I(X,Y) &= \sum_{i=1}^{K}\sum_{j=1}^{L} P_{XY}(x_i, y_j) \log \frac{P_{XY}(x_i, y_j)}{P_1(X = x_i) P_2(Y = y_j)} \\ &= \sum_{n=1}^{KL} P_{XY}(x_{\lceil n/L \rceil}, y_{n-(\lceil n/L \rceil -1)*L}) \\ &\quad \cdot \log \frac{P_{XY}(x_{\lceil n/L \rceil}, y_{n-(\lceil n/L \rceil -1)*L})}{P_1(X = x_{\lceil n/L \rceil}) P_2(Y = y_{n-(\lceil n/L \rceil -1)*L})}. \end{aligned} \tag{41}$$

Since

$$\sum_{n=1}^{KL} P_{XY}\left(x_{\lceil n/L\rceil}, y_{n-(\lceil n/L\rceil-1)*L}\right) = \sum_{i=1}^{K}\sum_{j=1}^{L} P_{XY}\left(x_i, y_j\right) = 1,$$

$$\sum_{n=1}^{KL} P_1\left(X = x_{\lceil n/L\rceil}\right) P_2\left(Y = y_{n-(\lceil n/L\rceil-1)*L}\right)$$

$$= \sum_{i=1}^{K}\sum_{j=1}^{L} P_1\left(X = x_i\right) P_2\left(Y = y_j\right) = 1, \quad (42)$$

we obtain

$$I(X,Y) = \sum_{n=1}^{KL} P_{XY}\left(x_{\lceil n/L\rceil}, y_{n-(\lceil n/L\rceil-1)*L}\right) \cdot \log \frac{P_{XY}\left(x_{\lceil n/L\rceil}, y_{n-(\lceil n/L\rceil-1)*L}\right)}{P_1\left(X = x_{\lceil n/L\rceil}\right) P_2\left(Y = y_{n-(\lceil n/L\rceil-1)*L}\right)}. \quad (43)$$

Property 2. Let X and Y be two discrete random variables. The mutual information between X and Y satisfies

$I(X, Y) \geq 0$ (44)

with equality if and only if X and Y are independent.

Proof. Let us use the mappings between two-dimensional indices and one-dimensional index in the proof of Property 1. By Lemma 24, $I(X, Y) \geq 0$ with equality if and only if $P_{XY}(x_{\lceil n/L\rceil}, y_{n-(\lceil n/L\rceil-1)*L}) = P_1(X = x_{\lceil n/L\rceil}) P_2(Y = y_{n-(\lceil n/L\rceil-1)*L})$ for $n = 1, 2, \ldots, L, L + 1, \ldots, 2L, \ldots, (K - 1)L + \cdots + K$, that is, $P_{XY}(x_i, y_j) = P_1(X = x_i) P_2(Y = y_j)$ for $i = 1, \ldots, K$ and $j = 1, 2, \ldots, L$, or X and Y are independent.

Corollary 26. If X is a constant random variable, that is, $K = 1$, then, for any random variable Y,

$$I(X, Y) = 0. \quad (45)$$

Proof. Suppose the range of X is a constant x and the sample space has only one point ω. Then, $P_1(X = x) = P_1(\omega) = 1$. For any $j = 1, 2, \ldots, L$,

$$P_{XY}\left(x, y_j\right) = \sum_{i=1}^{1} P_{XY}\left(x, y_j\right) = P_2\left(Y = y_j\right)$$

$$= P_1\left(X = x\right) P_2\left(Y = y_j\right). \quad (46)$$

Thus, X and Y are independent. By Property 2, $(X, Y) = 0$.

Lemma 27 (see [8]). Let be discrete random variables with K values. Then

$0 \leq H(X) \leq \log K$ (47)

with equality if and only if the values are equally probable.

Property 3. Let X and Y be two discrete random variables. Then the following relationships among mutual information, entropy, and conditional entry hold:

$$I(X,Y) = H(X) - H(X \mid Y)$$
$$= H(Y) - H(Y \mid X) = I(Y,X). \tag{48}$$

Proof. Consider

$$I(X,Y) = \sum_{i=1}^{K}\sum_{j=1}^{L} P_{XY}(x_i, y_j)\log\frac{P_{XY}(x_i, y_j)}{P_1(X = x_i)\,P_2(Y = y_j)}$$

$$= \sum_{i=1}^{K}\sum_{j=1}^{L} P_{XY}(x_i, y_j)\log\frac{P_{Y|X}(x_i, y_j)}{P_2(Y = y_j)}$$

$$= -\sum_{i=1}^{K}\sum_{j=1}^{L} P_{XY}(x_i, y_j)\log P_2(Y = y_j)$$

$$+\sum_{i=1}^{K}\sum_{j=1}^{L} P_{XY}(x_i, y_j)\log P_{Y|X}(x_i, y_j)$$

$$= -\sum_{i=1}^{K}\sum_{j=1}^{L} P_{XY}(x_i, y_j)\log P_2(Y = y_j)$$

$$-\left(-\sum_{i=1}^{K}\sum_{j=1}^{L} P_{XY}(x_i, y_j)\log P_{Y|X}(x_i, y_j)\right)$$

$$= -\sum_{j=1}^{L}\sum_{i=1}^{L} P_{XY}(x_i, y_j)\log P_2(Y = y_j)$$

$$-\left(-\sum_{i=1}^{K}\sum_{j=1}^{L} P_{XY}(x_i, y_j)\log P_{Y|X}(x_i, y_j)\right)$$

$$= -\sum_{j=1}^{L} P_2(Y = y_j)\log P_2(Y = y_j)$$

$$-\left(-\sum_{i=1}^{K}\sum_{j=1}^{L} P_{XY}(x_i, y_j)\log P_{Y|X}(x_i, y_j)\right)$$

$$= H(Y) - H(Y \mid X). \tag{49}$$

Combining the above properties and noting that $(X \mid Y)$ and $H(Y \mid X)$ are both nonnegative, we obtain the following properties.

Property 4. Let X and Y be 2 discrete random variables with and values, respectively. Then

$$0 \le I(X,Y) \le H(Y) \le \log L,$$
$$0 \le I(X,Y) \le H(X) \le \log K. \tag{50}$$

Moreover, if and only if and are independent.

NEWLY DEFINED MUTUAL INFORMATION IN MACHINE LEARNING

Machine learning is the science of getting machines (computers) to automatically learn from data. In a typical learning setting, a training set S contains N examples (also known as samples, observations, or records) from an input space $X = \{X_1, X_2, \ldots, _M\}$ and their associated output values y from an output space Y (i.e., dependent variable). Here, $X_1, X_2, \ldots, _M$ are called features, that is, independent variables. Hence, S can be expressed as

$$S = \{x_{i1}, x_{i2}, \ldots, x_{iM}, y_i\}, \quad i = 1, 2, \ldots, N, \tag{51}$$

where feature X_j has values $x_{1j}, x_{2j}, \ldots, x_{Nj}$ for $j = 1, 2, \ldots, M$.

A fundamental objective in machine learning is to find a functional relationship between input X and output Y. In general, there are a very large number of features, many of which are not needed. Sometimes, the output Y is not determined by the complete set of the input features $\{X_1, X_2, \ldots, _M\}$. Rather, it is decided by only a subset of them. This kind of reduction is called feature selection. Its purpose is to choose a subset of features to capture the relevant information. An easy and natural way for feature selection is as follows.

(1) Evaluate the relationship between each individual input feature x_i and the output Y.

(2) Select the best set of attributes according to some criterion.

Calculation of Newly Defined Mutual Information

Since mutual information measures dependency between random variables, we may use it to do feature selection in machine learning. Let us calculate mutual information between an input feature X and output Y. Assume X has K different values $\omega_1, \omega_2, \ldots, _K$. If X has missing values, we will use ω_1 to represent all the missing values. Assume Y has L different values $\rho_1, \rho_2, \ldots, _L$.

Let us build a two-way frequency or contingency table by making X as the row variable and Y as the column variable like in [8]. Let O_{ij} be the frequency (could be 0) of (ω_i, ρ_j) for i=1 to K and j=1 to L. Let the row and column marginal totals be $n_i \cdot$ and $n_{\cdot j}$, respectively. Then

$$\begin{aligned} n_{i\cdot} &= \sum_j O_{ij}, \\ n_{\cdot j} &= \sum_i O_{ij}, \\ N &= \sum_i \sum_j O_{ij} = \sum_i n_{i\cdot} = \sum_j n_{\cdot j}. \end{aligned} \tag{52}$$

Let us denote the relative frequency O_{ij}/N by p_{ij}. We have the two-way relative frequency table; see Table 2.

Since

$$\sum_{i=1}^{K}\sum_{j=1}^{L} p_{ij} = \sum_{i=1}^{K} p_{i\cdot} = \sum_{j=1}^{L} p_{\cdot j} = 1, \quad 53)$$

$\{p_{i\cdot}\}_{i=1}^{K}$, $\{p_{\cdot j}\}_{j=1}^{L}$, and $\{p_{ij}\}_{i=1}^{K}$ can each serve as a probability measure.

Now we can define random variables for X and Y as follows. For convenience, we will use the same names X and Y for the random variables

$$X : (\Omega_1, \mathscr{F}_1, P_X) \longrightarrow R \quad (54)$$

as $X(\omega_i)=xi$, where $\Omega_1 = \{\omega_1, \omega_2,...,\omega_K\}$ and $P_X(\{\omega_i\}) = n_i\cdot/N = p_i\cdot$ for $i = 1, 2, \ldots, K$. Note that $x_1, x_2,...,_K$ could be any real numbers as long as they are distinct to guarantee that X is a one-to-one mapping. In this case, $P(X = x_i)=P_X(\{\omega_i\})$.

Similarly,

$$Y : (\Omega_2, \mathscr{F}_2, P_2) \longrightarrow R \quad (55)$$

as $Y(\rho_j)=y_j$, where $\Omega_2 = \{\rho_1, \rho_2,...,\rho_L\}$ and $P_Y(\{\rho_j\}) = n_{\cdot j}/N = p\cdot_j$ for $j = 1, 2, \ldots, L$. Also, $y_1, y_2,...,_K$ could be any real numbers as long as they are distinct to guarantee that Y is a one-to-one mapping. In this case, $P(Y = y_j)=P_Y(\{\rho_j\})$.

Now define a mapping P_{XY} from $\Omega_1 \times \Omega_2$ to R as follows:

$$P_{XY}(\omega_i, \rho_j) = p_{ij} = \frac{O_{ij}}{N}. \quad (56)$$

Since

$$\sum_{i=1}^{K}\sum_{j=1}^{L} P_{XY}(\omega_i, \rho_j) = 1,$$

$$\sum_{j=1}^{L} P_{XY}(\omega_i, \rho_j) = \sum_{j=1}^{L} p_{ij} = p_{i\cdot} = P_X(\{\omega_i\}),$$

$$\sum_{i=1}^{K} P_{XY}(\omega_i, \rho_j) = \sum_{i=1}^{K} p_{ij} = p_{\cdot j} = P_Y(\{\rho_j\}), \quad (57)$$

$\{p_{ij}\}_{i=1}^{K}$ is a joint probability measure by Proposition 14. Finally, we can calculate mutual information as follows:

$$I(X,Y) = \sum_{i=1}^{K}\sum_{j=1}^{L} P_{XY}(\omega_i,\rho_j)\log\frac{P_{XY}(\omega_i,\rho_j)}{P_X(\omega_i)P_Y(\rho_j)}$$

$$= \sum_{i=1}^{K}\sum_{j=1}^{L} p_{ij}\log\frac{p_{ij}}{p_{i\cdot}p_{\cdot j}}. \qquad (58)$$

It follows from Corollary 26 that if X has only one value, then $(X, Y) = 0$. On the other hand, if X has all distinct values, the following result shows that mutual information will reach the maximum value.

Proposition 28. If all the values of X are distinct, then $(X, Y) = H(Y)$.

Proof. If all the values of X are distinct, then the number of different values of X equals the number of observations; that is, $K=N$. From Tables 1 and 2, we observe that

(1) $O_{ij} = 0$ or 1 for all $i = 1, 2, \ldots, K$ and $j = 1, 2, \ldots, L$,

(2) $p_{ij} = O_{ij}/N = 0$ or $1/N$ for all $i = 1, 2, \ldots, K$ and $j = 1, 2, \ldots, L$,

(3) for each $j = 1, 2, \ldots, L$, since $O_{1j}+O_{2j}+\cdots+O_{Kj} = n_{\cdot j}$, there are $n{\cdot}j$ nonzero O_{ij}'s or equivalently $n_{\cdot j}$ nonzero p_{ij}'s,

(4) $p_{i}\cdot = 1/N$, $i = 1, 2, \ldots, K$.

Using the above observations and the fact that 0 log 0=0 we have

$$I(X,Y) = \sum_{i=1}^{K}\sum_{j=1}^{L} p_{ij}\log\frac{p_{ij}}{p_{i\cdot}p_{\cdot j}}$$

$$= \sum_{i=1}^{K} p_{i1}\log\frac{p_{i1}}{p_{i\cdot}p_{\cdot 1}} + \sum_{i=1}^{K} p_{i2}\log\frac{p_{i2}}{p_{i\cdot}p_{\cdot 2}}$$

$$+\cdots+\sum_{i=1}^{K} p_{iL}\log\frac{p_{iL}}{p_{i\cdot}p_{\cdot L}}$$

$$= \sum_{p_{i1}\neq 0}\frac{1}{N}\log\frac{1/N}{p_{\cdot 1}/N} + \sum_{p_{i2}\neq 0}\frac{1}{N}\log\frac{1/N}{p_{\cdot 2}/N}$$

$$+\cdots+\sum_{p_{iL}\neq 0}\frac{1}{N}\log\frac{1/N}{p_{\cdot L}/N}$$

$$= \sum_{p_{i1}\neq 0}\frac{1}{N}\log\frac{1}{p_{\cdot 1}} + \sum_{p_{i2}\neq 0}\frac{1}{N}\log\frac{1}{p_{\cdot 2}}$$

$$+\cdots+\sum_{p_{iL}\neq 0}\frac{1}{N}\log\frac{1}{p_{\cdot L}}$$

$$= \frac{n_{\cdot 1}}{N}\log\frac{1}{p_{\cdot 1}} + \frac{n_{\cdot 2}}{N}\log\frac{1}{p_{\cdot 2}} +\cdots+ \frac{n_{\cdot L}}{N}\log\frac{1}{p_{\cdot L}}$$

$$= p_{\cdot 1} \log \frac{1}{p_{\cdot 1}} + p_{\cdot 2} \log \frac{1}{p_{\cdot 2}} + \cdots + p_{\cdot L} \log \frac{1}{p_{\cdot L}}$$

$$= H(Y). \quad (59)$$

Table 1: Frequency table

	ρ_1	ρ_2	$\cdots$	ρ_j	$\cdots$	ρ_L	Total
ω_1	O_{11}	O_{12}	$\cdots$	O_{1j}	$\cdots$	O_{1L}	$n_{1\cdot}$
ω_2	O_{21}	O_{22}	$\cdots$	O_{2j}	$\cdots$	O_{2L}	$n_{2\cdot}$
$\cdots$	$\cdots$	$\cdots$	$\cdots$	$\cdots$	$\cdots$	$\cdots$	$\cdots$
ω_i	O_{i1}	O_{i2}	$\cdots$	O_{ij}	$\cdots$	O_{iL}	$n_{i\cdot}$
$\cdots$	$\cdots$	$\cdots$	$\cdots$	$\cdots$	$\cdots$	$\cdots$	$\cdots$
ω_K	O_{K1}	O_{K2}	$\cdots$	O_{Kj}	$\cdots$	O_{KL}	$n_{K\cdot}$
Total	$n_{\cdot 1}$	$n_{\cdot 2}$	$\cdots$	$n_{\cdot j}$	$\cdots$	$n_{\cdot L}$	N

Table 2: Relative frequency table

	ρ_1	ρ_2	$\cdots$	ρ_j	$\cdots$	ρ_L	Total
ω_1	p_{11}	p_{12}	$\cdots$	p_{1j}	$\cdots$	p_{1L}	$p_{1\cdot}$
ω_2	p_{21}	p_{22}	$\cdots$	p_{2j}	$\cdots$	p_{2L}	$p_{2\cdot}$
$\cdots$	$\cdots$	$\cdots$	$\cdots$	$\cdots$	$\cdots$	$\cdots$	$\cdots$
ω_i	p_{i1}	p_{i2}	$\cdots$	p_{ij}	$\cdots$	p_{iL}	$p_{i\cdot}$
$\cdots$	$\cdots$	$\cdots$	$\cdots$	$\cdots$	$\cdots$	$\cdots$	$\cdots$
ω_K	p_{K1}	p_{K2}	$\cdots$	p_{Kj}	$\cdots$	p_{KL}	$p_{K\cdot}$
Total	$p_{\cdot 1}$	$p_{\cdot 2}$	$\cdots$	$p_{\cdot j}$	$\cdots$	$p_{\cdot L}$	1

Applications of Newly Defined Mutual Information in Credit Scoring

Credit scoring is used to describe the process of evaluating the risk a customer poses of defaulting on a financial obligation [15–19]. The objective is to assign customers to one of two groups: good and bad. Machine learning has been successfully used to build models for credit scoring [20]. In credit scoring, Y is a binary variable, good and bad, and may be represented by 0 and 1. To apply mutual information to credit scoring, we first calculate mutual information for every pair of(X, Y) and then do feature selection based on values of mutual information. We propose three ways.

Absolute Values Method

From Property 4, we see that mutual information (X, Y) is nonnegative and upper bounded by $\log(L)$ and that $I(X, Y) = 0$ if and only if X and Y are independent. In this sense, high mutual information indicates a large reduction in uncertainty, while low mutual information indicates a small reduction. In particular, zero mutual information means the two random variables are independent. Hence, we may select those features whose mutual information with Y is larger than some threshold based on needs.

Relative Values

From Property 4, we have $0 \leq (X,Y)/H(Y) \leq 1$. Note that $(X,Y)/H(Y)$ is relative mutual information, which measures how much information X catches from Y. Thus, we may select those features whose relative mutual information $(X, Y)/H(Y)$is larger than some threshold between 0 and 1 based on needs.

Chi-Square Test for Independency

For convenience, we will use the natural logarithm in mutual information. We first state an approximation formula for the natural logarithm function. It can be proved by the Taylor expansion like in Kullback's book [5].

Lemma 29. Let p and q be two positive numbers less than or equal to 1. Then

$$p \ln \frac{p}{q} \approx (p-q) + \frac{(p-q)^2}{2q}. \qquad (60)$$

The equality holds if and only if p=q. Moreover, the close p is to q, the better the approximation is.

Now let us denote $N \times I(X,Y)$ by $\hat{I}(X,Y)$. Then, applying Lemma 29, we obtain

$$2\hat{I}(X,Y) = 2N\sum_{i=1}^{K}\sum_{j=1}^{L} p_{ij} \ln \frac{p_{ij}}{p_{i\cdot}p_{\cdot j}}$$

$$= 2\sum_{i=1}^{K}\sum_{j=1}^{L} O_{ij} \ln \frac{O_{ij}/N}{(n_{i\cdot}/N)(n_{\cdot j}/N)}$$

$$= 2\sum_{i=1}^{K}\sum_{j=1}^{L} O_{ij} \ln \frac{O_{ij}}{n_{i\cdot}n_{\cdot j}/N}$$

$$\approx 2\sum_{i=1}^{K}\sum_{j=1}^{L}\left(O_{ij} - \frac{n_{i\cdot}n_{\cdot j}}{N}\right)$$

$$+\sum_{i=1}^{K}\sum_{j=1}^{L}\frac{(O_{ij} - n_{i\cdot}n_{\cdot j}/N)^2}{n_{i\cdot}n_{\cdot j}/N}$$

$$= 2\sum_{i=1}^{K}\sum_{j=1}^{L} O_{ij} - 2\frac{\sum_i n_{i\cdot}}{N}\frac{\sum_j n_{\cdot j}}{N}$$

$$+\sum_{i=1}^{K}\sum_{j=1}^{L}\frac{(O_{ij} - n_{i\cdot}n_{\cdot j}/N)^2}{n_{i\cdot}n_{\cdot j}/N}$$

$$= 2N - 2\frac{N}{N} + \sum_{i=1}^{K}\sum_{j=1}^{L}\frac{(O_{ij} - n_{i\cdot}n_{\cdot j}/N)^2}{n_{i\cdot}n_{\cdot j}/N}$$

$$= \sum_{i=1}^{K}\sum_{j=1}^{L}\frac{\left(O_{ij} - n_{i.}n_{.j}/N\right)^2}{n_{i.}n_{.j}/N}$$

$$= \chi^2. \tag{61}$$

The last equation means the previous expression $\sum_{i=1}^{K}\sum_{j=1}^{L}((O_{ij} - n_{i.}n_{.j}/N)^2/(n_{i.}n_{.j}/N))$ follows χ^2 distribution. According to [5], it follows χ^2 distribution with a degree of freedom of $(K - 1)(L - 1)$. Hence, $2N \times (X, Y)$ approximately follows χ^2 distribution with a degree of freedom of $(K - 1)(L - 1)$. This is the well-known Chi-square test for independence of two random variables. This allows using the Chi-square distribution to assign a significant level corresponding to the values of mutual information and $(K - 1)(L - 1)$.

The null and alternative hypotheses are as follows.

H_0: X and Y are independent (i.e., there is no relationship between them).

H_1: X and Y are dependent (i.e., there is a relationship between them).

The decision rule is to reject the null hypothesis at the α level of significance if the χ^2 statistic

$$\sum_{i=1}^{K}\sum_{j=1}^{L}\frac{\left(O_{ij} - n_{i.}n_{.j}/N\right)^2}{n_{i.}n_{.j}/N} \approx 2N \times I(X,Y) \tag{62}$$

is greater than χ_U^2, the upper-tail critical value from a Chisquare distribution with $(K - 1)(L - 1)$ degrees of freedom. That is,

$$\text{Select feature } X \text{ if } I(X,Y) > \frac{\chi_U^2}{2N}. \tag{63}$$

Take credit scoring, for example. In this case, L=2. Assume feature X has 10 different values; that is, $K = 10$. Using a level of significance of $\alpha = 0.05$, we find χ_U^2 to be 16.9 from a Chisquare table with $(K - 1)(L - 1) = 9$, and select this feature only if $I(X, Y) > 16.9/2N$.

Assume a training set has N examples. We can do feature selection by the following procedure.

(i) Step 1. Choose a level of significance of α, say, 0.05.

(ii) Step 2. Find K, the number of values of feature X.

(iii) Step 3. Build the contingency table for X and Y.

(iv) Step 4. Calculate (X, Y) from the contingency table.

(v) Step 5. Find χ^2_U with $(K-1)(L-1)$ degrees of freedom from a Chi-square table or any other sources such as SAS.

(vi) Step 6. Select X if $(X, Y) > 16.9/2N$ and discard it otherwise. (vii) Step 7. Repeat Steps 2–6 for all features.

If the number of features selected from the above procedure is smaller or larger than what you want, you may adjust the level of significant α and reselect features using the procedure.

Adjustment of Mutual Information in Feature Selection

In Section 5.2, we have proposed 3 ways to select feature based on mutual information. It seems that the larger the mutual information (X, Y), the more dependent X on Y. However, Proposition 28 says that if X has all distinct values, then (X, Y) will reach the maximum value $H(Y)$ and $I(X, Y)/H(Y)$ will reach the maximum value 1.

Therefore, if X has too many different values, one may bin or group these values first. Based on the binned values, mutual information is calculated again. For numerical variables, we may adopt a three-step process.

(i) Step 1: select features by removing those with small mutual information.

(ii) Step 2: do binning for the rest of numerical features.

(iii) Step 3: select features by mutual information.

Comparison with Existing Feature Selection Methods

There are many other feature selection methods in machine learning and credit scoring. An easy way is to build a logistic model for each feature with respect to the dependent variable and then select features with p values less than some specific values. However, this method does not apply to any nonlinear models in machine learning.

Another easy way of feature selection is to calculate the covariance of each feature with respect to the dependent variable and then select features whose values are larger than some specific value. Yet, mutual information is better than the covariance method [21] in that mutual information measures the general dependence of random variables without making any assumptions about the nature of their underlying relationships.

The most popular feature selection in credit scoring is done by information value [15–19]. To calculate information between an independent variable and the dependent variable, a binning algorithm is used to group similar attributes into a bin. Difference between the information of good accounts and that of bad accounts in each bin is then calculated. Finally, information value is calculated as the sum of information differences of all bins. Features with information value larger than 0.02 are believed to have strong predictive power. However, mutual information is a better measure than information value. Information value focuses only on the linear relationships of variables, whereas mutual information can potentially offer some advantages information value for nonlinear models such as gradient boosting model [22]. Moreover, information value depends on binning algorithms and the bin size. Different binning algorithms and/or different bin sizes will have different information value.

CONCLUSIONS

In this paper, we have presented a unified definition for mutual information using random variables with different probability spaces. Our idea is to define the joint distribution of two random variables by taking the marginal probabilities into consideration. With our new definition of mutual information, different joint distributions will result in different values of mutual information. After establishing some properties of the new defined mutual information, we proposed a method to calculate mutual information in machine learning. Finally, we applied our newly defined mutual information to credit scoring.

ACKNOWLEDGMENTS

The author has benefited from a brief discussion with Dr. Zhigang Zhou and Dr. Fuping Huang of Elevate about probability theory.

REFERENCES

1. G. D. Tourassi, E. D. Frederick, M. K. Markey, and C. E. Floyd Jr., "Application of the mutual information criterion for feature selection in computer-aided diagnosis," Medical Physics, vol. 28, no. 12, pp. 2394–2402, 2001.
2. I. Guyon and A. Elisseeff, "An introduction to variable and feature selection," Journal of Machine Learning Research, vol. 3, pp. 1157–1182, 2003.
3. A. Navot, On the role of feature selection in machine learning [Ph.D. thesis], Hebrew University, 2006.
4. C. E. Shannon, "A mathematical theory of communication," The Bell System Technical Journal, vol. 27, no. 3, pp. 379–423, 1948.
5. S. Kullback, Information Theory and Statistics, John Wiley & Sons, New York, NY, USA, 1959.
6. M. S. Pinsker, Information and Information Stability of Random Variables and Processes, Academy of Science, USSR, 1960, (English Translation by A. Feinstein in 1964 and published by Holden-Day, San Francisco, USA).
7. R. B. Ash, Information Theory, Interscience Publishers, New York, NY, USA, 1965.
8. T. M. Cover and J. A. Thomas, Elements of Information Theory, John Wiley & Sons, New York, NY, USA, 2nd edition, 2006.
9. R. M. Fano, Transmission of Information, MIT Press, Cambridge, Mass, USA; John Wiley & Sons, New York, NY, USA, 1961.
10. N. Abramson, Information Theory and Coding, McGraw-Hill, New York, NY, USA, 1963.
11. R. G. Gallager, Information Theory and Reliable Communication, John Wiley & Sons, New York, NY, USA, 1968.
12. R. B. Ash and C. A. Doleans-Dade, Probability & Measure Theory, Academic Press, San Diego, Calif, USA, 2nd edition, 2000.
13. I. Braga, "A constructive density-ratio approach to mutual information estimation: experiments in feature selection," Journal of Information and Data Management, vol. 5, no. 1, pp. 134–143, 2014.
14. L. Paninski, "Estimation of entropy and mutual information," Neural Computation, vol. 15, no. 6, pp. 1191–1253, 2003.

15. M. Refaat, Credit Risk Scorecards: Development and Implementation Using SAS, Lulu.com, New York, NY, USA, 2011.
16. N. Siddiqi, Credit Risk Scorecards: Developing and Implementing Intelligent Credit Scoring, John Wiley & Sons, New York, NY, USA, 2006.
17. G. Zeng, "Metric divergence measures and information value in credit scoring," Journal of Mathematics, vol. 2013, Article ID 848271, 10 pages, 2013.
18. G. Zeng, "A rule of thumb for reject inference in credit scoring," Mathematical Finance Letters, vol. 2014, article 2, 2014.
19. G. Zeng, "A necessary condition for a good binning algorithm in credit scoring," Applied Mathematical Sciences, vol. 8, no. 65, pp. 3229–3242, 2014.
20. K. Kennedy, Credit scoring using machine learning [Ph.D. thesis], School of Computing, Dublin Institute of Technology, Dublin, Ireland, 2013.
21. R. J. McEliece, The Theory of Information and Coding, Cambridge University Press, Cambridge, UK, Student edition, 2004.
22. J. H. Friedman, "Greedy function approximation: a gradient boosting machine," The Annals of Statistics, vol. 29, no. 5, pp. 1189–1232, 2001.

6

Memory Effects can make the Transmission Capability of a Communication Channel Uncomputable

David Elkouss [1] & David Pérez-García[2,3]

[1] QuTech, Delft University of Technology, Lorentzweg 1, 2628 CJ Delft, The Netherlands
[2]Departamento de Análisis Matemático and Instituto de Matemática Interdisciplinar, Universidad Complutense de Madrid, 28040 Madrid, Spain
[3] ICMAT, c/ Nicolás Cabrera, Campus de Cantoblanco, 28049 Madrid, Spain

ABSTRACT

Most communication channels are subjected to noise. One of the goals of information theory is to add redundancy in the transmission of information so that the information is transmitted reliably and the amount of information

Citation: Elkouss, D., Pérez-García, D. "Memory effects can make the transmission capability of a communication channel uncomputable". Nat Commun 9, 1149 (2018). https://doi.org/10.1038/s41467-018-03428-0

transmitted through the channel is as large as possible. The maximum rate at which reliable transmission is possible is called the capacity. If the channel does not keep memory of its past, the capacity is given by a simple optimization problem and can be efficiently computed. The situation of channels with memory is less clear. Here we show that for channels with memory the capacity cannot be computed to within precision 1/5. Our result holds even if we consider one of the simplest families of such channels—information-stable finite state machine channels—restrict the input and output of the channel to 4 and 1 bit respectively and allow 6 bits of memory.

INTRODUCTION

The need to manipulate large amounts of information is one of the main characteristics of our society. It is crucial to protect the information against noise and errors in order to ensure its reliable transmission and long-term storage. It is important also to do so in the optimal way so that communication channels transmit and memories store trustworthily as much information as possible. This problem motivated Shannon[1], already in 1948, to develop the theory of communications. The natural problem that Shannon posed is, given a noisy communication channel, find the maximum rate of information it can transmit with an arbitrarily small error.

In an ingenuity tour de force, he proved that for channels that keep no memory of their past uses (called memoryless), this quantity—the capacity of the channel—defined in such operational way, has a simple entropic expression. It coincides with the maximization, on all inputs to the channel, of the so-called mutual information between input and output in one single use of the channel. This coding result was complemented years later by the Blahut-Arimoto (BA) algorithm[2,3], which allows to efficiently approximate the capacity of any memoryless channel within any desired precision.

The situation for channels with memory is less clear. Regarding coding theorems, more and more general classes of channels were successfully dealt with[4,5,6,7] culminating in the generalized capacity formula[8]. In this last work, Verdu and Han derived a generalization of Shannon's coding theorem which essentially makes no assumption regarding the structure of the channel. When it comes to algorithms that approximate the capacity, despite considerable effort, the situation is nowadays less successful. Even if we restrict to the simplest case of channels with memory, the so-called finite state machine channels (FSMCs), the problem remains open. There is a rich literature dealing with particular cases (see e.g. refs. 9·10·11·12·13·14·15·16·17·18·19). However, these results do not address FSMCs in full generality or cannot guarantee the precision of the result.

It is the main aim of this work to show that an algorithm that computes approximately the capacity of an arbitrary FSMC cannot exist. Since computable functions are exactly those that can be computed by an algorithm, this is equivalent to show that any function that approximates sufficiently the capacity of any FSMC must necessarily be uncomputable.

RESULTS

Main Statements

Aiming at an impossibility result, the simpler the family of channels we consider, the stronger the result. This is why we consider FSMCs. The same result hence holds true for any more general family of channels with memory.

In order to be precise, an FSMC with n possible input symbols (the number of possible output symbols will be always 2) and m possible states in the memory is determined by9 a set of conditional probability assignments. The set of conditional probability assignments p(y,s|x,s′) describes the probability of output symbol y and transition to state s in the memory if the FSMC is in state s' and gets x as input. Moreover, we will only consider FSMCs in which the initial state is fixed and known to the sender and receiver. We denote the initial state by s_0.

To avoid problems of approximating $p(y, s|x, s')$ we will only consider FSMCs for which the probability assignments $p(y, s|x, s')$ are rational numbers. Moreover, we will only consider FSMCs for which $p(y, s|x, s')$ are in product form $p(y|x, s')\ p(s|x, s')$ and which are information stable. Information stable channels are one of the simplest classes of channels with memory. For these channels the capacity is given by the limit of the mutual information rate20 and it is not necessary to consider the most general capacity formula8.

Our main result can then be stated as

Main Result 1: Any function that on input the set of probability assignments $\{p(y|x, s'), p(s|x, s')\}_{s,y,x,s'}$ of an information stable FSMC **N** with

10 input symbols and 62 states, outputs a rational number c so that the capacity of **N** verifies

$$|C(\mathbf{N}) - c| \leq \frac{1}{5}, \qquad (1)$$

must be uncomputable.

It is obvious that the same result then holds for n input symbols and m states as long as $n \geq 10$ and $m \geq 62$. For example, taking $n=16$ and $m=64$ we get a channel with 4 bits of input, 1 bit of output and 6 bits of memory.

Indeed, we will prove something slightly stronger. Let us recall that a decision problem can be cast as a function with values in {0, 1}, where 1 stands for accept and 0 for reject. When the associated function is uncomputable, the decision problem is called undecidable.

Fix a rational number $\lambda \in (0, 1]$. We will give explicitly a subfamily S_λ of FSMCs (information stable and with rational conditional probability assignments in product form) with 10 input symbols and 62 states, with the additional property that all channels $N \in S_\lambda$ have capacity $\geq \lambda$ or $\leq \lambda/2$.

Main Result 2: It is undecidable to know whether $N \in S_\lambda$, given by its set of probability assignments $\{p(y|x, s'), p(s|x, s')\}_{s, y, x, s'}$, has capacity $\geq \lambda$ or $\leq \lambda/2$.

It is clear that if we consider our Main Result 2 for S_1 we get Main Result 1. That is, if we could approximate the capacity within error 1/5 then, given a channel from S_1 for which we know its capacity is $\leq 1/2$ or ≥ 1, we could decide which is the case. However, we know by Main Result 2 that the problem is undecidable.

Proof Sketch

The idea behind our proof is to construct a family of channels such that the capacity of a channel in the family is related to some property of a probabilistic finite automaton (PFA). Our construction is indirect, we first give a map form PFAs to FSMCs; then we define the channel family as the set of FSMCs that are the image of a PFA via this map. The important property of this map, proved in Theorem 1, is that the capacity of a channel in the image set is given by the value of its preimage PFA (see the PFA section). We now sketch the structure of the proof and point to the appropriate sections for further detail.

FSMCs, are controlled by a finite state machine. The state of the finite state machine determines the (memoryless) channel that is applied to the input. Then depending both on the input and the current state it transitions probabilistically to the next state. A PFA is a finite state machine that transitions probabilistically from state to state depending on the current state and the input (see Fig. 1). Hence, it is possible to identify the finite state machine controlling an FSMC with a PFA.

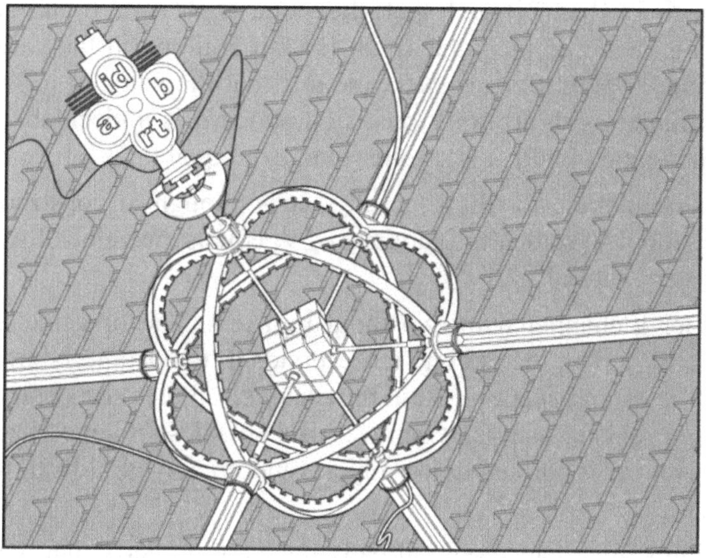

Figure 1: A noisy Rubik cube solver as an example of a probabilistic finite automaton (PFA). This PFA has as many states as different Rubik cube configurations. It begins in some predefined state and can be manipulated with four different buttons or input alphabet symbols: $\{a, b, \text{id}, \text{rt}\}$. A Rubik cube can be solved by combinations of only two sequences of rotations36. The press of the buttons a, b will, with some probability, implement one of these two sequences and with the complementary probability apply a random rotation. The buttons id, rt will make the state of the Rubik cube either stay idle or bring it back to the initial state. The accepting state is the solved configuration of the cube. The value of this automaton would be the maximum probability of taking the initial configuration to the solved configuration by pressing a sequence of buttons (Credit: Francisco García Moro).

A concrete input into a PFA, that is a sequence of input symbols, is accepted if after reading the input the PFA ends in a subset of the states called accepting states, otherwise the input is rejected. Informally, the value of a PFA is the maximum probability of ending in an accepting state. It turns out that many decision problems related to the value cannot be solved. Notably, given some value $\lambda \in (0, 1)$ and a PFA A it is undecidable to know if the value of A is greater than λ21,22,23. Here, we use a recent proof of this result by Hirvensalo24. In order to prove a result about approximations, we amplify this result about decision problems with a very original PFA construction by Gimbert and Oualhadj25, which is the key ingredient in our proof. With this construction, it is possible to embed any PFA A into a larger PFA B_λ (with $\lambda \in [0, 1/2]$) such that: the value of B_λ is $\leq\lambda$ if and only if the value of A is $\leq 1/2$ and the value of B_λ is 2λ if and only if the value of A is $>1/2$. Joining both arguments, we conclude that the value of a PFA cannot be approximated with arbitrary precision, since it is undecidable to know whether the value of a PFA is smaller than λ or equal to 2λ.

To go from there to Main Result 2 it is enough to construct for any PFA A a channel V_A so that the capacity of V_A equals the value of A. The idea for that is very natural:

Consider a channel with two input registers. The first one is used to control the PFA. The second one corresponds to the data to be transmitted. If the PFA is in an accepting state the channel outputs the contents of the second input register, that is, it behaves as a noiseless channel. Otherwise it is only noise, i.e. it outputs uniformly at random a symbol from the output alphabet.

Intuitively, this map should already have the property that the capacity of V_A equals the value of A. However, without an additional gadget, we cannot conclude this. Let us see with an example why it does not suffice. Consider for instance a PFA that transitions from the initial state to an accepting state with probability 1/2 and with probability 1/2 to some other state. Moreover, suppose that these two states are final in the sense that the PFA cannot leave them once reached. Such a PFA would have value 1/2. However, the capacity of the associated channel would be zero because the error probability of any code would always be greater than 1/4. In order to solve this problem we concatenate the map with a function $\gamma(\cdot)$ from PFAs to PFAs that adds to the PFA a reset and a freeze symbols. The reset symbol takes the PFA back to the initial state while the freeze symbol keeps the state of the PFA unchanged. That the value of an automaton A does not change under this map, i.e. the value of A equals the value of γ(A). But, for PFAs with the additional reset and freeze symbols we can show the desired result that the capacity of the channel V_γ(A) equals the value of the automaton A. This is our main technical result, proved in Theorem 1.

The intuition between the equality of the capacity of channel V_A and the value of A, val_A, is as follows. For any $\delta > 0$ there exists a word with value greater than $val_A - \delta$. By feeding this word into the control register, the channel will transition into a final state with probability at least $val_A - \delta$. The state of the channel can then be frozen making the mutual information rate tend to $val_A - \delta$. However, this rate might not be achievable. In order to show achievability then, we induce a memoryless channel by choosing for the control input a periodic sequence that ends with a reset symbol. More concretely, for $\delta > 0$, the sequence consists of: a word with a value larger than $val_A - \delta$, a number of freeze symbols that guarantee an information rate larger than $val_A - 2$ and a reset symbol. In the other direction, one would not expect a capacity larger than val_A. The reason is that the channel outputs a symbol uniformly at random when it is in a non-final state and this happens with probability at least $1 - val_A$.

Finally, note that at this point we do not know yet that the channel is information stable. Indeed, the proof of this fact will use crucially Theorem 1.

Formal Statements of the Main Results

So far we have introduced the notion of uncomputable functions as those that cannot be computed with an algorithm (similarly the notion of undecidable problems). In order to make this definition, and hence the Main Results, mathematically rigorous, we have to recall the definition of a Turing Machine (TM) as the formal definition of what an algorithm is. For more details one can consult for instance26,27.

A TM represents a machine with a finite set of states that can read from and write to an infinitely long memory in the form of a tape. The tape is divided into cells that can hold a single symbol from a finite alphabet. Initially, the tape contains some arbitrary but finite string that we call the input followed by an infinite sequence of blank symbols. The operation of the machine is controlled by a head that sits on top of a cell of the tape. The head operates as follows: it reads the symbol below it; then, depending on the symbol and the current state it writes a symbol, moves left or right and transitions to a new state. The set of states includes the halting state. The TM halts after it transitions to the halting state. The output of the TM consists of the, possibly empty, string of symbols starting from the leftmost non-blank symbol to the rightmost non-blank symbol.

Formally, a TM is defined by a triple $M=(Q,\Sigma,\delta)$ where Q represents the finite set of states including an initial and a halting state, Σ is the finite set of symbols that a cell may contain and it includes the blank symbol and $\delta:(Q\times\Sigma)\mapsto(Q\times\Sigma\times\{L,R\})$ is the transition function.

A configuration is a complete description of the status of a TM. It consists of the current state, the contents of the tape and the position of the head. In the initial configuration, the tape contains the input string and the head of the TM is in the initial state and situated on top of the leftmost cell of the input. Once the initial configuration is fixed a TM evolves deterministically and may or may not eventually halt.

Let us fix $n=10$, $m=62$. In order to specify an FSMC with n input symbols and m states, it is enough to give $N=nm(2+m)=39{,}680$ rational numbers corresponding to the conditional probability assignments. It is very easy to construct an injective map $\sigma(\cdot)$ from vectors of N positive rational numbers to the natural numbers, which then can be transformed into a valid input of a TM. For instance, it would be transformed into a string of zeroes and ones if $\Sigma=\{0, 1, \#\}$. Main Results 1 and 2 can be then respectively restated as:

Main Result 1: There does not exist any TM that halts on all inputs of the form $\sigma(\mathbf{N})$ for $\mathbf{N}\in S_1$ and outputs a rational number c such that the capacity of $\mathbf{N}$ verifies

$$|C(\mathbf{N}) - c| \le \frac{1}{5}. \tag{2}$$

Main Result 2: There does not exist any TM that halts on all inputs of the form $\sigma(\mathbf{N})$ for $\mathbf{N}\in S_\lambda$ and outputs 1 if the capacity of $\mathbf{N} \ge \lambda$ and 0 if the capacity of $\mathbf{N} \le \lambda/2$.

DISCUSSION

We have proven that no algorithm can exist that approximates the capacity for all information stable FSMCs to any desired precision.

Our construction builds directly on top of several strong undecidability results of PFAs. Recent developments underlying these results suggest that it should be possible to reduce the dimensions of our construction28. It is an interesting problem to find the minimal dimensions for which uncomputability holds.

It is important to notice also that the channels appearing in our construction have long-term memory. Combined with the known results for memoryless channels, this suggests the existence of a tradeoff between the time-scale of the memory of a channel and the efficiency to compute its capacity. Giving precise quantitative bounds in this direction is an interesting open question.

It is also worth exploring other problems that could be attacked with similar techniques. The proof technique can be extended to the capacities of quantum channels with memory implying an even stronger inapproximability result in that case. We will make the explicit analysis in a forthcoming paper. Similar long-term memory effects appear in other interesting situations, associated with other entropic quantities. One paradigmatic example is cryptography, where in order to analyse the security of the sequential use of a device, one needs to assume the worst case-scenario in which the adversary keeps memory of its past uses. Both in the classical and in the quantum case, the techniques of this paper could provide insights on the difficulty to provide optimal results in cryptographic settings.

Furthermore, our result connects with recent work regarding the different capacities of memoryless quantum channels[29,30], showing some evidence that these capacities might be uncomputable. Also, memoryless zero error capacities, both classical and quantum, are known to have highly non-trivial behaviour[31,32,33,34,35]. Unfortunately, the techniques used here exploit directly the memory of the channel and hence cannot be directly applied to the memoryless capacities. The question is however of unquestionable interest.

METHODS

Notation

We denote random variables by capital letters X, Y,..., sets and PFA—see below for the definition—by calligraphic capital letters X,Y,... channels by capital bold face letters **X**, **Y**,..., and instances of random variables by lower case letters x, y,.... We denote vectors with the same convention, whenever confusion might arise a superscript indicates the number of components of the vector and a subscript the concrete component: $X^n=(X_1, X_2,..., X_n)$ or $x^n=(x_1, x_2,..., x_n)$. We indicate a consecutive subset of n components of the vector with subscript notation $[a, a+n-1]$: $x_{[a, a+n-1]}=(x_a, x_{a+1},..., x_{a+n-2}, x_{a+n-1})$.

A vector is called a probability vector if all its entries are non-negative and add up to one. A matrix is called a stochastic matrix if all its columns are probability vectors. A stochastic matrix takes probability vectors to probability vectors.

Probabilistic Finite Automata

A PFA consists of a finite set of inputs and a finite set of states. One of these states is the initial state and a subset of the states are accepting states.

The action of the PFA is defined by the transition probabilities from one state to another as a function of the input symbols. A word is a sequence of symbols. After a word is fed to a PFA in the initial state, the PFA will transition from state to state and will end up in an accepting state with some probability. We call this probability the accepting probability of a word. Intuitively, we can understand a PFA as a machine with noisy knobs, the input symbols, and the input word is a sequence of knobs that tries to steer the machine into some desired state. See Fig. 1 for an example for a formal definition.

Given some PFA, we denote by val_A the supremum of the acceptance probabilities over all input words:

$$\mathrm{val}_{\mathcal{A}} = \sup_{\mathbf{w}} \mathrm{val}(\mathcal{A}, \mathbf{w}), \tag{3}$$

where val(A, w) denotes the value of w when input into the PFA A and the optimization runs over all words of finite length.

We consider two types of PFAs that we name as freezable and resettable.

We call a PFA a freezable PFA if one of the transition matrices is equal to the identity matrix X_{id}. The reason is that for such a PFA reading the symbol corresponding to the identity leaves the state probabilities unchanged. Let u be any probability vector, then

$$u = \mathcal{X}_{\mathrm{id}} u. \tag{4}$$

We call a PFA a resettable PFA if one of the transition matrices, X_{rt}, takes the state back to the initial state. Let u be any probability vector, then

$$v = \mathcal{X}_{\mathrm{rt}} u. \tag{5}$$

We let γ be a map from PFAs to PFAs such that for all PFA A, γ(A) is freezable and resettable. More concretely:

Definition 1

Given a PFA A={Q,W,X,v,F},we define γ(A)={Q,W∪{id, rt}, X∪{X_{id}, X_{rt}}, v, F}γ as an automaton that extends AA with the two additional input symbols id and rtand the corresponding matrices X_{id} and X_{rt} as given by (4) and (5).

The key lemma we will need about PFAs, essentially due to Gimbert and Oualhadj25, is the fact that their value cannot be approximated within a constant error. Let us give the precise statement. Fix a rational number $\lambda \in (0, 1]$.

Lemma 1

One can give explicitly a subfamily T_λ *of rational freezable and resettable PFA with alphabet size* 5 *and* 62 *states with the following properties:*

- val_A *is either* $\geq\lambda$ *or* $\leq\lambda/2$ *for all* $A \in T_\lambda$.
- *It is undecidable to know which is the case.*

The Family S_λ and the Proof of Main Result 2

Given a freezable and resettable PFA A we define the channel V_A as follows. The input alphabet of the channel takes values in {0,1}×W, which we identify with two different input registers: a data input and a control input. The data input is transmitted to the output: noiselessly if A is in an accepting state or, if A in any other state, the channel outputs uniformly at random an element of the output alphabet. More concretely, the output of the channel is defined by the following conditional probability:

$$p(y_n|x_n, s_{n-1}) = \begin{cases} \frac{1}{2} & \text{if } s_{n-1} \notin \mathcal{F} \\ 1 & \text{if } s_{n-1} \in \mathcal{F} \text{ and } y_n = x_n \\ 0 & \text{else} \end{cases} \quad (6)$$

The control input is fed to A, which begins in the initial state, and the state transition probabilities are dictated by the PFA:

$$p(s_n|c_n, s_{n-1}) = \langle \pi_{s_n}, \mathcal{X}_{c_n} \pi_{s_{n-1}} \rangle. \quad (7)$$

We connect the properties of PFA with the capacity of FSMCs in the next Theorem:

Theorem 1

The capacity of V_A *is given by*

$$C(\mathbf{V}_{\mathcal{A}}) = \mathrm{val}_{\mathcal{A}}. \qquad (8)$$

We defer the proof.

The family S_λ in Main Result 2 is defined simply as

$$\mathcal{S}_\lambda = \{\mathbf{V}_{\mathcal{A}} : \mathcal{A} \in \mathcal{T}_\lambda\}.$$

with T_λ the family introduced in Lemma 1 and defined.

Main Result 2 is then a trivial consequence of Lemma 1 and Theorem 1.

Furthermore one can leverage Theorem 1 to show that all the channels in S_λ are information stable:

Corollary 1

Given $V_A \in S_\lambda$, V_A *is information stable.*

ACKNOWLEDGEMENTS

The authors are grateful to Toby Cubitt, Jeremy Ribeiro, Eddie Schoute, Henry Pfister, Marco Tomamichel and Michael Wolf for stimulating discussions and feedback. We thank Kenneth Goodenough for pointing us to the upper bound on the Riemann zeta function and David Reeb and Tobias Koch for extremely valuable comments that highly improve the presentation of the paper. DPG acknowledges support from MINECO (grant MTM2014-54240-P), from Comunidad de Madrid (grant QUITEMAD+-CM, ref. S2013/ICE-2801), and the European Research Council (ERC) under the European Union's Horizon 2020 research and innovation programme (grant agreement No 648913). This work has been partially supported by ICMAT Severo Ochoa project SEV-2015-0554 (MINECO). DE acknowledges support from TU Delft Open Access Fund.

REFERENCES

1. Shannon, C. E. A mathematical theory of communication. *Bell Syst. Tech. J.* 27, 623–656 (1948).
2. Blahut, R. E. Computation of channel capacity and rate-distortion functions. *IEEE Trans. Inf. Theory* 18, 460–473 (1972).
3. Arimoto, S. An algorithm for computing the capacity of arbitrary discrete memoryless channels. *IEEE Trans. Inf. Theory* 18, 14–20 (1972).
4. Dobrushin, R. L. General formulation of Shannon's main theorem in information theory. *Am. Math. Soc. Trans.* 33, 323–438 (1963).
5. Ahlswede, R. The weak capacity of averaged channels. *Z. Wahrscheinlichkeit.* 11, 61–73 (1968).
6. Winkelbauer, K. On the coding theorem for decomposable discrete information channels I. *Kybernetika* 7, 109–124 (1971).
7. Kieffer, J. C. A general formula for the capacity of stationary nonanticipatory channels. *Inf. Control* 26, 381–391 (1974).
8. Verdu, S. & Han, T. A general formula for channel capacity. *IEEE Trans. Inf. Theory* 40, 1147–1157 (1994).
9. Gallager, R. G. *Information Theory and Reliable Communication* Vol. 2 (Springer, New York, 1968)
10. Mushkin, M. & Bar-David, I. Capacity and coding for the Gilbert-Elliott channels. *IEEE Trans. Inf. Theory* 35, 1277–1290 (1989).
11. Goldsmith, A. J. & Varaiya, P. P. Capacity, mutual information, and coding for finite-state Markov channels. *IEEE Trans. Inf. Theory* 42, 868–886 (1996).
12. Pfister, H. D., Soriaga, J. B., & Siegel, P. H. On the achievable information rates of finite state ISI channels. *Global Telecommunications Conference, 2001. GLOBECOM '01. IEEE* 5, 2992–2996 (2001)
13. Arnold, D. M., Loeliger, H.-A., Vontobel, P. O., Kavčić, A. & Zeng, W. Simulation-based computation of information rates for channels with memory. *IEEE Trans. Inf. Theory* 52, 3498–3508 (2006).
14. Sharma, V. & Singh, S. K. Entropy and channel capacity in the regenerative setup with applications to Markov channels. In *Proc. 2001 IEEE International Symposium on Information Theory*, 283 (IEEE, 2001)

15. Kavčić, A. On the capacity of Markov sources over noisy channels. *Global Telecommunications Conference, 2001. GLOBECOM '01. IEEE* 5, 2997–3001 (2001).
16. Holliday, T., Goldsmith, A. & Glynn, P. Capacity of finite state channels based on Lyapunov exponents of random matrices. *IEEE Trans. Inf. Theory* 52, 3509–3532 (2006).
17. Pfister, H. D. The capacity of finite-state channels in the high-noise regime. Preprint at https://arxiv.org/abs/1001.1214 (2010).
18. Vontobel, P. O., Kavcic, A., Arnold, D.-M. & Loeliger, H.-A. A generalization of the Blahut–Arimoto algorithm to finite-state channels. *IEEE Trans. Inf. Theory* 54, 1887–1918 (2008).
19. Han, G. A randomized algorithm for the capacity of finite-state channels. *IEEE Trans. Inf. Theory* 61, 3651–3669 (2015).
20. Hu, G.-D. On Shannon theorem and its converse for sequence of communication schemes in the case of abstract random variables. In *Trans. 3rd Prague Conference on Information Theory, Statistical Decision Functions, Random Processes* 285–333 (Czechslovak Academy of Sciences, Prague, 1964)
21. Ginsburg, S. *The Mathematical Theory of Context Free Languages [Mit Fig.]* (McGraw-Hill Book Company, 1966)
22. Paz, A. *Introduction to Probabilistic Automata* (Academic Press, Inc., Orlando, FL, 1971).
23. Condon, A. & Lipton, R. J. On the complexity of space bounded interactive proofs. In *30th Annual Symposium on Foundations of Computer Science* 462–467 (IEEE, 1989).
24. Hirvensalo, M. Improved undecidability results on the emptiness problem of probabilistic and quantum cut-point languages. In *SOFSEM 2007: Theory and Practice of Computer Science*309–319 (Springer, 2007).
25. Gimbert, H. & Oualhadj, Y. In *Automata, Languages and Programming. ICALP 2010. Lecture Notes in Computer Science*, Vol. 6199 (eds Abramsky, S. et al.) 527–538 (Springer, Berlin-Heidelberg, 2010).
26. Michael, S. *Introduction to the Theory of Computation*, Vol. 2 (Thomson Course Technology, Boston, 2006).
27. Arora, S. & Barak, B. *Computational Complexity: A Modern Approach* (Cambridge University Press, New York, 2009).

28. Neary, T. & Ollinger, N. Undecidability in binary tag systems and the post correspondence problem for five pairs of words. In *32nd International Symposium on Theoretical Aspects of Computer Science* Vol. 30, 649–661 (Schloss Dagstuhl–Leibniz-Zentrum fuer Informatik, 2015).
29. Cubitt, T., et al. Unbounded number of channel uses are required to see quantum capacity. *Nat. Commun.* 6, 6739 (2015).
30. Elkouss, D. & Strelchuk, S. Superadditivity of private information for any number of uses of the channel. *Phys. Rev. Lett.* 115, 040501 (2015).
31. Noga, A. The Shannon capacity of a union. *Combinatorica* 18, 301–310 (1998).
32. Chen, J., Cubitt, T. S., Harrow, A. W. & Smith, G. Entanglement can completely defeat quantum noise. *Phys. Rev. Lett.* 107, 250504 (2011).
33. Cubitt, T. S., Chen, J. & Harrow, A. W. Superactivation of the asymptotic zero-error classical capacity of a quantum channel. *IEEE Trans. Inf. Theory* 57, 8114–8126 (2011).
34. Cubitt, T. S. & Smith, G. An extreme form of superactivation for quantum zero-error capacities. *IEEE Trans. Inf. Theory* 58, 1953–1961 (2012).
35. Shirokov, M. E. On channels with positive quantum zero-error capacity having vanishing *n*-shot capacity. *Quant. Inf. Process* 14, 3057–3074 (2015)
36. Singmaster, D. *Notes on Rubik's magic cube* (Enslow Pub Inc., Hillside, 1981)

SECTION 3:
SIGNALING OVER AWGN CHANNELS

7

An M-QAM Signal Modulation Recognition Algorithm in AWGN Channel

Ahmed K. Ali and Ergun Erçelebi

Department of Electric and Electronic Engineering, Gaziantep University, Gaziantep 27310, Turkey

ABSTRACT

Computing the distinct features from input data, before the classification, is a part of complexity to the methods of automatic modulation classification (AMC) which deals with modulation classification and is a pattern recognition problem. However, the algorithms that focus on multilevel quadrature amplitude modulation (M-QAM) which underneath different channel scenarios is well detailed. A search of the literature revealed that few studies were performed on the classification of high-order M-QAM modulation schemes such as 128-QAM, 256-QAM, 512-QAM, and 1024-QAM. This work focuses on the investigation of the powerful capability of the natural logarithmic properties and the possibility of extracting higher

Citation: Ahmed K. Ali and Ergun Erçelebi. "An M-QAM Signal Modulation Recognition Algorithm in AWGN Channel" Hindawi Scientific Programming Volume 2019, Article ID 6752694, 17 pages https://doi.org/10.1155/2019/6752694

order cumulant's (HOC) features from input data received raw. The HOC signals were extracted under the additive white Gaussian noise (AWGN) channel with four effective parameters which were defined to distinguish the types of modulation from the set: 4-QAM~1024-QAM. This approach makes the classifier more intelligent and improves the success rate of classification. The simulation results manifest that a very good classification rate is achieved at a low SNR of 5 dB, which was performed under conditions of statistical noisy channel models. This shows the potential of the logarithmic classifier model for the application of M-QAM signal classification. Furthermore, most results were promising and showed that the logarithmic classifier works well under both AWGN and different fading channels, as well as it can achieve a reliable recognition rate even at a lower signal-to-noise ratio (less than zero). It can be considered as an integrated automatic modulation classification (AMC) system in order to identify the higher order of M-QAM signals that has a unique logarithmic classifier to represent higher versatility. Hence, it has a superior performance in all previous works in automatic modulation identification systems.

INTRODUCTION

Efficacious information transmission can be realized very clearly in trendy communication systems, and the transmitted signals are typically modulated by using various modulation ways. Modulation recognition is an intermediate way that must usually be achieved before signal demodulation and information detection, and it represents the substantial feature in modern radio systems to give knowledge on modulation signals rather than signal demodulation and can be used in decoding both civilian and military applications such as cognitive radio, signal identification, menace assessment, spectrum senses, and management, which allows to more efficiently use the available spectrum and increase the speed of data transfer. Furthermore, the unknown signal classification is a decisive weapon in electronic warfare scenarios. The electronic support management system plays a paramount role as a source of information is required to conduct electronic counter repression, threat analysis, warning, and target acquisition.

Particular recognition to modulation is the identification of types of the transmitted signals that lie in noncooperative channel environment groups which are significant for following up the signal demodulation and data extraction, and this was considered as the major turn to automatic modulation classification (AMC) which became an attractive subject for researchers, yet there is a major challenging for engineers who deal with the design of software-defined radio systems (SDRSs) and transmission deviation. The implementation of sophisticated information services and systems to military applications under a congested electromagnetic spectrum is a major concern issue to communication engineers. Friendly signals must be safely transmitted and received, whilst enemy signals should be found and jammed [1].

The first base of AMC theory is provided in [2], the essential ideas proposed at Stanford University in domestic project and was documented since 1969, and nowadays, the AMC is extensively used and well known for communication engineers as a significant part in the internal design of intelligent radio systems [3, 4].

Currently, digital modulation identification algorithms can be split into two techniques: a maximum likelihood hypothesis method which is based on decision theory [5] and statistical pattern recognition which relies upon the feature extraction ideas [6, 7], in uncooperative channel environment. The algorithms of the pattern recognition technique are usually used in practical application, and that is due to the absence of prior knowledge from received modulation signals [8]. The discrimination method has particularly beneficial features for pattern recognition which are initially extracted from the vectors of data and the identification of the signal modulation mode that was completed upon the coverage between characteristic parameters and limits of the extracted features which was considered as the known mode of the modulation type [9].

In the last five years, a new modulation classification technique has been proposed at the same time with the evolution of the automatic digital signal modulation recognition algorithms research, and this new modulation offers a new trend of complicated signal sets and higher order modulation signals. The most commonly used features are high-order cumulants (HOC) and high-order moment (HOM) [10]. The HOC-based techniques show superior performance to classify the digital modulation signals at a low signal-to-noise ratio (SNR); these algorithms were documented in literature [11–13]. Nevertheless, they were not appropriate to classify higher-order M-QAM modulation formats. Constellation shape varies from one modulation signal type to another in order to be considered as a robust signature.

The constellation shape was used in the algorithms in references [14–16] to recognize various types of higher order modulation including 8-PSK and 16-QAM. The recognition of success rate probabilities is around 95%, but it still required more SNR which may reach 4 dB. However, the constellation shape can be classified into M-PSK and M-QAM, and at the same time, they are sensitive to several wireless channels that could seriously make confusions in work; these comprise frequency offset, phase rotation, and the application of raised cosine roll-off filters. Under this situation, the symbol rate and frequency offset setting must be more accurate during the periods of presentment, and at the same time, the most widely used features are a cyclic spectrum and cyclic frequency that could help multiple unwanted signals to be detected with each of temporal and frequency-based overlaps [17–20].

In 2012, Dobber and his coworkers [21] suggested two novel algorithms for the recognition of 4-QAM, 16-QAM, 64-QAM, and QAMV.29 modulation schemes; also other [22, 23] authors suggested to use higher order cyclic cumulants (CCs) which were obtained from received signals as features for modulation classification. Also, this had been put forward in [24] which presents a method to classify a number of mixed modulation schemes with an assortment of *N*-class problems, and this technique also achieved a notable performance at low ranges of SNR. Due to the development of algorithms in machine learning, the researchers had started to design classification algorithms based on some aspects of machine learning which improved the classification ability and gives further pros to the classifier in terms of distinguished types of modulation signals.

Some investigations which were achieved on the behaviors of hidden naive Bayes (HNB) [25] and on combination of naive Bayes (NB) and other types of classifiers to create a new classifier named multiple classifier [26] certified that such classifier types were highly effective compared to conventional classifiers in terms of a small number of training iterations.

Supervised learning technique was combined with a modified K-means algorithm that is based on four famous optimization algorithms. This modification presented a new generation of the AMC technique. Likewise, artificial neural networks (ANNs) with genetic algorithms were studied and considered by Norouz and his coworkers [27], and also support vector machine (SVM) classifier in [28, 29].

The ways based on ANN and SVM show a superior performance although there is less prior information of signal features. However, multiple training samples and a long training period are necessary to accomplish sufficient learning which increases the computational complexity and makes the processing time longer.

In the last five years, researches widely explored the new techniques in order to reduce the required SNR and make it more efficient of recognition capability through focusing on robust features and classifier designs. One of the weak points of the previous algorithms is that the nature of the decision-tree is that it requires fixed threshold values due to the features that had been proposed by the authors, and these features are highly sensitive to any changes in SNR that can make the threshold values be valid for small ranges of SNR above to 10 dB. However, there had been no work until now that concentrates on the recognition of higher order QAM signals in terms of the shape of feature distribution curves.

In fact, this assumption gives a good understanding of the behavior of systems and reflects their major trends; therefore, under the circumstance of a channel corrupted by AWGN, this paper derived a relationship among of the higher order cumulant as features and threshold levels, in order to evaluate the performance of M-QAM modulation recognition technique.

MATHEMATICAL ANALYSES

Logarithmic Calculation

Below are some theoretical analyses that assumes two logarithmic functions denoted $f_1(x_1)$ and $f_2(x_2)$ carrying different variable "x_1, x_2" but have the same base value "n" which can be written as follows:

$$f_1(x_1) = \log_n[x_1],$$
$$f_1(x_2) = \log_n[x_2], \quad (1)$$

where the range of variables $x_1 \neq x_2$ and the base value is considered to be the same. The ratio between $f_1(x_1)$ and $f_2(x_2)$ can be expressed as

$$W(x_1, x_2) = \frac{\log_n[x_1]}{\log_n[x_2]} = \frac{\ln[x_1]/\ln(n)}{\ln[x_2]/\ln(n)}. \quad (2)$$

In general, the logarithmic equations can be expressed into another equivalent form, written as

$$\log_n[x_1] = \frac{\ln[x_1]}{\ln(n)}. \quad (3)$$

Since [x] is a log function, it can also be expressed as $\log_r[x]$,, where "r" indicates the base value of natural logarithms. Therefore, also equation (2) can be calculated as

$$W(x_1, x_2) = \frac{\ln[x_1]/\ln(n)}{\ln[x_2]/\ln(n)}. \tag{4}$$

Due to the equality between numerator and denominator, the part has been eliminated. This gives the result expressed as below:

$$W(x_1, x_2) = \frac{\ln[x_1]}{\ln[x_2]}, \tag{5}$$

where x_2 is the constant value "10" that makes s $w(x_1, x_2)$ a constant as well. The logarithmic functions are defined based on the higher order cumulant. In Section 3, the feature distribution curves with logarithmic properties are obtained, and the plot depicts that the modulation scheme is less sensitive to the variation of SNR.

PARAMETER EXTRACTION BASED ON LOGARITHMIC PROPERTIES

Parameter Extraction Based on HOC

It is well notable that a random variable under complex-stationary process always has a zero mean value. Therefore, the following transformation formula which is based on higher order moments can be accomplished:

$$C_{11} = \mathrm{cum}(y(n)^*, y(n)^*) = M_{11},$$

$$C_{22} = \mathrm{cum}(y(n)^*, y(n)^*) = M_{22} - M_{20}^2 - 2M_{11}^2,$$

$$C_{33} = \mathrm{cum}(y(n)^*, y(n)^*, y(n)^*) = M_{33} - 6M_{20}M_{31}$$
$$- 9M_{22}M_{11} + 18(M_{20})^2 M_{11} + 12M_{11}^3,$$

$$C_{44} = \mathrm{cum}(y(n)^*, y(n)^*, y(n)^*, y(n)^*) = M_{44} - M_{40}^2$$
$$- 18M_{22}^2 - 54M_{20}^4 - 144M_{11}^4 - 432M_{20}^2 M_{11}^2$$
$$+ 12M_{40}M_{20}^2 + 192M_{31}M_{11}M_{20} + 144M_{22}M_{11}^2$$
$$+ 72M_{22}M_{20}^2, \tag{6}$$

where the sign (..) represents the cumulant operation, while the superscript "*" represents the operation of complex conjugate [30].

Improved Higher Order Cumulant Feature

The logarithmic expression formulas in (7)–(10) below show the modified higher order cumulant in terms of logarithm. The pros of modification not only make these features insensitive to noise but also help classifiers to recognize the increase in the higher order modulation signals. The simulation condition tests of 10,000 signal realizations from {4~1024}-QAM each consisting signal length N=4096 with a phase offset of "π/6" with the statistical average value are shown in Figures 1–4, the distribution curve of each feature with varying values of the SNR. Through the simulation results, it is clear that the feature distribution curve is not significantly affected by variation of noise. However, this modification provides a better improvement in the achieved classification activity:

$$f_a = \log_{10}\left[\left|C_{11}\right|\right], \quad (7)$$

$$f_b = \log_{10}\left[\left|C_{22}\right|\right], \quad (8)$$

$$f_c = \log_{10}\left[\left|C_{33}\right|\right], \quad (9)$$

$$f_d = \log_{10}\left[\left|C_{44}\right|\right]. \quad (10)$$

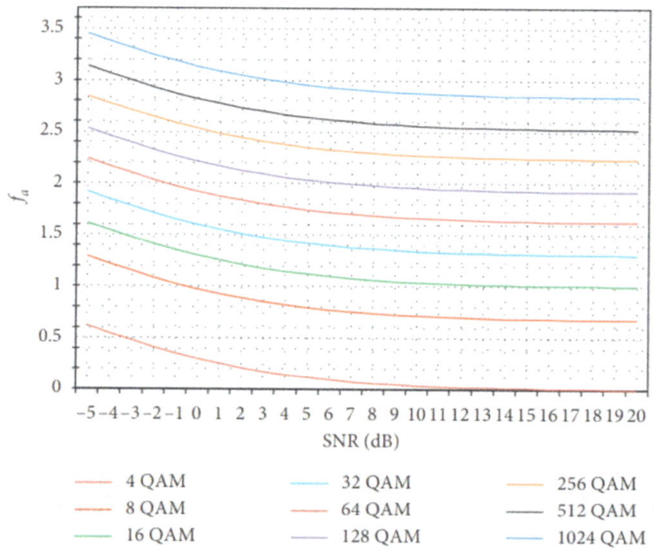

Figure 1: Feature "f_a" distribution curve vs SNR with the test of 10,000 signal realizations from {4~1024}-QAM each consisting signal length N=4096 with a phase offset of "π/6."

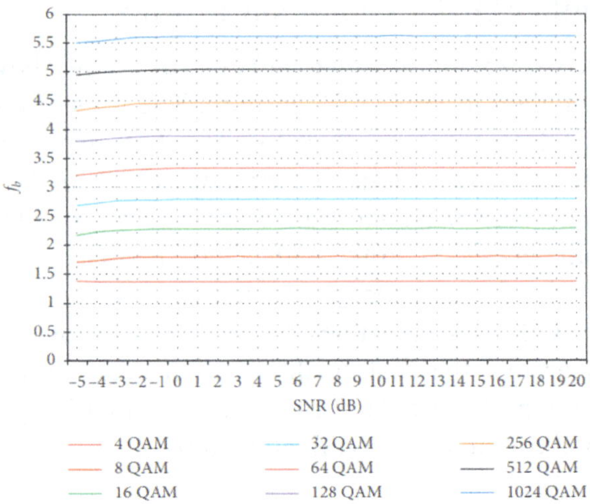

Figure 2: Feature "f_b" distribution curve vs SNR with the test of 10,000 signal realizations from {4~1024}-QAM each consisting signal length N=4096 with a phase offset of "π/6".

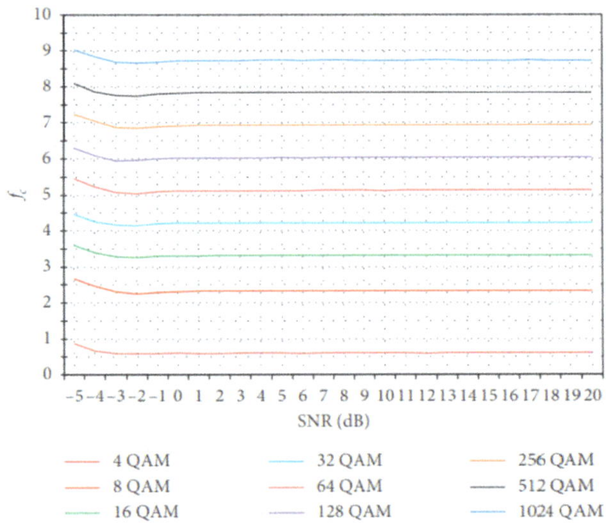

Figure 3: Feature "f_c" distribution curve vs SNR with the test of 10,000 signal realizations from {4~1024}-QAM each consisting signal length N=4096 with a phase offset of "π/6".

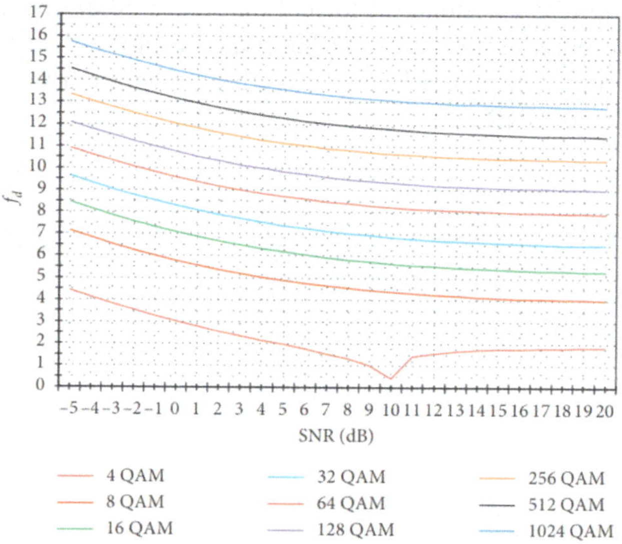

Figure 4: Feature "f_d" distribution curve vs SNR with the test of 10,000 signal realizations from {4~1024}-QAM each consisting signal length N=4096 with a phase offset of "π/6".

DETAILS ON ALGORITHM AND SIMULATION PERFORMANCE

The above four HOC parameters have been derived and modified by using the logarithmic function to get optimum quality to recognize the M-QAM signals in the AWGN channel. However, the basic procedure of the automatic modulation classification system is shown in Figure 5 [31], while the actual decision procedure of the algorithm that has been proposed is demonstrated in Figure 6, yet the logarithmic modulation recognition method is effective with nine sets of M-QAM signals. The feature distribution curves in Figures 1–4 show the extracted features of modulation formats. Feature extraction is a primary step in the pattern recognition method. Threshold settings can then be determined depending on the distribution conditions of the extracted features. The threshold levels have been determined based on the projection of two lines upward and downward with effective space to avoid misclassification. The lines are parallel to the SNR axis (within a certain range of SNR). The created cluster region is used to separate required class distribution samples (modulation format that required discrimination). The points are computed empirically on the *y*-axis (feature distribution level), and this criterion will give a highly effective recognition rate even with a minimum value of SNR.

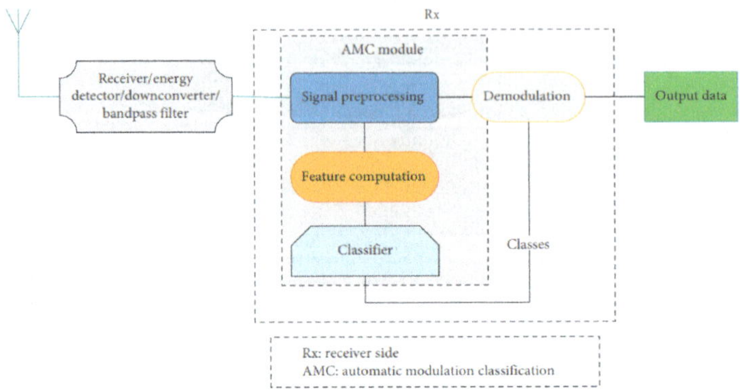

Figure 5: Procedure for automatic modulation classification.

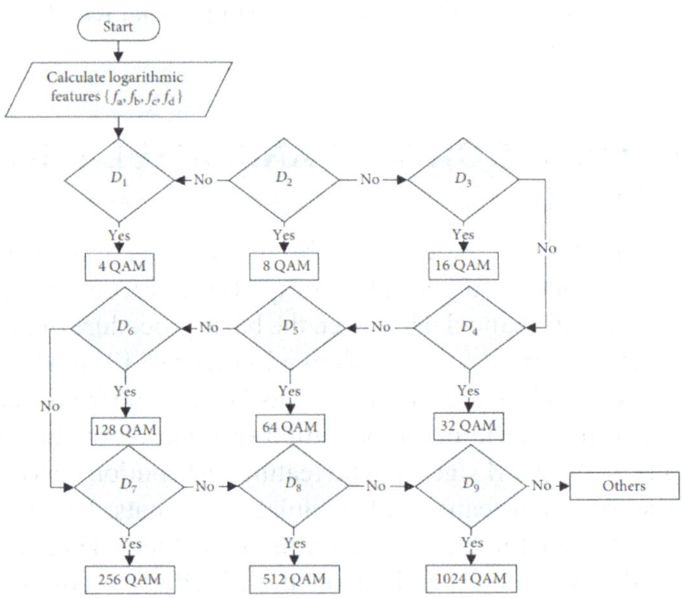

Figure 6: Flowchart of the M-QAM recognition algorithm.

Table 1 represents the feature threshold values which were set empirically. The Boolean equations (12)–(16) illustrated the logical decision process, while Figure 7 shows the block diagram of the M-QAM logarithmic recognition algorithm, which is posterior to easily implement an extension of the proposed method in real time.

Table 1: Optimum threshold values to each logarithmic feature Nature.

I	f_a	f_b	f_C	f_d
0	0.5	1.5	1.5	3
1	0.95	2.2	3	5
2	1.3	2.7	4	6.4
3	1.52	3.1	5	7.6
4	1.8	3.7	6.3	9
5	2.2	4.3	6.6	10
6	2.5	4.8	7.5	11
7	3	5.7	9	12
8	4.5	7.5	12.5	15.5

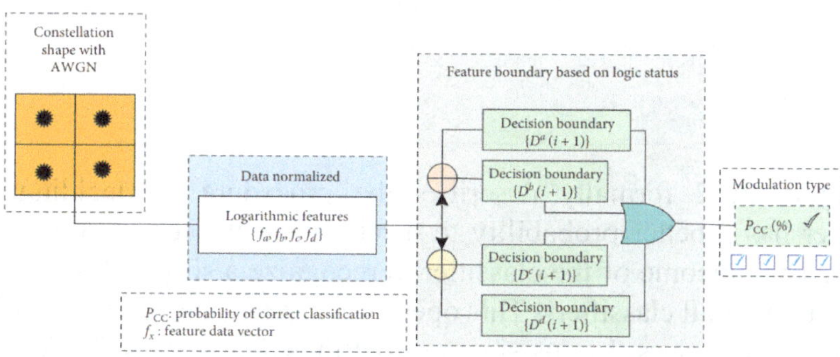

Figure 7: The block diagram of our logarithmic classification system.

Algorithm 1 formally describes the procedures for the feature extraction and the decision for modulation recognition. On the contrary, in the feature extracting procedure, the input is represented by the data row after modulated with M-QAM. However, $s_i = d_x$ signal passes through a noisy channel

$s_i = d'_x$, and the first step in extracting is fetching [C_{11}, C_{22}, C_{33}, C_{44}]; after computing the cumulant vector of signals, the process of the logarithmic classifier is begun and tested $\{f_{a_0} < f_x(\varphi_0)\}$, if so $\text{index}_1 = 1$ for the 4-QAM signal.

Else if $\{f_{x_i} \geq f_x(\varphi_i)\ \&\&\ f_{x_i} \leq f_x(\varphi_i)\}$ $\text{index}_i = i'$, where $2^{n'}$ QAM, $n' > 2,\ i > 1, i' > 1$.

The same procedure is carried out for $\{f_{a_i}, f_{b_i}, f_{c_i}, f_{d_i}\}$ with the increase in the index i, until $i=8$.

Algorithm 1: Logarithmic classifier of higher order QAM modulation signals.

Input: generate digital M-QAM modulation signals; $Dx = d_1, d_2, d_3, \ldots, d_i$, and $x \in \{a, b, c, d\}$, $M = \{2^n\}$, $n = 2, 3, 4, 5, 6, 7, 8, 9, 10$ $s_i = d_x$; s_i row of data after modulated by M-QAM signal Output: $v \in$ index $\{f_{a_i}, f_{b_i}, f_{c_i}, f_{d_i}\}$; output boundary with index_i Step 1: $s_i = d'_x$; corrupted by Gaussian random noise Calculate cumulants C_{11}, C_{22}, C_{33}, and C_{44} Step 2: $f_{a_i} = \log_{10}[\lvert C_{11_i}(s_i)\rvert]$; $f_{b_i} = \log_{10}[\lvert C_{22_i}(s_i)\rvert]$; $f_{c_i} = \log_{10}[\lvert C_{33_i}(s_i)\rvert]$; $f_{d_i} = \log_{10}[\lvert C_{44_i}(s_i)\rvert]$; Step 3: If $\{f_{a_0} < f_x(\varphi_0)\}$; 4-QAM $\text{index}_i = 1$; Step 4: Else If $\{f_{x_i} \geq f_x(\varphi_i)\ \&\&\ f_{x_i} \leq f_x(\varphi_i)\}$; where $2^{n'}$ QAM; $n' > 2$, $i > 1$, $i' > 1$ $\text{index}_i = \text{i}'$; Step 5: Return v;

Algorithm 2 formally describes the procedure of deciding which classifier has a better probability to discriminate the required modulation format. The outcome of the classifier to recognize a single class is boosted by one, two, or all classifiers. This operation depends on feature distribution curves for each class. Let us suppose that their feature distribution curves have no overlapping parts and the curves are approximately parallel to the SNR axis (within a certain range of SNR). All classifiers set "true." The input is $\{f_{a_i}, f_{b_i}, f_{c_i}, f_{d_i}\}$, with i index.

Algorithm 2: Choose a higher and better classifier.

Input: $v \in \text{index}\{f_{a_i}, f_{b_i}, f_{c_i}, f_{d_i}\}_i$; input boundary with index_i
Output: P_{CC_i}; probability of correct recognition
Step 1: If $(\text{index} f_{a_i} == i')\{||(\text{index} f_{b_i} == i')||(\text{index} f_{c_i} == i')||(\text{index} f_{d_i} == i')\}$;
$CC_{\text{index SNR}} = CC_i + 1$; correct rate counts
End
Step 2: Return P_{cc};

If $(\text{index} f_{a_i} == i')||(\text{index} f_{b_i} == i')||(\text{index} f_{c_i} == i')||(\text{index} f_{c_i} == i')$ is true, then increase the index $CC_{\text{index SNR}}$. This procedure continues until $\{f_{a_i}, f_{b_i}, f_{c_i}, f_{d_i}\}_{i=8}$. Algorithm 2 has been developed to shorten the execution time of signal recognition and classifier speed increment and, moreover, to improve the accuracy of the classifier. Algorithm 2 can be summarized as follows: the comparison is achieved by logical operation of the sequential status $\{f_{a_i}, f_{b_i}, f_{c_i}, f_{d_i}\}_i$. As result if either present statue could not recognize the candidate class, in both cases of signal/noise power ratio is greater or less than 0 dB. To overcome that situation, other classifiers strengthened the present status and boosted the probability of the correct recognition rate before deciding the correct modulation.

The mathematical expression of the decision boundary of the logarithmic classifier algorithm of the higher order QAM signals is expressed below. For instance, for an input feature $D^x_{(1)}$, the threshold levels are concluded from the results in Table 1 at i = 0 and expressed as logical status:

$$D^x_{(1)} = [f_a \geq f_a(\varphi_0) \,\&\, f_a < f_a(\varphi_2)]. \tag{11}$$

Remark. $i' = (i+1)$ for 4-QAM, $i' = 1$; likewise the expression of $D^x_{(i)>1}$ is as follows:

$$D^a_{(i+1)} = [f_a \geq f_a(\varphi_{i-1}) \,\&\, f_a < f_a(\varphi_{i+1})], \tag{12}$$

$$D^b_{(i+1)} = [f_b \geq f_b(\varphi_{i-1}) \,\&\, f_b < f_b(\varphi_{i+1})], \tag{13}$$

$$D^c_{(i+1)} = [f_c \geq f_c(\varphi_{i-1}) \,\&\, f_c < f_c(\varphi_{i+1})], \tag{14}$$

$$D^d_{(i+1)} = [f_d \geq f_d(\varphi_{i-1}) \,\&\, f_d < f_d(\varphi_{i+1})], \tag{15}$$

$$D_i = D^a_{(i+1)} \Big|\Big| D^b_{(i+1)} \Big|\Big| D^c_{(i+1)} \Big|\Big| D^d_{(i+1)}, \tag{16}$$

where $f_x(\varphi_i)$ is the threshold value to each of the logarithmic features, in which $x \in \{a, b, c, d\}$ and $i = 0, 2, 3, 4, \ldots, 8$, where the threshold count here is 8, and also $i' = 1, 2, 3, 4, \ldots, 9$ is a class pattern number of QAM signals that is being recognized.

Referring to the flowchart in Figure 7, and equations (12)–(16), under the effect of SNR, the Monte Carlo simulation was carried 10000 times realization for each one of the nine M-QAM signals and then it was tested by using the proposed logarithmic classifier, and the phase was offset by using three degrees starting from m 1/6π as the reference point.

This work emerges some promising results, and the simulation results are arranged in Figures 8–15. The most striking result that formed in Figures 8–11 shows that when SNR ≥5, the recognition rate of M-QAM signals can reach over 99% even with variation in both sample length and phase offset; therefore, the algorithm proposed in this paper is very efficient in recognizing these signals, and in order to evaluate the performance of the proposed algorithm, a comparison is conducted in terms of the probability of correct classification (PCC) versus SNR which can be formulated as

$$P_{CC} = \sum_{k=1}^{j} p(\omega_k \mid \omega_k) P(\omega_k),$$

$$\text{probability of correct classification rate \%} = P_{CC} \cdot \%100. \qquad (17)$$

In the same direction, "J" is the modulation signal candidate classification, $(\omega_1, \omega_2, \omega_3, \omega_j)$, where $P(\omega_k)$ is the probability of the modulation scheme when (ω_k) occur, while $P(\omega_k \mid \omega_k)$ is the probability of correct classification when an (ω_k) constellation is sent [32].

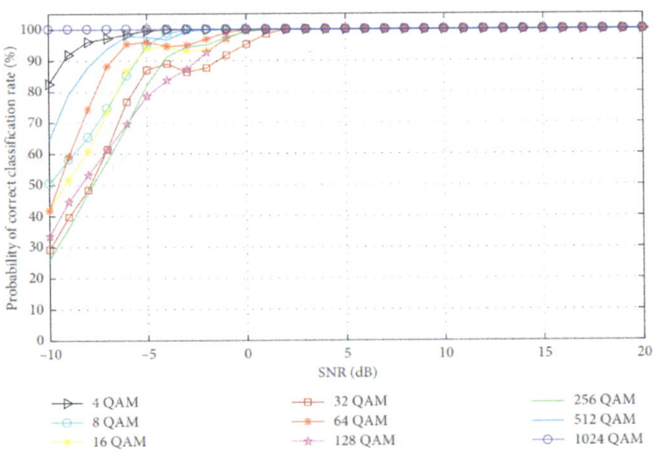

Figure 8: The probability of correct classification rate vs SNR with 10,000 signal realizations from {4~1024}-QAM each consisting signal length N=4096 and phase offset "π/6"

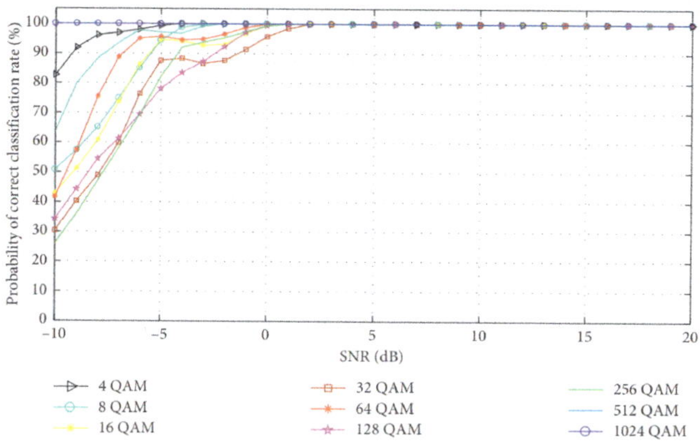

Figure 9: The probability of correct classification rate vs SNR with 10,000 signal realizations from {4~1024}-QAM each consisting signal length N=4096 and phase offset "π/4"

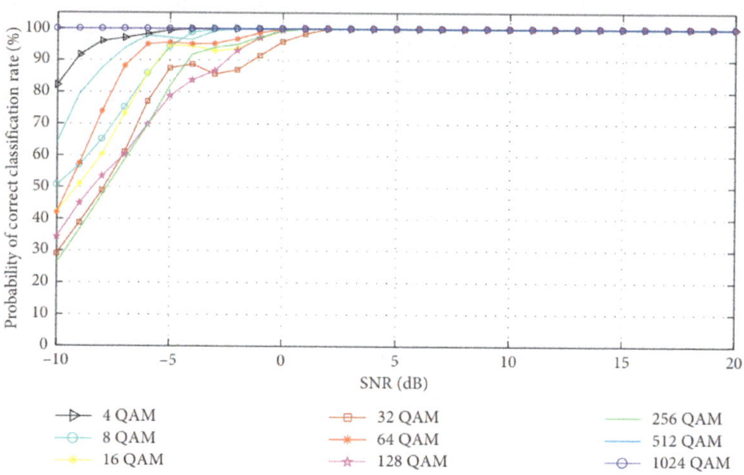

Figure 10: The probability of correct classification rate vs SNR with 10,000 signal realizations from {4~1024}-QAM consisting signal length N=4096 and phase offset "π/3".

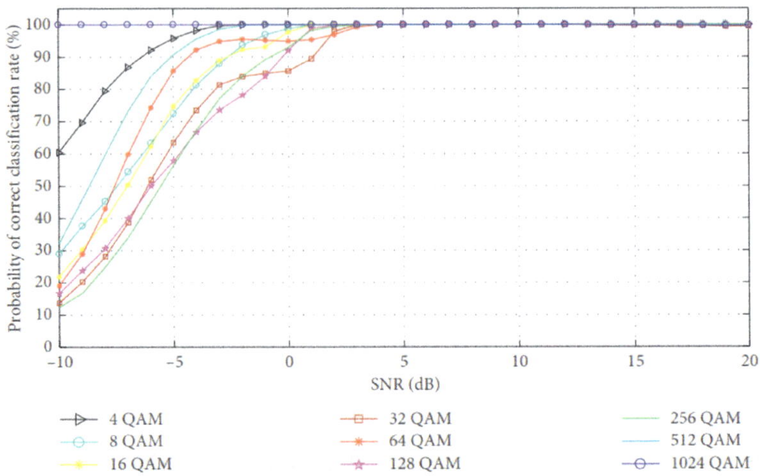

Figure 11: The probability of correct classification rate vs SNR with 10,000 signal realizations from {4~1024}-QAM each consisting variable signal length from N={64 128 265 512 1024 2048 4096} and phase offset "π/6".

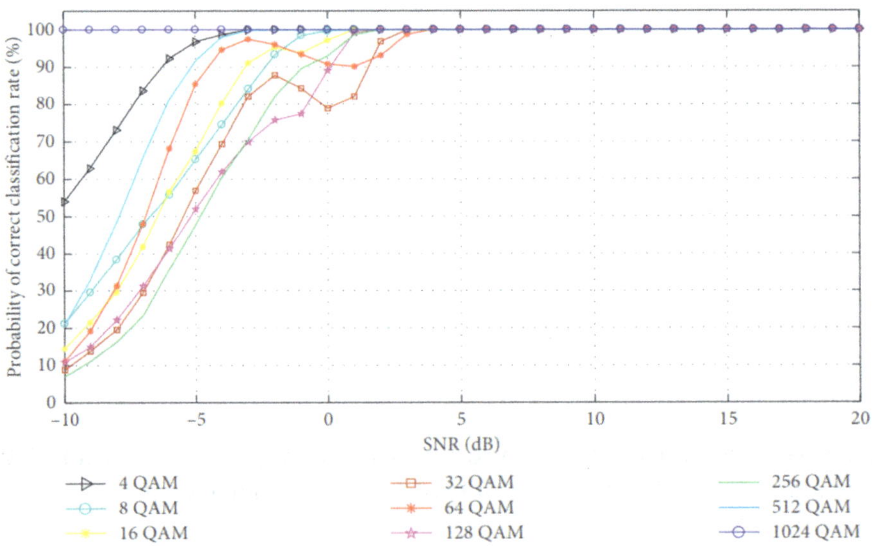

Figure 12: The probability of correct classification rate vs SNR with 10,000 signal realizations {4~1024}-QAM each consisting signal length N=256 and phase offset "π/6".

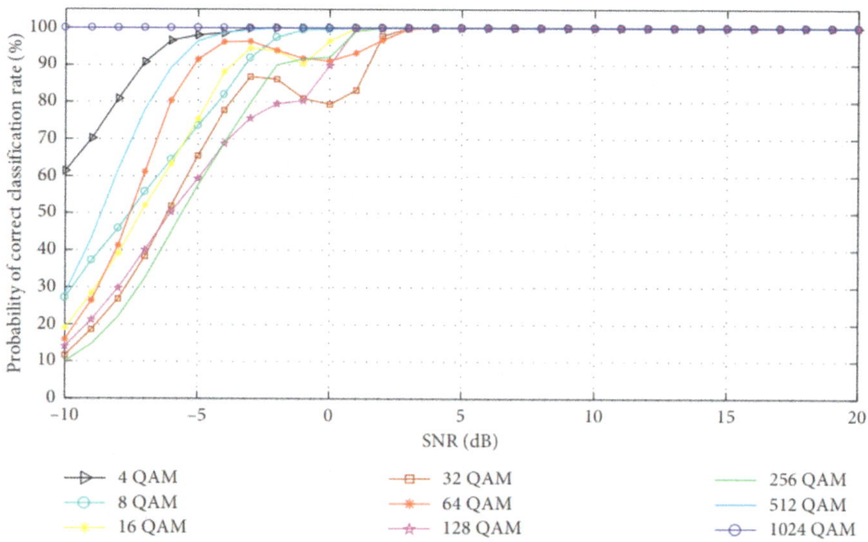

Figure 13: The probability of correct classification rate vs SNR with 10,000 signal realizations from {4~1024}-QAM each consisting signal length N=512 and phase offset "π/6"

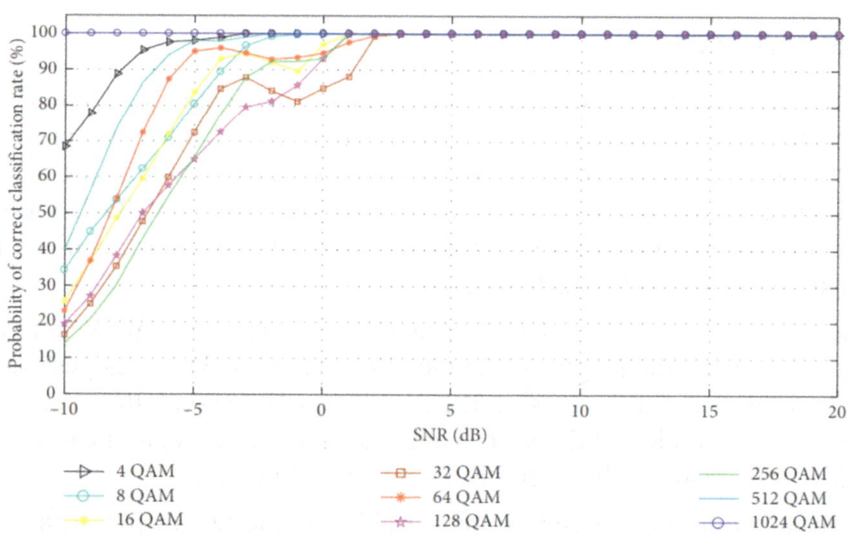

Figure 14: The probability of correct classification rate vs SNR with 10,000 signal realizations from {4-1024}-QAM each consisting signal length N=1024 and phase offset "π/6"

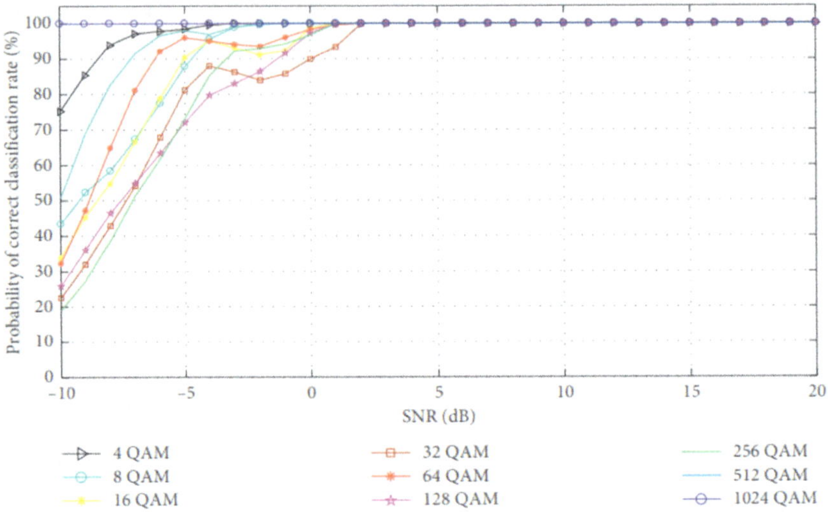

Figure 15: The Probability of correct classification rate vs SNR with 10,000 signal realizations {4-1024}-QAM each consisting signal length N=2048 and phase offset "π/6"

DISCUSSION

All the simulations have been done under the computer simulation environment. The experiments were carried out in three stages in the presence of AWGN: first, extract the HOC features and then pass through the feature modifier, the latter will be a logarithmic classifier based on HOC. Second, make a decision based on the threshold algorithm, and third, achieve the evaluation of PCC % for each modulation type.

It appears from the simulation results that the best performance curve is achieved in using the length of sample; 4096 samples. This was performed based on the fact that increasing the data set can overcome the overfitting problem. Nevertheless, this performance has the highest classification rate at a low SNR from –3 dB to 20 dB, as shown in Figure 8.

On the contrary, the curves for variable sample length [64, 128, 256, 512, 1024, 2048, and 4096] indicate a converged accuracy in the same test range of the SNR, and this can be clearly observed in Figure 11.

The accuracy of classifications is lower when reducing the number of samples transmitted, while in the proposed algorithm, the sample length has been reduced to [2048, 1024, 512, 256]. The probability of the correct classification ratio ranges from about 80% to 100% in the SNR range of 0 to 20 dB, as shown in Figures 12–15, respectively. This is mainly due to the reason that we proposed a method which considered that the HOC's feature is based on logarithmic modification; this will create a new classifier known as a logarithmic classifier. However when , the classifier accuracy deteriorates but does not collapse the correct classification rate that becomes below 75% and the phase offset will be considered as shown in Figure 9 in which the phase offset were , while in Figure 10, the phase offset is . The proposed algorithm gives a consistent classification accuracy of about 99.7%, in the range of SNR from 2 dB to 20 dB.

It was observed when the phase offset changed; the probability of the correct classification rate is almost equal, and this was due to the high robustness of the proposed classifier. Eventually, the superior performance of the proposed classifier changed the phase offset, and also variability of the transmitted signal length is attributed to the efficient of the proposed logarithmic classifier.

Evaluation Accuracy and Complexity with Other Methods

Because of the complexity of the classification algorithm, this research suggests a different method of classifications to conventional modulation [10, 11], which requires a different logarithmic operations. Yet, still it is crucial in terms of high probabilities to correct the recognition rate and less process is required to calculate the cumulant orders; at the same time, the other studies do not highlight these promising lines [21, 23, 33–35].

Although cumulant-based classifier methods were convenient to cope with M-QAM modulation recognition, yet it does not provide a sufficient classification rate M-QAM modes under fading channel, and it can even produce missing classification in some circumstances, while the logarithmic classifier of M-QAM signals has a better probability of correct recognition rate at SNR range between −5 dB and 20 dB.

The performance of the logarithmic classifier of M-QAM is superior in what was proposed in terms of its ability to distinguish between high order and very high order of QAM signals and thus can be considered as one of the pillars for the modulation techniques in the future due to unprecedented demand for the wireless communication technology expecting to reach an estimated 7.6 billion by 2020 [36].

According to Table 2, the fourth-order cumulant classifier has the lowest complexity, while the combination between forth-, sixth-, and eighth-order cumulants based classifier has the highest complications due to the use of exponential operations; however, it is worth to clarify that the logarithmic classifier in this research work depends on unique combination between lowest and highest orders of cumulant, and this combination will create an integrity and a less complexity classifier system. The complexity of several methods proposed that occurred in previous work was compared in detail as shown in Table 2.

Table 2: Comparison between proposed classifier with other systems in the literature

Reference	Classifier structure		Modulation type		System performance		Simulation tool
	Features	Complexity	M-QAM	M-PSK	SNR (dB)	Accuracy (%)	
[33]	$C_{40}, C_{41}, C_{42}, C_{60}, C_{61}, C_{62}, C_{63}$	High	16, 64	4	−3	Not covering	MATLAB functions are invoked to implement and evaluate the performance
					0	Not covering	
					≥+5	78.4–100	
[34]	$C_{40}, C_{41}, C_{42}, C_{60}, C_{61}, C_{62}, C_{63}$	High	16, 64	2, 4	−3	Not covering	
					0	Not covering	
					≥+5	89.8–100	
[11]	$C_{11}, C_{22}, C_{33}, C_{44}$	Medium	16~256	No	−3	98.33	
					0	100	
					≥+5	100	
[35]	$C_{20}, C_{21}, C_{40}, C_{41}, C_{42}, C_{60}, C_{61}, C_{62}, C_{63}$	High	16, 64, 256	2, 4, 8	−3	Not covering	
					0	77.6	
					≥+5	99.96	
[23]	$C_{20}, C_{21}, C_{40}, C_{41}, C_{42}, C_{60}, C_{61}, C_{62}, C_{63}, C_{80}, C_{81}, C_{82}, C_{83}, C_{84}$	High	16, 64, 256	2, 4, 8, and others	−3	Not covering	
					0	Not covering	
					≥+5	Unknown	
[21]	$C_{20}, C_{21}, C_{41}, C_{42}, C_{63}$	Medium	16, 64, V.29	4	−3	60	
					0	81	
					≥+5	90	
[10]	$C_{20}, C_{21}, C_{40}, C_{41}, C_{42}$	Medium	8, 16, 32, 64	4, 8, 16, 32, 64	−3	26	
					0	45	
					≥+5	95	
Proposed algorithm	Logarithmic cumulant	Medium	(8~1024)	4	−3	75–95	
					0	80–98	
					≥+5	100	

"~" refers values from M-QAM to M-QAM and "," refers values M-QAM and M-QAM; note that signal length $N = 4096$.

The results shown in Figures 16 and 17 indicate the performance degradation of classification for QAM signals versus their performance in the AWGN channel. It is not difficult to understand that the cause of the performance degradation comes from the multipath channel fading and Doppler fading effects model. The weakness performance of cumulant-based classifiers with fading channels has been documented in [22]. The only exception in the logarithmic classifier that maintains the same level of classification accuracy is typically at SNR >5 dB despite the different fading channel models used.

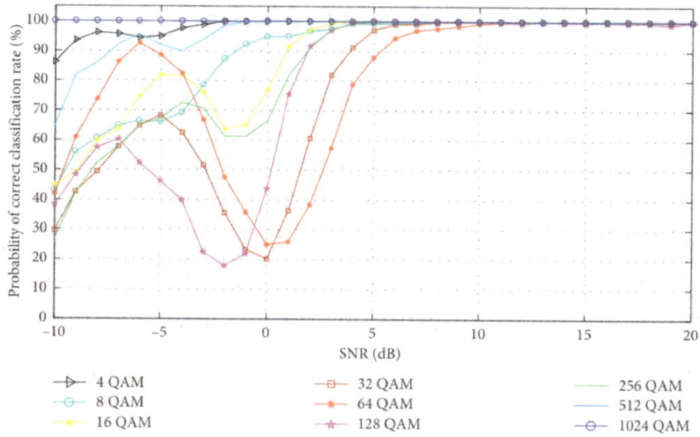

Figure 16: The probability of correct classification rate vs SNR with 10,000 signal realizations from {4~1024}-QAM each consisting signal length N=4096 and phase offset "π/6" under AWGN plus fast, frequency selective Ricean fading channel with $f_D = 5\,\text{kHz}$.

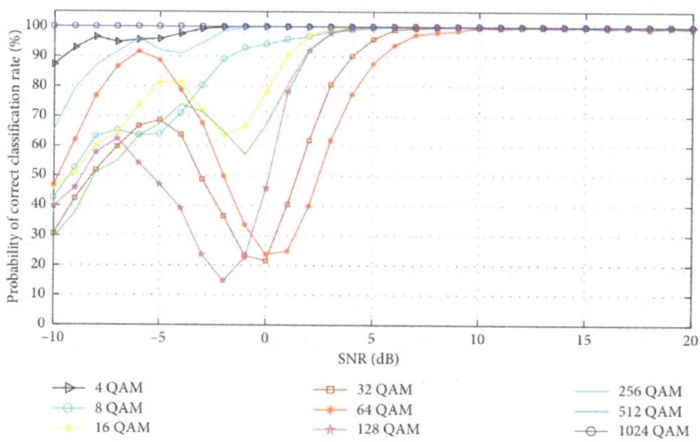

Figure 17: The probability of correct classification rate vs SNR with 10,000 signal realizations from {4~1024}-QAM each consisting signal length N=4096 and phase offset "π/6" under AWGN plus fast, frequency selective Rayleigh fading channel with $f_D = 5\,\text{kHz}$.

ROBUSTNESS EVALUATION CASES

Under Different Channel Models

Research in the modulation classification of digital signals approximately started since forty years ago, and the robustness evaluation of conventional AMC methods under the AWGN channel has been well established in the literature [37–39]. Similarly, the use of AWGN as the channel model is a common practice for communication engineers and researchers [19, 40].

In information theory, the basic communication channel model AWGN that represents the effect of several random processes occurring in nature is in the same direction, by which AWGN is used as a channel model where the communication impairment is an addition of white noise with a constant spectral density and normally distributed amplitude. In this work, the proposed logarithmic classifier has shown superior performance under the AWGN channel even with low SNR, also with variable lengths of the transmitted signal.

Table 2 shows the robustness analysis among previous works which were investigated in the development of the modulation classification algorithm. It appears very clearly that the logarithmic classifier has higher classification accuracy of modulation schemes {4-1024}-QAM, which also achieves a high percentage when the.

Although AWGN alone could be a proper model to the degradation of signals in free space, yet the existence of physical obstructions such as building and towers can cause multipath propagation losses. It implies that the signal reaches the receiver site via multiple propagation paths. Respecting to the relative lengths of the paths, this could be either constructive or destructive interference. Furthermore, relative motion between the transmitter and receiver may cause the Doppler shift. The performance of the proposed classifier is also investigated in the AWGN channel with fast frequency-selective fading also with the effect of Rayleigh and Rician magnitude channel fading, while the phase was randomly distributed and was assumed to be constant through the symbol period, which indicates no loss in signals. Rician factor is defined as the ratio between the power of the direct path and the power of the reflected paths, which is a constant value. Here, it is assumed to be 3 dB.

The specification used was based on Molisch and his coworker research [41], with a symbol rate of symbols per second, average path gains dB, path delays seconds, and maximum Doppler shift of 5 kHz.

Figures 16 and 17 depict the probability of correct recognition rate at signal length N=4096 in the AWGN plus fast, frequency selective with 50 kHz of maximum Doppler shifts with Rician and Rayleigh fading channels, respectively. Furthermore, Figures 18 and 19 show the probability of correct recognition rate with variable signal lengths in AWGN plus fast, frequency selective with 50 kHz of maximum Doppler shifts in Rician and Rayleigh fading channels, respectively.

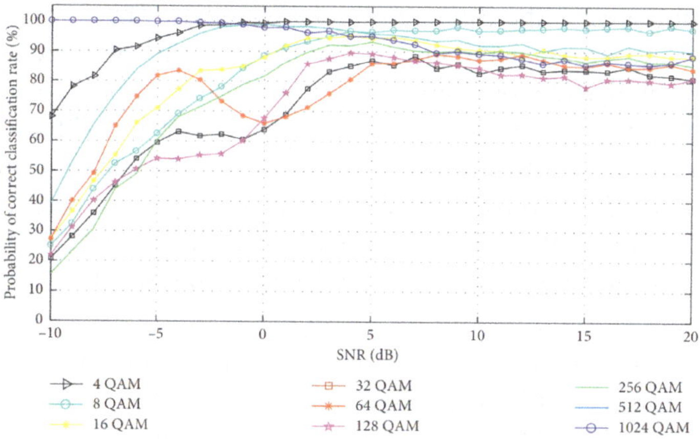

Figure 18: The probability of correct classification rate vs SNR with 10,000 signal realizations from {4~1024}-QAM each consisting variable signal length from N={64 128 265 512 1024 2048 4096} and phase offset "π/6" under AWGN plus fast, frequency selective Ricean fading channel with f_D = 5 kHz.

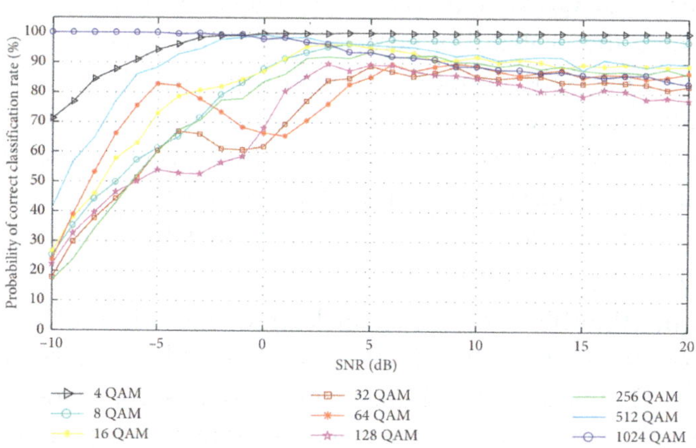

Figure 19: The probability of correct classification rate vs SNR with 10,000 signal realizations from {4~1024}-QAM each consisting variable signal length

from $N=\{64\ 128\ 265\ 512\ 1024\ 2048\ 4096\}$ and phase offset "π/6" under AWGN plus fast, frequency selective Rayleigh fading channel with $f_D = 5\,\text{kHz}$.

As could be observed in Figures 16 and 17, respectively, under signal propagation over multipath, it gives an accepted classification rate when SNR is not lower than 4 dB, although with presence of effect of the Doppler shift. Also, it could be observed in Figures 16 and 17. The probability of the correct classification rate at SNR > 4 dB is higher for the Rician fading channels compared to the Rayleigh fading channels. This is because of the existence of the strong signal due to the line of sight component. It is well noted that a higher Doppler shift results in a lower probability of correct classification [41]. As well as contrast to that, the logarithmic classifier provides unexpected performance even with the existence of the Doppler shift, and signal is transmitted with a variable length between {64-4096}. It is also important to underline that the proposed classifier vouchsafes satisfactory results under both fading channel environments Rician and Rayleigh.

As could also be observed in Figures 16 and 17, mainly, at SNR lower than −2 dB, the correct classification rate has fallen below 80%, especially to {32, 64, 128, 256} QAM signals. This could be noted when the SNR is lower than 0 dB.

The intraclass recognition of the modulation order using the logarithmic classifier gives different results depending on the modulation type. For example, the simulations in this work illustrated that this recognition will be better for 1024-QAM and 512-QAM schemes than other modulation schemes, where the probability percent is high to correct the classification rate. Similarly, the SNR is not lower than 0 dB, and the results show that the probability of the recognition rate gives the lowest percentage for 32-QAM and 64-QAM signals, which can be observed in Figures 16–19. It is also important to mention that the fluctuations in performance in the correct recognition rate is in higher order QAM signals, and that is due to the internal structures of the logarithmic classifier.

It seems that the effect of variation in the frequency offset of the transmitted signals almost insignificantly influences the logarithmic classifier although under SNR=−2 and random signal lengths. Figures 20(a) and 20(b) depict the probability of the correct classification rate vs frequency offset and signal length N=4096. Likewise Figure 20(b) depicts the probability of the correct classification rate vs frequency offset with a variable signal length and the signals corrupted by AWGN. This end could be an important point to the proposed classifier among the AMC methods that have tested the influence of the offset frequency to the transmitted signal. In the same direction, Figure 21 shows a change in the number of transmitted symbols vs probability of the correct classification rate. It also well-observed that the proposed method has a better classification rate than the proposed algorithm does [23]. Even with this, it is a very difficult circumstance with SNR=−2 dB in the AWGN channel. Eventually, the logarithmic classification algorithm is not the best one. On the contrary, it facilitated the process of classification of higher order QAM signals. Nevertheless, degradation points of the logarithmic classifier are that it is a channel-dependent model. Any variations in the type of noise (colored vs. AWGN), the number of transmitted signals present in the channel, and the type of possible modulated signals require an overhaul of all four tree threshold levels.

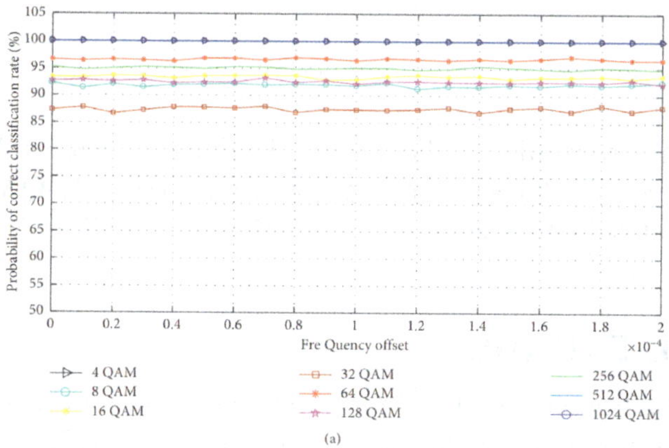

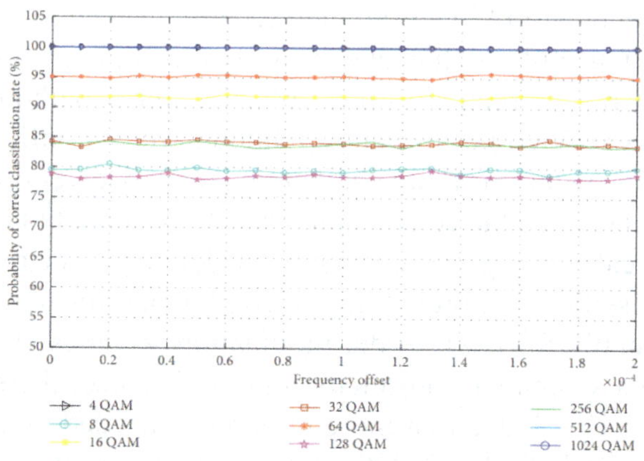

Figure 20: (a) Probability of the correct classification rate vs frequency offset and signal length N=4096. (b) Probability of the correct classification rate vs

frequency offset with variable signal lengths from $N = \{64\ 128\ 265\ 512\ 1024\ 2048\ 4096\}$ with a phase offset of "π/6" {4~1024}-QAM with 10,000 signal under AWGN—only SNR = −2 dB.

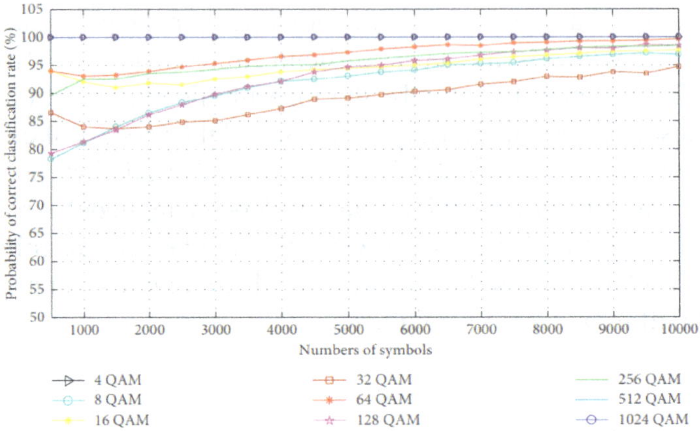

Figure 21: The probability of the correct classification rate vs the number of symbols SNR with 10,000 signal realizations from {4~1024}-QAM, phase offset "π/6", and signal degradation using AWGN—only SNR = −2 dB.

GENERAL TRENDS

Unlike the traditional AMC algorithms, the logarithmic classifier has functional features, which make them the preferable classifier for the medium and higher order of M-QAM modulation schemes. Likewise, the logarithmic classifier is very useful for channels that only receive corruption by AWGN. The prominent results related to the AWGN channel without fading effect have shown unexpected performance. Although the classification of higher order modulation signals is more challenging than to lower order signals due to the distance between the constellation points is small, but on the contrary, it appears nonreal convergence during the evaluation of the robustness of the logarithmic classifier. The probability of correct classification suffered from disparity especially for lower order modulation schemes at the expense of schemes that have smaller constellation point distance; therefore from this end, it might be concluded that more research is needed on the logarithmic classifier particularly SNR = 0. Although cumulant-based classifier schemes with many variations is the convenient way to cope with M-QAM modulation schemes, signal cumulant features, or uncoordinated combination between cumulants and moments does not provide sufficient classification under multipath fading channels and can even produce contrast in the classification rate. It is also notable that this classifier technique is not unique in terms of fixed threshold and there exist various ways to determine threshold levels automatically in order to improve the classification sense to the desired modulation schemes. However, the comprehensive review reveals that the logarithmic classifier with fixed threshold has overcome all weaknesses points even with a severe noisy environment, moreover, under multipath propagation channels. On the contrary, this work has provided references that help researchers make progress toward in the direction of developed new generation of AMC systems, which is called as the logarithmic classifier.

CONCLUSION

This paper provides a method that transacts with the modulation recognition of higher order M-QAM signals under the effect of AWGN on a channel environment through the usage of classifier which is cumulant-based natural logarithmic characteristics with fixed thresholds.

The combination of higher order cumulant and properties of the logarithmic functions created a new generation of the cumulant-based classifier, who provides a superior performance to classify the higher order M-QAM signals even with a low range of SNR.

The simulation results indicate that the correct classification rate can reach over 92% at an SNR of 4 dB under the AWGN, even as it seems moderate over different fading channels.

According to a comparison with the recently proposed algorithms in the literature, the performance of the logarithmic classifier was efficient and also less complex.

REFERENCES

1. O. A. Dobre, A. Abdi, Y. Bar-Ness, and W. Su, "Survey of automatic modulation classification techniques: classical approaches and new trends," *IET Communications*, vol. 1, no. 2, pp. 137–156, 2007.
2. Z. Xing and Y. Gao, "Method to reduce the signal-to-noise ratio required for modulation recognition based on logarithmic properties," *IET Communications*, vol. 12, no. 11, pp. 1360–1366, 2018.
3. J. A. Sills, "Maximum-likelihood modulation classification for PSK/QAM," in *Proceedings of the IEEE Military Communications Conference Proceedings MILCOM 1999*, vol. 1, pp. 217–220, Atlantic City, NJ, USA, October-November 1999.
4. F. Hameed, O. Dobre, and D. Popescu, "On the likelihood-based approach to modulation classification," *IEEE Transactions on Wireless Communications*, vol. 8, no. 12, pp. 5884–5892, 2009.
5. W. Wei and J. M. Mendel, "A new maximum-likelihood method for modulation classification," in *Proceedings of the 1995 Conference Record of the Twenty-Ninth Asilomar Conference on Signals, Systems and Computers*, vol. 2, pp. 1132–1136, Pacific Grove, CA, USA, October 1995.
6. E. E. Azzouz and A. K. Nandi, "Automatic identification of digital modulation types," *Signal Processing*, vol. 47, no. 1, pp. 55–69, 1995.
7. M. Abdelbar, W. H. Tranter, L. Fellow, T. Bose, and S. Member, "Cooperative cumulants-based modulation classification in distributed networks," *IEEE Transactions on Cognitive Communications and Networking*, vol. 4, no. 3, pp. 446–461, 2018.
8. J. L. Xu, W. Su, and M. Zhou, "Likelihood-ratio approaches to automatic modulation classification," *IEEE Transactions on Systems, Man, and Cybernetics, Part C (Applications and Reviews)*, vol. 41, no. 4, pp. 455–469, 2011.
9. H. Hosseinzadeh, F. Razzazi, and A. Haghbin, "A self training approach to automatic modulation classification based on semi-supervised online passive aggressive algorithm," *Wireless Personal Communications*, vol. 82, no. 3, pp. 1303–1319, 2015.
10. A. Swami and B. M. Sadler, "Hierarchical digital modulation classification using cumulants," *IEEE Transactions on Communications*, vol. 48, no. 3, pp. 416–429, 2000.
11. I. A. Hashim, J. W. Abdul Sadah, T. R. Saeed, and J. K. Ali,

"Recognition of QAM signals with low SNR using a combined threshold algorithm," *IETE Journal of Research*, vol. 61, no. 1, pp. 65–71, 2015.

12. M. L. Dukiü and G. B. Markoviü, "Automatic modulation classification using cumulants with repeated classification attempts," in *20th Telecommunications Forum (TELFOR)*, vol. 4, pp. 424–427, 2012.
13. D. S. Chirov, "Application of the decision trees to recognize the types of digital modulation of radio signals in cognitive systems of HF communication," in *Proceedings of the 2018 Systems of Signal Synchronization, Generating and Processing in Telecommunications*, pp. 1–6, Minsk, Belarus, July 2018.
14. B. G. Mobasseri, "Constellation shape as a robust signature for digital modulation recognition," in *Proceedings of the IEEE Conference on Military Communications MILCOM 1999*, pp. 442–446, Piscataway, NJ, USA, October-November 1999.
15. B. G. Mobasseri, "Digital modulation classification using constellation shape," *Signal Processing*, vol. 80, no. 2, pp. 251–277, 2000.
16. C. Yin, B. Li, Y. Li, and B. Lan, "Modulation classification of MQAM signals based on density spectrum of the constellations," in *Proceedings of the 2010 2nd International Conference on Future Computer and Communication ICFCC 2010*, vol. 3, pp. 57–61, Wuhan, China, May 2010.
17. A. Fehske, J. Gaeddert, and J. H. Reed, "A new approach to signal classification using spectral correlation and neural networks," in *Proceedings of the First IEEE International Symposium on New Frontiers in Dynamic Spectrum Access Networks DySPAN 2005*, pp. 144–150, Baltimore, MD, USA, November 2005.
18. W. C. Headley, J. D. Reed, and C. R. C. Da Silva, "Distributed cyclic spectrum feature-based modulation classification," in *Proceedings of the 2008 IEEE Wireless Communications and Networking Conference*, pp. 1200–1204, Las Vegas, NV, USA, March-April 2008.
19. Y. Yuan, P. Zhao, B. Wang, and B. Wu, "Hybrid maximum likelihood modulation classification for continuous phase modulations," *IEEE Communications Letters*, vol. 20, no. 3, pp. 450–453, 2016.
20. H. T. Fu, Q. Wan, and R. Shi, "Modulation classification based on cyclic spectral features for co-channel time-frequency overlapped two-signal," in *Proceedings of the 2009 Pacific-Asia Conference*

on Circuits, Communications and Systems PACCS 2009, pp. 31–34, Chengdu, China, May 2009.

21. O. A. Dobre, M. Oner, S. Rajan, and R. Inkol, "Cyclostationarity-based robust algorithms for QAM signal identification," *IEEE Communications Letters*, vol. 16, no. 1, pp. 12–15, 2012.
22. O. A. Dobre, A. Abdi, Y. Bar-Ness, and W. Su, "Cyclostationarity-based modulation classification of linear digital modulations in flat fading channels," *Wireless Personal Communications*, vol. 54, no. 4, pp. 699–717, 2010.
23. O. A. Dobre, Y. Bar-Ness, and W. Su, "Higher-order cyclic cumulants for high order modulation classification," in *Proceedings of the 2003 IEEE conference on Military communications MILCOM 2003*, pp. 112–117, Boston, MA, USA, October 2003.
24. C. M. Spooner, "On the utility of sixth-order cyclic cumulants for RF signal classification," in *Proceedings of the Conference Record-Asilomar Conference on Signals, Systems and Computers*, vol. 1, pp. 890–897, Pacific Grove, CA, USA, November 2001.
25. F. Ghofrani, A. Jamshidi, and A. Keshavarz-Haddad, "Internet traffic classification using hidden naive Bayes model," in *Proceedings of the ICEE 2015 23rd Iranian Conference on Electrical Engineering*, vol. 10, pp. 235–240, Tehran, Iran, May 2015.
26. F. Ghofrani, A. Keshavarz-Haddad, and A. Jamshidi, "Internet traffic classification using multiple classifiers," in *Proceedings of the 2015 7th International Conference on Knowledge and Smart Technology (KST) IKT 2015*, Chonburi, Thailand, January 2015.
27. S. Norouzi, A. Jamshidi, and A. R. Zolghadrasli, "Adaptive modulation recognition based on the evolutionary algorithms," *Applied Soft Computing*, vol. 43, pp. 312–319, 2016.
28. H. Agirman-Tosun, Y. Liu, A. M. Haimovich et al., "Modulation classification of MIMO-OFDM signals by independent component analysis and support vector machines," in *Proceedings of the 2011 Conference Record of the Forty Fifth Asilomar Conference on Signals, Systems and Computers (ASILOMAR)*, pp. 1903–1907, Pacific Grove, CA, USA, November 2011.
29. D. Boutte and B. Santhanam, "A hybrid ICA-SVM approach to continuous phase modulation recognition," *IEEE Signal Processing Letters*, vol. 16, no. 5, pp. 402–405, 2009.

30. B. Ramkumar, "Automatic modulation classification for cognitive radios using cyclic feature detection," *IEEE Circuits and Systems Magazine*, vol. 9, no. 2, pp. 27–45, 2009.
31. A. Ali, F. Yangyu, and S. Liu, "Automatic modulation classification of digital modulation signals with stacked autoencoders," *Digital Signal Processing*, vol. 71, pp. 108–116, 2017.
32. S. Xi and H.-C. Wu, "Robust automatic modulation classification using cumulant features in the presence of fading channels," in *Proceedings of theIEEE Wireless Communications and Networking Conference, 2006. WCNC 2006*, pp. 2094–2099, Las Vegas, NV, USA, April 2006.
33. Z. Zhu, M. Waqar Aslam, and A. K. Nandi, "Genetic algorithm optimized distribution sampling test for M-QAM modulation classification," *Signal Processing*, vol. 94, no. 1, pp. 264–277, 2014.
34. N. Ahmadi and R. Berangi, "Modulation classification of QAM and PSK from their constellation using Genetic Algorithm and hierarchical clustering," in *Proceedings of the 2008 3rd International Conference on Information & Communication Technologies: from Theory to Applications*, vol. 11670, Damascus, Syria, April 2008.
35. A. Abdelmutalab, K. Assaleh, and M. El-Tarhuni, "Automatic modulation classification based on high order cumulants and hierarchical polynomial classifiers," *Physical Communication*, vol. 21, pp. 10–18, 2016.
36. M. H. Alsharif and R. Nordin, "Evolution towards fifth generation (5G) wireless networks: current trends and challenges in the deployment of millimetre wave, massive MIMO, and small cells," *Telecommunication Systems*, vol. 64, no. 4, pp. 617–637, 2017.
37. C.-S. Park, J.-H. Choi, S.-P. Nah, W. Jang, and D. Y. Kim, "Automatic modulation recognition of digital signals using wavelet features and SVM," in *Proceedings of the 2008 10th International Conference on Advanced Communication Technology*, vol. 1, pp. 387–390, Gangwon-Do, South Korea, February 2008.
38. N. Ahmadi, "Using fuzzy clustering and TTSAS algorithm for modulation classification based on constellation diagram," *Engineering Applications of Artificial Intelligence*, vol. 23, no. 3, pp. 357–370, 2010.
39. L. Zhou, Z. Sun, and W. Wang, "Learning to short-time Fourier transform in spectrum sensing," *Physical Communication*, vol. 25, pp. 420–425, 2017.

40. M. H. Valipour, M. M. Homayounpour, and M. A. Mehralian, "Automatic digital modulation recognition in presence of noise using SVM and PSO," in *Proceedings of the 6th International Symposium on Telecommunications IST 2012*, pp. 378–382, Tehran, Iran, November 2012.

41. A. F. Molisch, K. Balakrishnan, D. Cassioli, and C.-C. Chong, "IEEE 802.15.4a channel model-final report," *Environments*, pp. 1–40, 2005.

8

Mathematical Analysis for Detection Probability in Cognitive Radio Networks over Wireless Communication Channels

Rohitha Ujjinimatad[1] and Siddarama R. Patil[2]

[1]Department of Electronics and Communication Engineering , Ballari Institute of Technology and Management , Bellary 583 104 , Karnataka , India
[2]Electronics and Communication Engineering Department , PDA College of Engineering , Gulbarga , Karnataka , India

ABSTRACT

In this study, the authors consider the problem of spectrum sensing based on energy detection method in cognitive radio over wireless communication channels when users experience fading and non-fading effects. The closed-form analytical expressions for the detection probability are derived over non-fading additive white Gaussian noise channel and Rayleigh and log-normal shadowing fading channels. The detection probability involves Marcum-Q function, summations and integrations in the early research papers, which are replaced by closed-form expressions in this study. The

Citation: Ujjinimatad, Rohitha; Patil, Siddarama R.: "Mathematical analysis for detection probability in cognitive radio networks over wireless communication channels", The Journal of Engineering, 2014, 2014, (8), p. 445-449, https://doi.org/10.1049/joe.2014.0173.

probability distribution function of fading channels is used to obtain the expressions for detection probability. The new derived numerical results are simulated under various parameters. The performance of the derived theoretical expressions closely matches with the simulated results.

INTRODUCTION

Owing to the increasing popularity of wireless devices in recent years, the radio spectrum has been an extremely scarce resource. However, all bands of the spectrum are not fully utilised at specific times or at particular geographic locations according to the report of Federal Communications Commission (FCC) [1]. The underutilisation of spectrum in many of the frequency bands is because of the conventional fixed spectrum allocation policy. To overcome the problem of spectrum scarcity, the license exempted secondary users (SUs) are allowed to exploit the unused spectrum holes over some frequency ranges by using the cognitive radio (CR) technology [2]. The challenge in implementing the CR technology is that the SUs must accurately monitor and be aware of the presence of the primary users (PUs) over a particular spectrum. Several efficient methods have been proposed to address this challenge [3 – 6]. The detection of unknown deterministic signals with energy detection (ED) is the simplest and most popular method since it does not require any a priori knowledge of the PUs signal and has much lower computational and implementation complexity. The performance of ED over non-fading additive white Gaussian noise (AWGN) channel was first addressed by Urkowitz and comprehensively derived analytic expressions for the probability of detection and the probability of false alarm [7]. Recently, the detection problem over fading channels has been revised in [8 , 9]. The expressions derived for the detection probability over fading channels using energy detector in [9 – 14] depend on the number of terms in the summation and other functions. The implementations of these expressions are susceptible to truncation errors. We summarise our contributions as follows. We provide approximate expressions for the detection probability over non-fading AWGN channel and Rayleigh and log-normal shadowing fading channels. The expressions reported in [9 – 14] for AWGN channel involves the Marcum-Q function, which is replaced by closed-form expression in our derivation (16). It has much less computational complexity. The derived analytical expressions over fading channels in the recent papers consist of summations and integrations with the limitations. These are susceptible to truncation errors. We compare our new derived expressions (16), (22) and (30) with the expressions reported in [9 – 14] and we show that our new derived expressions do not involve summations and integrations and they can be used with no limitations. The simulation results of the considered ED method over both AWGN and fading channels are compared with the theoretical evaluation of the derived expressions. The rest of this paper is organised as follows. In Section 2, the system model is discussed. We derive expressions for the detection probability over fading channels in Section 3. The performance of the derived expressions and simulation results are presented in Section 4. Conclusions are drawn in Section 5.

MODEL

The energy-sensing model is as shown in Fig. 1 . The received signal r(t) is filtered to the required bandwidth B to reject the noise and adjacent signals. The filtered signal is sampled and quantised using analog-to-digital converter. Next, a square-law device and an integrator measure the received signal energy. The output of the integrator, represented by the decision statistic Y, is compared with a predetermined threshold γ to determine the presence (H_1) or absence (H_0) of a PU. The presence or absence of a PU signal can be modelled as binary hypotheses problem as originally proposed by Urkowitz [7] and Pridham and Urkowitz [15] and latter followed by most researchers [16 , 17]. The received signal samples under two hypotheses defined as

$$r(n) = \begin{cases} w(n) & : H_0 \\ hs(n) + w(n) & : H_1 \end{cases} \tag{1}$$

where n = 1, 2, …, N. s(n) is the PU signal which is assumed to be an unknown deterministic signal and w(n) is the AWGN with zero mean and variance σ_n^2. h is the channel gain between PU and SU. The energy detector is defined as

$$Y = \sum_{n=1}^{N} |r(n)|^2 \tag{2}$$

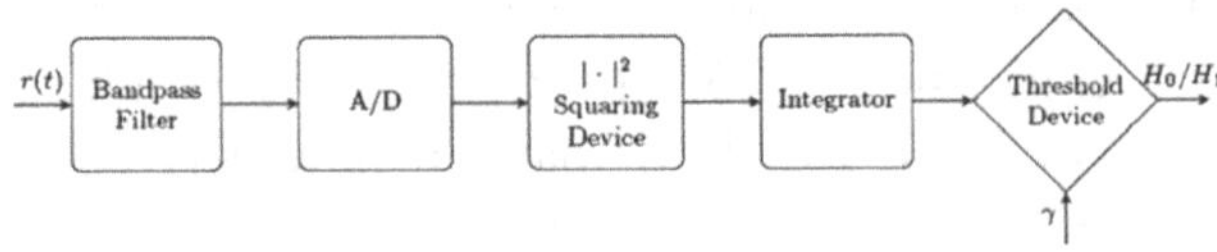

Figure 1: Energy detector model.

The distribution of the decision variable Y will be central chi-square X_N^2 under H $_0$ and non-central chi-square $\tilde{X}_N^2$ with N degrees of freedom under H $_1$. The distribution can be expressed as [18]

$$Y \sim \begin{cases} X_N^2 & : H_0 \\ \tilde{X}_N^2 & : H_1 \end{cases} \tag{3}$$

Its probability density function can be written as

$$f_Y(y) = \begin{cases} \frac{1}{\sigma_n^N 2^{N/2}\Gamma(N/2)} y^{(N/2)-1} \exp\left(\frac{-y}{2\sigma_n^2}\right) & : H_0 \\ \frac{1}{2\sigma_n^2}\left(\frac{y}{\zeta}\right)^{(N-2)/4} \exp\left[\frac{-1}{2\sigma_n^2}(\zeta + y)\right] I_{(N/2)-1}\left(\frac{\sqrt{\zeta y}}{\sigma_n^2}\right) & : H_1 \end{cases} \tag{4}$$

where the non-centrality parameter $\zeta = \sum_{i=1}^{N} \mu_i^2$ and μ_i is the mean of the ith Gaussian random variable of test Y. I $_M$ (·) is the Mth modified Bessel function of the first kind, which has a series representation [19]

$$I_M(Z) = \frac{\sum_{k=0}^{\infty} (Z/2)^{2k+M}}{k!\Gamma(M + k + 1)} \tag{5}$$

and Γ(·) is the gamma function $\Gamma(z) = \int_0^\infty t^{z-1} e^{-t}\, dt.$. The test statistic Y is compared with a predetermined threshold γ, to decide on one of the two hypotheses H_0 and H_1. The probability of detection (P_d) and probability of false alarm (P_{fa}) can be computed as

$$P_d = P(Y > \gamma : H_1) = \int_\gamma^\infty f_Y(y)\, dy \tag{6}$$

$$P_{fa} = P(Y > \gamma : H_0) = \int_\gamma^\infty f_Y(y)\, dy \tag{7}$$

Probability of False Alarm under AWGN Channel

To derive the probability of false alarm over non-fading AWGN channel, the probability density function defined under H_0 in (4) is inserted in (7). We define signal-to-noise ratio (SNR) (ρ) as $\rho = h^2\zeta/\sigma_n^2$

$$P_{\text{fa}} = \int_{\gamma}^{\infty} \frac{1}{\sigma_n^N 2^{N/2}\Gamma(N/2)} y^{(N/2)-1} \exp\left(\frac{-y}{2\sigma_n^2}\right) \mathrm{d}y \tag{8}$$

Substituting $t = y/\sigma_n^2$ in (8) and further integrating the P_{fa} results in

$$P_{\text{fa}} = \int_{\gamma/\sigma_n^2}^{\infty} \frac{t^{(N/2)-1}}{2^{N/2}\Gamma(N/2)} \exp\left(\frac{-t}{2}\right) \mathrm{d}t \tag{9}$$

The right-hand side of (9) is the chi-squared density [20, 26.4.2] which can be written as $P_{\text{fa}} = Q_{\chi_N^2}(\gamma/\sigma_n^2)$. In the energy detector, the threshold γ can be computed using $\gamma = \sigma_n^2\ [\sqrt{2/N}Q^{-1}(P_{\text{fa}}) + 1]$, where $Q(x) = (1/\sqrt{2\pi}) \int_x^{\infty} \mathrm{e}^{-u^2/2}\,\mathrm{d}u$ is the standard Gaussian tail probability function.

Probability of Detection under AWGN Channel

To derive the probability of detection over non-fading AWGN channel, the probability density function defined under H_1 in (4) is inserted in (6). By substituting $t = y/\sigma_n^2$ and using definition of ρ, the probability of detection is given by

$$P_{\text{d}} = \int_{\gamma/\sigma_n^2}^{\infty} \frac{1}{2}\left(\frac{t}{\rho}\right)^{(N-2)/4} \exp\left(\frac{-(t+\rho)}{2}\right) I_{(N/2)-1}\left(\sqrt{t\rho}\right) \mathrm{d}t \tag{10}$$

Substituting $t = x^2$, $\rho = a^2$ and with the further simplification, P_d can be written as

$$P_{\text{d}} = \frac{1}{a^{(N/2)-1}} \int_{\sqrt{\gamma/\sigma_n^2}}^{\infty} x^{N/2} \exp\left(\frac{-(x^2+a^2)}{2}\right) I_{(N/2)-1}(ax)\, \mathrm{d}x \tag{11}$$

We can rewrite (11) using the definition of generalised Marcum-Q function [21, 1.2] as

$$P_{\text{d}} = Q_{N/2}\left(\sqrt{\rho}, \sqrt{\frac{\gamma}{\sigma_n^2}}\right) \tag{12}$$

Since the precise computation of Marcum-Q function is quite difficult, it is represented in the new form as [21, 2.6]

$$P_{\text{d}} = 1 - \sum_{n\geq 0} (-1)^n \exp\left(\frac{-\rho}{2}\right) \frac{L_n^{(N/2)-1}(\rho/2)}{\Gamma((N/2)+n+1)} \left(\frac{\gamma}{2\sigma_n^2}\right)^{(N/2)+n} \tag{13}$$

where $L_n^k(x)$ is the generalised Laguerre polynomial of degree n and order k. The absolute convergence of the series in (13) can be easily shown [21 , 22] as

$$\left|\sum_{n\geq 0}(-1)^n \exp\left(\frac{-\rho}{2}\right)\frac{L_n^{\frac{N}{2}-1}\left(\frac{\rho}{2}\right)}{\Gamma\left(\frac{N}{2}+n+1\right)}\left(\frac{\gamma}{2\sigma_n^2}\right)^{\frac{N}{2}+n}\right|$$

$$\leq \exp\left(\frac{-\rho}{2}\right)\sum_{n\geq 0}\frac{1}{\Gamma((N/2)+n+1)}\left(\frac{\gamma}{2\sigma_n^2}\right)^{(N/2)+n}\left|L_n^{(N/2)-1}\left(\frac{\rho}{2}\right)\right| \tag{14}$$

$$= \exp\left(\frac{-\rho}{4}\right)\frac{1}{\Gamma(N/2)}\left(\frac{\gamma}{2\sigma_n^2}\right)^{(N/2)-1}\exp\left(\frac{\gamma}{2\sigma_n^2}-1\right) \tag{15}$$

Inserting (15) in (13) the P_d for AWGN channel is approximated as

$$P_{\mathrm{d}} \cong 1-\exp\left(\frac{-\rho}{4}\right)\frac{1}{\Gamma(N/2)}\left(\frac{\gamma}{2\sigma_n^2}\right)^{(N/2)-1}\exp\left(\frac{\gamma}{2\sigma_n^2}-1\right) \tag{16}$$

PROBABILITY OF DETECTION UNDER FADING CHANNELS

We define the average SNR $\bar{\rho}$ as $\bar{\rho}=E[h^2]\zeta/\sigma_n^2$, where $E(\cdot)$ denotes the expectation operator.

Rayleigh Fading Channel

The probability distribution function (PDF) of ρ in case of Rayleigh fading channel is given as

$$f(\rho)=\frac{1}{\bar{\rho}}\exp\left(\frac{-\rho}{\bar{\rho}}\right) \tag{17}$$

Using (16), we obtain the probability of detection as

$$P_{\mathrm{d_{Ray}}}=\int_0^{\infty}\left[1-\exp\left(\frac{-\rho}{4}\right)\frac{1}{\Gamma(N/2)}\left(\frac{\gamma}{2\sigma_n^2}\right)^{(N/2)-1}\exp\left(\frac{\gamma}{2\sigma_n^2}-1\right)\right]\frac{1}{\bar{\rho}}\exp\left(\frac{-\rho}{\bar{\rho}}\right)\,\mathrm{d}\rho \tag{18}$$

$$P_{\mathrm{d_{Ray}}}=\underbrace{\int_0^{\infty}\frac{1}{\bar{\rho}}\exp\left(\frac{-\rho}{\bar{\rho}}\right)\mathrm{d}\rho}_{\text{first term}}-\frac{1}{\bar{\rho}}\frac{1}{\Gamma(N/2)}\left(\frac{\gamma}{2\sigma_n^2}\right)^{(N/2)-1}\exp\left(\frac{\gamma}{2\sigma_n^2}-1\right)\underbrace{\int_0^{\infty}\exp\left(\frac{-\rho}{4}\right)\exp\left(\frac{-\rho}{\bar{\rho}}\right)\mathrm{d}\rho}_{\text{second term}} \tag{19}$$

The first term of (19) is simplified to

$$\int_0^{\infty}\frac{1}{\bar{\rho}}\exp\left(\frac{-\rho}{\bar{\rho}}\right)\mathrm{d}\rho=1 \tag{20}$$

The second term of (19) is simplified to

$$\int_0^{\infty}\exp\left(\frac{-\rho}{4}\right)\exp\left(\frac{-\rho}{\bar{\rho}}\right)\mathrm{d}\rho=\frac{4\bar{\rho}}{4+\bar{\rho}} \tag{21}$$

After substituting (20) and (21) in (19), the approximate probability of detection for Rayleigh channel is

$$P_{d_{Ray}} \cong 1 - \frac{4}{4+\bar{\rho}} \frac{1}{\Gamma(N/2)} \left(\frac{\gamma}{2\sigma_n^2}\right)^{(N/2)-1} \exp\left(\frac{\gamma}{2\sigma_n^2} - 1\right) \tag{22}$$

Log-normal Shadowing Channel

The link quality in terrestrial and satellite land-mobile systems is affected by slow variation of the mean signal level because of the shadowing from terrain, buildings and trees. The performance of a communication system will depend only on the shadowing if the radio receiver is able to average out the fast multipath fading or if an efficient micro-diversity system is used to eliminate the effects of multipath. Empirical measurements reveal that shadowing can be modelled by a log-normal distribution for various outdoor and indoor environments, in which case the probability distribution of ρ is given by [23]

$$f(\rho) = \frac{4.34}{\sqrt{2\pi}\sigma\rho} \exp\left(-\frac{(10\log_{10}\rho - \mu)^2}{2\sigma^2}\right) \tag{23}$$

where μ(d B) and σ(d B) are the mean and the standard deviation of 10 $\log_{10} \rho$, respectively. To the best of our knowledge, closed-form expression cannot be obtained for the detection probability using (16) and (23). The log-normal distribution can be closely approximated by the Wald distribution [10]. Its PDF is given by

$$f(\rho) = \sqrt{\frac{\eta}{2\pi}} \rho^{-3/2} \exp\left(-\frac{\eta(\rho-\theta)^2}{2\theta^2\rho}\right) \tag{24}$$

where $\theta = E(\rho)$ denotes the expectation of ρ and η is the shape parameter. These parameters are related with μ and σ as given below

$$\theta = \exp\left(\frac{\mu}{4.34} + \frac{\sigma^2}{37.67}\right)$$

$$\eta = \frac{\theta}{\exp(\sigma^2/18.84) - 1} \tag{25}$$

Using (16) and (24), the approximate probability of detection for log-normal shadowing channel is given as

$$P_{d_{Log-Normal}} \cong \int_0^\infty \left[1 - \exp\left(\frac{-\rho}{4}\right) \frac{1}{\Gamma(N/2)} \left(\frac{\gamma}{2\sigma_n^2}\right)^{(N/2)-1} \exp\left(\frac{\gamma}{2\sigma_n^2}\right) - 1\right] \sqrt{\frac{\eta}{2\pi}} \rho^{-3/2} \exp\left(-\frac{\eta(\rho-\theta)^2}{2\theta^2\rho}\right) d\rho \tag{26}$$

$$P_{d_{\text{Log-Normal}}} \cong \sqrt{\frac{\eta}{2\pi}}\left[\underbrace{\int_0^{\infty} \rho^{-3/2}\exp\left(-\frac{\eta(\rho-\theta)^2}{2\theta^2\rho}\right) \mathrm{d}\rho}_{\text{Firstterm}} - \frac{1}{\Gamma(N/2)}\left(\frac{\gamma}{2\sigma_n^2}\right)^{(N/2)-1} \exp\left(\frac{\gamma}{2\sigma_n^2}-1\right)\underbrace{\int_0^{\infty} \exp\left(\frac{-\rho}{4}\right)\rho^{-3/2}\exp\left(-\frac{\eta(\rho-\theta)^2}{2\theta^2\rho}\right) \mathrm{d}\rho}_{\text{Secondterm}}\right] \tag{27}$$

Using [24, 3.471-9], the first and second terms of (27) are reduced to

$$\int_0^{\infty} \rho^{-3/2}\exp\left(-\frac{\eta(\rho-\theta)^2}{2\theta^2\rho}\right) \mathrm{d}\rho = \sqrt{\frac{\eta}{2\pi}}\exp\left(\frac{\eta}{\theta}\right)\int_0^{\infty} \rho^{-(1/2)-1}\exp\left(-\frac{\eta}{2\rho}-\frac{\eta\rho}{2\theta^2}\right) \mathrm{d}\rho = \exp\left(\frac{\eta}{\theta}\right)\sqrt{\frac{2\eta}{\pi\theta}}K_{-1/2}\left(\frac{\eta}{\theta}\right) \tag{28}$$

$$\begin{aligned}\int_0^{\infty} \rho^{-3/2}\exp\left(\frac{-\rho}{4}-\frac{\eta(\rho-\theta)^2}{2\theta^2\rho}\right)\mathrm{d}\rho &= \sqrt{\frac{\eta}{2\pi}}\exp\left(\frac{\eta}{\theta}\right)\int_0^{\infty} \rho^{-(1/2)-1}\exp\left(-\frac{\eta}{2\rho}-\left(\frac{\eta}{2\theta^2}+\frac{1}{4}\right)\rho\right) \mathrm{d}\rho \\ &= \sqrt{\frac{\eta}{2\pi}}\exp\left(\frac{\eta}{\theta}\right)2\left(\frac{\eta}{2((\eta/2\theta^2)+(1/4))}\right)^{-1/4}K_{-1/2}\left[2\sqrt{\frac{\eta}{2}\left(\frac{\eta}{2\theta^2}+\frac{1}{4}\right)}\right]\end{aligned} \tag{29}$$

where $K_v(x)$ is a modified Bessel function of second kind. By substituting (28) and (29) in (27), the approximate probability of detection over log-normal shadowing channel is given by

$$P_{d_{\text{Log-Normal}}} \cong \exp\left(\frac{\eta}{\theta}\right)\left\{\sqrt{\frac{2\eta}{\pi\theta}}K_{-1/2}\left(\frac{\eta}{\theta}\right) - \frac{1}{\Gamma(N/2)}\left(\frac{\gamma}{2\sigma_n^2}\right)^{(N/2)-1}\exp\left(\frac{\gamma}{2\sigma_n^2}-1\right)\sqrt{\frac{2\eta}{\pi}}\left(\frac{\eta}{2((\eta/2\theta^2)+(1/4))}\right)^{-1/4}K_{-1/2}\left[2\sqrt{\frac{\eta}{2}\left(\frac{\eta}{2\theta^2}+\frac{1}{4}\right)}\right]\right\} \tag{30}$$

RESULTS AND DISCUSSION

In this section, we present the performance of the derived expressions for AWGN and fading channels through simulation. Receiver operating characteristic (ROC) has been widely used in the signal detection theory. It is an ideal technique to quantify the trade-off between the probability of detection and probability of false alarm. In the simulation, we use complementary ROC curves (P_{md} against P_{fa}) to show the detection performance of ED over non-fading AWGN and fading channels. The noise variance factor (σ2n)σn2 is considered in each simulation result.

In Fig. 2 , we generate the complementary ROC curves for AWGN channel for different values of ρ with N = 10 and noise variance $\sigma_n^2 = 0.5$. In this figure, the deviation between (12) which involves generalised Marcum-Q function and the derived closed-form expression (16) is shown. The deviation between (12) and (16) is too small. It is clear that increasing SNR decreases the miss-detection probability and improvement in the detection probability diminishes at low SNR.

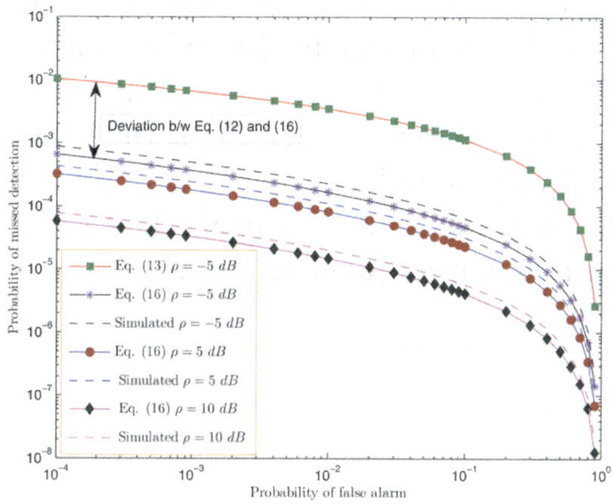

Figure 2: Complementary ROC curves for AWGN channel for different values of ρ with N = 10 and $\sigma_n^2 = 0.5$.

Fig. 3 shows the complementary ROC curves for different values of N in AWGN channel for ρ = −10 dB and $\sigma_n^2 = 0.75$. Clearly, if the value of N is greater then the network can achieve better performance. From this figure, we can see that the miss-detection probability decreases greatly with a small increase in value of N, which shows the improvement in the sensing performance.

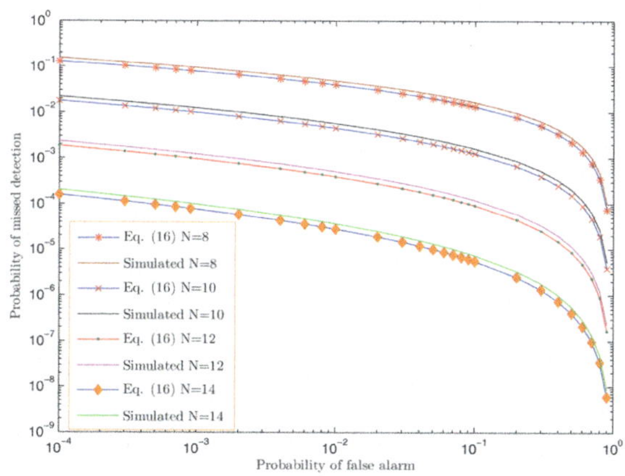

Figure 3: Complementary ROC curves for AWGN channel for different values of N with ρ = −10 dB and $\sigma_n^2 = 0.75$

In Fig. 4 , we have presented the performance evaluation of ED over Rayleigh channel based on both analytical and simulation for different values of N with $\bar{\rho} = -5$ and $\sigma_n^2 = 0.5$. As can be observed, the simulation results well confirm the analytical expression (22). Here, increasing the value of N improves the detection probability. Fig. 5 shows the complementary ROC curves for different values of $\bar{\rho}$ with N = 8 and $\sigma_n^2 = 0.5$. We note from the figure that there is an improvement in the performance with several dB's increment in $\bar{\rho}$.

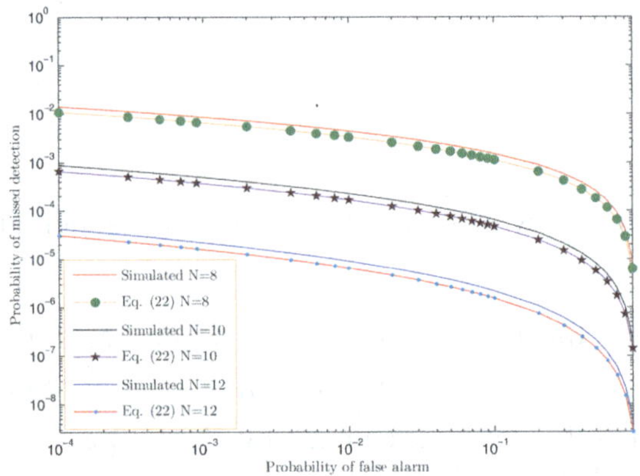

Figure 4: Complementary ROC curves for Rayleigh channel for different values of N with $\bar{\rho} = -5\ dB\ and\ \sigma_n^2 = 0.5$

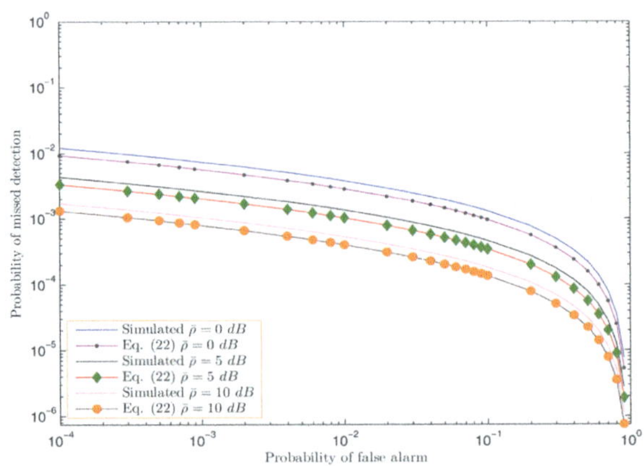

Figure 5: Complementary ROC curves for Rayleigh channel for different values of $\bar{\rho}$ with N = 8 and $\sigma_n^2 = 0.5$.

In Fig. 6 , we plot complementary ROC curves for log-normal shadowing channel for different values of mean (μ) with N = 8, σ = 4 and $\sigma_n^2 = 0.5$. There is a great improvement in the detection probability when μ is increased from 6 to 12 dB. Fig. 7 shows the complementary ROC curves for log-normal shadowing channel for different values of N with μ = 5 dB, σ = 3 and $\sigma_n^2 = 0.6$. We note that detection probability is low when N = 6 and we can achieve very high detection probability for N = 12. Here simulation results closely match with analytical results. Finally, we say that all the derived expressions for the detection probability over non-fading and fading channels are simulated in MATLAB under various parameters which concur with the simulation results.

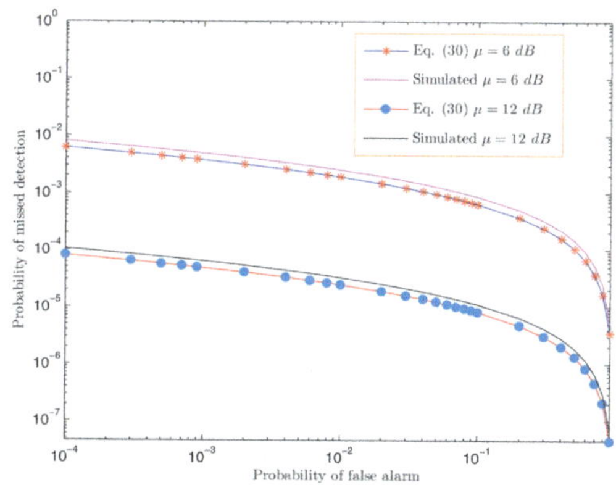

Figure 6: Complementary ROC curves for log-normal shadowing channel for different values of μ with N = 8, σ = 4 and $\sigma_n^2 = 0.5$.

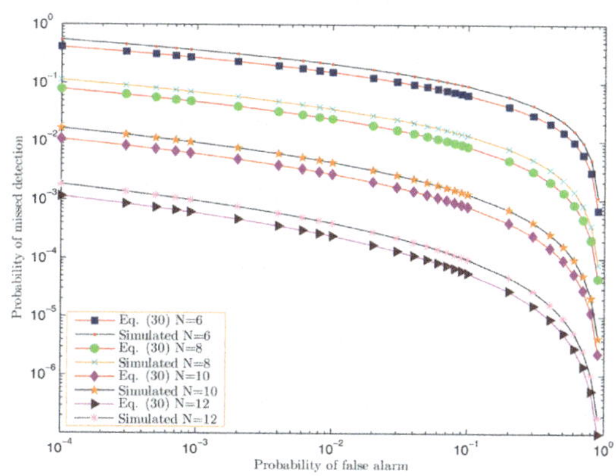

Figure 7: Complementary ROC curves for log-normal shadowing channel for different values of N with μ = 5 dB, σ = 3 and $\sigma_n^2 = 0.6$.

CONCLUSIONS

In this study, we derived the closed-form expressions for the probability of detection over non-fading AWGN and fading channels. To the best of our knowledge, (16), (22) and (30) are the new numerical results for the detection probability over considered channels. These theoretical expressions are evaluated under various parameters and verified by the simulation results. Our numerical results in Section 3 provide new mathematical analysis for the detection probability over fading channels.

REFERENCES

1. KolodzyP. AvoidanceI. ModelsS.: ' Spectrum policy task force'. Federal Communications Commission, Washington, DC, Report ET Docket , 2002 .
2. HaykinS.: ' Cognitive radio: brain-empowered wireless communications', IEEE Trans. Commun., 2005, **23**, pp. 201–220 .
3. TaherpourA.NorouziY.Nasiri-KenariM.JamshidiA.Zeinalpour-YazdiZ.: ' Asymptotically optimum detection of primary user in cognitive radio networks', IET Commun., 2007, **1**, pp. 1138–1145 (doi: 10.1049/iet-com:20060645) .
4. GanesanG.LiY.: ' Cooperative spectrum sensing in cognitive radio part I: two user networks', IEEE Trans. Wirel. Commun., 2007, **6**, pp. 2204–2213 (doi: 10.1109/TWC.2007.05775) .
5. KimK.AkbarI.A.BaeK.K.UmJ.S.SpoonerC.M.ReedJ.H.: ' Cyclostationary approaches to signal detection and classification in cognitive radio'. DySPAN2007, April 2007, pp. 212–215.
6. GhasemiA.SousaE.: ' Opportunistic spectrum access in fading channels through collaborative sensing', J. Commun., 2007, **2**, pp. 71–82 (doi: 10.4304/jcm.2.2.71-82) .
7. UrkowitzH.: ' Energy detection of unknown deterministic signals', Proc. IEEE, 1967, **55**, pp. 523–531 (doi: 10.1109/PROC.1967.5573) .
8. KostylevV.I.: ' Energy detection of a signal with random amplitude'. Proc. IEEE Int. Conf. Commun., May 2002, pp. 1606–1610.
9. DighamF.F.AlouiniM.S.SimonM.K.: ' On the energy detection of unknown signals over fading channels', IEEE Trans. Commun., 2007, **55**, (1), pp. 21–24 (doi: 10.1109/TCOMM.2006.887483) .
10. SunH.LaurensonD.I.WangC.X.: ' Computationally tractable model of energy detection performance over slow fading channels', IEEE Commun. Lett., 2010, **14**, (10), pp. 924–926 (doi: 10.1109/LCOMM.2010.090710.100934) .
11. RaoA.AlouiniM.S.: ' Performance of cooperative spectrum sensing over non identical fading environments', IEEE Trans. Commun., 2011, **59**, (12), pp. 3249–3253 (doi: 10.1109/TCOMM.2011.082911.100222) .
12. RuttikK.KoufosK.JanttiR.: ' Detection of unknown signals in a fading environment', IEEE Commun. Lett., 2009, **13**, (7), pp. 498–500 (doi: 10.1109/LCOMM.2009.090169) .

13. SofotasiosP.C.RebeizE.ZhangL.TsiftisT.A.FreearS.CabricD.: ' Energy detection based spectrum sensing over k– μ and k– μ extreme fading channels', IEEE Trans. Veh. Technol., 2013, **62**, (3), pp. 1031–1040 (doi: 10.1109/TVT.2012.2228680) .
14. BanjadeV.R.S.RajathevaN.TellamburaC.: ' Performance analysis of energy detection with multiple correlated antenna cognitive radio in Nakagami- m fading', IEEE Commun. Lett., 2012, **16**, (4), pp. 502–505 (doi: 10.1109/LCOMM.2012.020212.112541) .
15. PridhamR.UrkowitzH.: ' Comment on energy detection of unknown deterministic signals', Proc. IEEE, 1968, **56**, (8), pp. 1379–1380 (doi: 10.1109/PROC.1968.6600) .
16. AxellE.LarssonE.: ' Optimal and sub-optimal spectrum sensing of OFDM signals in known and unknown noise variance', IEEE J. Sel. Areas Commun., 2011, **29**, (2), pp. 290–304 (doi: 10.1109/JSAC.2011.110203) .
17. GismallaE.AlsusaE.: ' On the performance of energy detection using Bartlett's estimate for spectrum sensing in cognitive radio systems', IEEE Trans. Signal Process., 2012, **60**, (7), pp. 3394–3404 (doi: 10.1109/TSP.2012.2191544) .
18. ShnidmanD.: ' The calculation of the probability of detection and the generalized Marcum Q-function', IEEE Trans. Inf. Theory, 1989, **35**, (2), pp. 389–400 (doi: 10.1109/18.32133) .
19. ProakisJ.G.SalehiM.: ' Digital communications' (McGraw-Hill, New York, 2008), pp. 46–47 .
20. AbramowitzM.StegunI.A.: ' Handbook of mathematical functions with formulas, graphs, and mathematical tables' (Dover, New York, 1974) .
21. AndrásS.BariczÁ.SunY.: ' The generalized Marcum-Q function: an orthogonal polynomial approach', Acta Univ. Sapientiae Math., 2011, **3**, (1), pp. 60–76 .
22. AltradO.MuhaidatS.: ' A new mathematical analysis of the probability of detection in cognitive radio over fading 'channels', EURASIP J. Wirel. Commun. Netw. 2013, **2013**, (159), doi: 10.1186/1687-1499-2013-159 .
23. SimonM.K.AlouiniM.-S.: ' Digital communications over fading channels' (John Wiley and Sons, Inc., 2004, 2nd edn.) .
24. GradshteynI.S.RyzhikI.M.: ' Tables of integrals, series, and products', 7th edn. (Elsevier Academic Press, 2007) .

9

Simple Adaptive Single Differential Coherence Detection of BPSK Signals in IEEE 802.15.4 Wireless Sensor Networks

Gaoyuan Zhang[1,2], Hong Wen[2], Longye Wang[2,3], Ping Xie[1], Liang Song[1], Jie Tang[2] and Runfa Liao[2]

[1]School of Electronic and Information Engineering, Henan University of Science and Technology, Luoyang 471023, China
[2]National Key Laboratory of Science and Technology on Communications, University of Electronic Science and Technology of China, Chengdu 611731, China
[3]School of Engineering and Technology, Tibet University, Lhasa 850000, China

ABSTRACT

In this paper, we propose an adaptive single differential coherent detection (SDCD) scheme for the binary phase shift keying (BPSK) signals in IEEE 802.15.4 Wireless Sensor Networks (WSNs). In particular, the residual carrier frequency offset effect (CFOE) for differential detection is adaptively estimated, with only linear operation, according to the changing channel conditions. It was found that the carrier frequency offset (CFO) and chip

Citation: Zhang G, Wen H, Wang L, et al. "Simple Adaptive Single Differential Coherence Detection of BPSK Signals in IEEE 802.15.4 Wireless Sensor Networks". Sensors (Basel, Switzerland). 2017 Dec;18(1) https://doi.org/10.3390/s18010052

signal-to-noise ratio (SNR) conditions do not need a priori knowledge. This partly benefits from that the combination of the trigonometric approximation sin−1(x)≈x and a useful assumption, namely, the asymptotic or high chip SNR, is considered for simplification of the full estimation scheme. Simulation results demonstrate that the proposed algorithm can achieve an accurate estimation and the detection performance can completely meet the requirement of the IEEE 802.15.4 standard, although with a little loss of reliability and robustness as compared with the conventional optimal single-symbol detector.

Keywords: single-symbol noncoherent detection; adaptive estimation; IEEE 802.15.4; wireless sensor networks

INTRODUCTION

Even since its introduction, the IEEE 802.15.4 low rate wireless personal area network (LR-WPAN) standard has found wide application in low complexity, low cost, low power consumption, and low data rate wireless connectivity among inexpensive devices [1], e.g., the smart sensor node in pervasive Wireless Sensor Networks (WSNs) [2]. WSNs can be applied to home automation, precision agriculture, health care, consumer electronic, industrial wireless control, environmental monitoring, and data collecting in battlefield awareness [3].

The specifications of the physical (PHY) layer and the medium access control (MAC) layer for IEEE 802.15.4 are defined in [1]. Much focus on its MAC protocol and capacity analysis has been witnessed in recent years [3,4,5]. Specially, Piyush et al. detailedly analyzed the capacity of WSNs, where *n* nodes are located in a region of area 1 m2. However, as we know, there is relatively less attention on the reliability issue in PHY layer, which is concerned in this paper. Specifically, the noncoherent signal detection algorithm for IEEE 802.15.4 WSNs is studied.

A binary phase shift keying (BPSK) direct sequence spread spectrum (DSSS) scheme is required in 868/915/950 MHz bands as indicated in [1]. A noncoherent receiver is preferred as compared to a coherent one in IEEE 802.15.4 BPSK receivers. This is because carrier acquisition and tracking is not required at the receiver in the battery-operated sensor node, although the complexity simplification is achieved at the cost of a compromise in detection performance. This degradation is often directly manifested in terms of the required signal-to-noise ratio (SNR) for a given bit error rate (BER) or packet-error rate (PER) [6,7].

Bit–level single differential coherent detection (SDCD) is known to be more popular in IEEE 802.15.4 BPSK receivers because of its attractive performance [2,8]. This follows that the chip–level SDCD suffers from serious performance loss due to chip-error propagation [9], while the bit-level SDCD combines soft decision values of the chip samples for detection and compensates for the constant carrier frequency offset effect (CFOE) in statistics after differential decoding (DD) [2].

To the best of our knowledge, all available detectors classified into this category [2,8,10] employ a two-step approach referred to here as the bit-level SDCD (BSDCD) approach: Step 1 (postcompensation step) estimates the residual CFOE based on a measurement derived from the whole preamble samples at the autocorrelator output and then feeds back a coherent reference to the following autocorrelation sample of the useful data; and Step 2 (decision step) detects the data bit-by-bit. The idea in the first step seems to be economically attractive for WSNs because of its simple implementation. Furthermore, the second step combines the simplicity of the approach without any compensation, with the improved performance due to including some degree of simple phase rotation of the detection statistic after DD, i.e., the differential detector output. Hence, this unified *estimation-detection* framework appears to offer a very reasonable trade-off between error performance and detection complexity.

This unified heuristic configuration can be considered an inspiration from the idea of the Generalized Likelihood Ratio Test (GLRT) [11], yet the receiver has partial prior knowledge of the nuisance parameter (e.g., an unknown CFOE) from known symbols (e.g., the preamble symbols). In principle, it can be interpreted as an approximation to the optimal coherent detectors [12]. The creative approximation is imperative for IEEE 802.15.4 WSNs where computation of the optimal detector is too difficult and uneconomical. It can also be classified into the data-aided (DA) noncoherent demodulation [13].

It is important to note that a similar approach was also introduced by Liu et al. in [14] for *M*-ary phase shift keying (*M*-PSK) with single differential (SD) coding. However, in this case, the CFO embedded in samples at differential detector input is considered to be recovered from some unknown modulated symbols. Therefore, the nonlinear operation (e.g., *M*th power for *M*-PSK) [15] is required to eliminate the dependence of these datas as indicated in Equation (7) of [14]. This is because the undesired parameter CFO is now a parameter of interest, whereas the data symbols are now just nuisance parameters.

For the aforementioned post-compensation step, several CFOE estimation approaches have been studied [2,8,10]. The full estimation scheme is powerful yet always not affordable, since the inverse tangent operation is considered to be the most complex to implement in WSNs [2]. The work of Lee et al. [8] first successfully simplifies this implementation; however, a considerable performance compromise cannot be avoided. This is because the mathematical approximation $\tan^{-1}(x) \approx 0$ is considered therein and thus the CFOE is quantified into four fixed values (in radians), i.e., $0, \pi/2, -\pi$, and $-\pi/2$ [10]. These phases clearly seems to be independent of the channel conditions, i.e., SNRs. Considering the more advisable trigonometric approximation $\tan^{-1}(x) \approx x$, we recently introduced an adaptive offset-term adjustment algorithm in [10]. It has the advantage of enabling the receiver to intelligently adjust the four constant phases above according to the time-varying channel conditions with respect to each packet transmission. However, a division operation is required to perform this process. This needs a complex implementation and is energy-consuming in floating-point form [16]. An efficient symbol-timing algorithm was also deeply discussed by Bloch et al. in [2], where the observation is necessary to be extended to three bit intervals. In this scheme, the most recent bit detection result is required in practice as shown in Equation (8) of [2]. Under this decision-feedback mechanism, error propagation may take place. For more details about these eatimators, the reader is referred to [10] and the reference therein.

The main contribution of this work can be highlighted as follows. The above so-called BSDCD technique is first deduced in detail, and its receiver structure is given. Then, for the compensation processor, some particular estimators are summarized and compared. Finally, an efficient estimator with only addition operation is provided for the compensation processor. It is worth noting that the high-order modulation [17], e.g., offset quadrature phase-shift keying (O-QPSK) with half-sine pulse shape, is also employed in IEEE 802.15.4 WSNs [1]. A simple double correlation based detector has been derived in [18,19], which however is out of the scope of this work.

To the best of our knowledge, this is the first work investigating the combination of the trigonometric approximation sin−1(x) ≈ x and the high chip SNR to simplify the optimal detector. The remainder of this paper is organized as follows. Section 2 concentrates on the signal model over an additive white Gaussian noise (AWGN) channel. The generalized BSDCD algorithm is explicitly proposed in Section 3, while the optimal BSDCD scheme is described in Section 4. Section 5 indicates several particular CFOE estimation schemes for the postcompensation process, and a new estimator is proposed in Section 6. Numerical results are addressed in Section 7. Finally, some conclusions and future work are offered in Section 8.

SYSTEM MODEL

Consider transmission over an AWGN channel and ideal timing synchronisation at the receiver in sensor node. The CFO and carrier phase offset (CPO) are assumed to be unknown and random but constant with respect to a packet transmission [2]. Without loss of generality, we follow the discrete-time signal model in [2] but with some changes. In particular, the baseband equivalent chip sequence for the *m*th bit E[m] is

$$r_{m,k} = s_{m,k}e^{j(2\pi kfT_c+\theta)} + \eta_{m,k}, \quad 1 \leq k \leq K. \tag{1}$$

Here, $s_{m,k}$ is the *k*th bipolar BPSK modulated chip, *f* is the CFO in Hz, and θ is arbitrary modulo-2π reduced CPO in radians. Tc is the chip duration, $\eta_{m,k}$ is a discrete-time, circularly symmetric, zero-mean complex AWGN with variance $\sigma^2_{m,k}$, and K=15 is the length of the pseudorandom number (PN) code [1]. We assume that *f* and θ are statistically independent of each other, and of the AWGN {ηm,k}.

Notice that the channel is supposed to only introduce CPO, but otherwise is assumed to be perfectly equalized. That is to say, a phase noncoherent channel is considered. Signal distortion caused by multipath fading is not considered for simplicity in describing detection principle, i.e., the indoor short-range propagation model is used, where the complex channel gain equals to 1 [19]. The CFO is due to instabilities associated with the inexpensive carrier oscillators in transmitting node and receiving node. The bandwidth of the receiving filter *B* is wide enough for all of the signal energy spread by CPO and CFO to pass [20]. In this case, all of the modeling and analysis in this work can be performed using the discrete-time complex baseband model in [8].

GENERALIZED BSDCD SCHEME

The receiver performs BSDCD of the *m*th bit E[m] by forming a complex-represented statistic from sample autocorrelation operation [8]:

$$A[m] = \sum_{k=1}^{L_1} r_{m,k} r^*_{m-1,k} = L_1 e^{j\phi_m} e^{jK\omega T_c} + \eta(0), \tag{2}$$

where L1 is the sample number, $1 \leq L_1 \leq K$, the superscript $*$ is complex conjugate, $\phi_m \in \{0,\pi\}$ is the actual information phase corresponding to the *m*th transmitted bit E[m], $\omega=2\pi f$, and $\eta(0)$ is an integrated noise sample.

Obviously, first-order differential modulation is a means of handling only one unknown invariable parameter in two adjacent bit intervals [14]. Thus, the CFOE $\varphi \triangleq K\omega T_c$ embedded in autocorrelator output A[m] is a residual nuisance parameter, generating a multiplicative phase distortion $e^{j\varphi}$ as indicated in Equation (2). It is by no means evident that this impairment, referred to as phase noise, can cause a significant performance loss if it is not wiped out. The avenue considered here is to equip the autocorrelator with a postcompensation processor. It not only is simple to implement but has the potential to be almost without performance degradation as the frequency shift increases. It is important to note that the exponential term $e^{jK\omega T_c}$ is referred as the CFOE in [2], but here is $K\omega T_c$.

Specifically, the CFOE is first recovered based on averaged autocorrelation samples of the preamble [8]:

$$Y = \frac{1}{(L_2-1)L_3} \sum_{m=2}^{L_2} \sum_{k=1}^{L_3} r_{m,k} r^*_{m-1,k} = e^{jK\omega T_c} + \eta(1). \tag{3}$$

Here, L2 is the observation length in bit intervals, and $2 \leq L_2 \leq J$, where J=32 is the bit number in preamble field [1]. L_3 is referred to as the number of the autocorrelation sample, $1 \leq L_3 \leq K$, and $\eta(1)$ is another complex-valued noise term.

Then, follow with the result reported in Equation (18) of [6], a decision rule can be intuitively written as:

$$\widehat{\phi}_m = \arg\max_{\phi_i} \Xi\left(A[m], \phi_i, Y\right), \tag{4}$$

where "argmax" denotes the argument that maximizes the following function with respect to the variable of interest, i.e., ϕ_i. Here, $\phi_i \in \{0,\pi\}$ is the hypothetical information phase and $\Xi(A[m],\phi_i, Y)$ denotes the detection metric, given by [6]

$$\Xi\left(A[m],\phi_i,Y\right)=Re\left\{A[m]q(Y)e^{-j\phi_i}\right\}. \tag{5}$$

In Equation (5), q(Y) is the so-called quantization function [2], which can be considered as a phase coherent reference to compensate the effect of the CFO on A[m]. Re{x} denotes the real part of *x*. Clearly, a bit detection error is generated when $\widehat{\phi}_m \neq \phi_m$.

Observe from Equations (2) and (5) that the quantization function q(Y) is used to undo rotation induced by CFOE in autocorrelator output A[m]. Bearing in mind that the BPSK alphabet is {+1,−1}, we can immediately arrive at another detection rule from Equation (4) in the following form:

$$\widehat{E}[m]=\begin{cases}0, & \text{if} \quad Re\{A[m]q(Y)\}\geq 0,\\ 1, & \text{if} \quad Re\{A[m]q(Y)\}<0.\end{cases} \tag{6}$$

This clearly is the receiver strategy followed in [2,8,10]. In fact, equip Equation (4) with respective quantization functions, and let truncationfactors $L_1=L_3=K$, $L_2=J$; then, we can arrive at the conventional detection approaches.

The details of Generalized BSDCD (GBSDCD) are presented in Algorithm 1. Note that the bits of the start-of-frame delimiter (SFD) in the first field and PHY header (PHR) in the second field of the physical layers protocol data unit (PPDU) are not considered in Algorithm 1. This simplicity is considered to make the ideas behind the detection principle apparent. In addition, the preamble field of the PPDU is composed of 32 binary zeros [1], therefore the transmitted bits E[m]=0, for $1\leq m\leq 32$. It is just based on these known bits that the postcompensation process can be performed at the receiver. This prior preamble field is primitively used for automatic gain control (AGC) convergence, diversity selection, timing acquisition, and coarse frequency acquisition in the IEEE 802.15.4 receivers [1,2].

The detector given here falls into those that use ad hoc methods [13], such as the decision-feedback-based method [21], the frequency-offset estimation-based method in [14] and the postcompensation-based method in this work. These algorithms can provide a wide range of trade-offs among detection complexity, detection speed, and PER performance.

Algorithm 1 Framework of GBSDCD Algorithm.

Input: rm,k: baseband sample for the *m*th bit E[m]; L1: sample number for each bit of the actual data; L2: observation length of the preamble in bit intervals; L3: number of the autocorrelation sample *m*th bit of the preamble; *J*: bit number in preamble field; *L*: payload length of the PPDU.

Output: E^[m]: detection of the *m*th bit of the actual data.

```
initial Y=0, A[m]=0, and J=32
for m=2; m≤L2; m++do
for k=1; k≤L3; k++ do
  A[m]←A[m]+rm,kr*m−1,k
  end for
  Y←Y+A[m]
end for
compute the quantization function q(Y);
for m=J+2; m≤J+L; m++do
   for k=1; k≤L1; k++ do
   A[m]←A[m]+rm,kr*m−1,k
   end for
   Ξ←A[m]q(Y)
   if Ξ<0 then
   E^[m]←1
   else
   E^[m]←0
   end if
end for
return E^[m]
```

The role of the differential coding is to facilitate CPO estimation and effectively translate the noncoherent channel to a coherent AWGN channel. Furthermore, the role of the preamble symbols is to facilitate CFOE estimation and effectively wipe out the instabilities associated with the inexpensive carrier oscillators in transmitting node. It is such exact knowledge of these symbols that reduces the degree of randomness and facilitates the estimation of the undesired parameter φ of interest. Note that, in the postcompensation process, the undesired parameter φ is now the parameter of interest that we would like to estimate, whereas the "parameters" of interest in the detection process, namely the data symbols, are now just nuisance (or undesired) parameters (cf. [22], p. 64).

THE AVAILABLE OPTIMAL QUANTIZATION FUNCTION

As described in the preceding section, the quantization function q(Y) can be considered as an estimator of the phasor $e^{-j\varphi}$ to eliminate the dependence of the residual nuisance parameter φ in autocorrelator output A[m]. It exhibits an important role not only in detection performance but in robustness to the CFO. Thus, in this section, we turn our attention towards developing q(Y) to achieve a benchmark that can be theoretically achieved.

In this case, q(Y) in Equation (5) can be expressed as:

$$q(Y) = Y^* = |Y|e^{-j\angle Y}. \tag{7}$$

Here, ∠Y is the principal value of the argument of the measurement *Y*, and |Y| denotes the magnitude of *Y*. The positive nature of magnitude |Y| means it does not affect the decision result in Equation (4). Hence, Equation (7) can be simplified as:

$$q(Y) = e^{-j\widehat{\varphi}}, \tag{8}$$

where φ^ must be set equal to the principal value of the argument of the measurement *Y*, i.e.,

$$\widehat{\varphi} = \begin{cases} \tan^{-1}\left(\frac{Im(Y)}{Re(Y)}\right), & \text{if } Re(Y) > 0, \\ \frac{\pi}{2}, & \text{if } Re(Y) = 0, Im(Y) \geq 0, \\ -\pi + \tan^{-1}\left(\frac{Im(Y)}{Re(Y)}\right), & \text{if } Re(Y) < 0, \\ -\frac{\pi}{2}, & \text{if } Re(Y) = 0, Im(Y) < 0. \end{cases} \triangleq \widehat{\varphi}_1, \tag{9}$$

Here, Im(Y) denotes the imaginary part of *Y*. Equation (9) is referred to here as the first old estimation model, and the receiver with it is the well-known conventional optimal receiver [2].

The noise term η(1) in Equation (3) is not Gaussian but can be approximated as Gaussian when chip signal-to-noise ratio (SNR) is reasonably high [23]. As a consequence, we can think of the measurement *Y* in Equation (3) as the equivalent received signal at time *m* and η(1) as an equivalent AWGN channel. In such condition, Equation (9) can be interpreted as data−aided maximum likelihood (DAML) estimates of the phase φ and Equation (8) a DAML estimate of the phasor e−jφ (cf. [24], p. 166). In this case, the receiver structure is given in Figure 1 [25]. In Figure 1, the switch 1 (SW 1) should open in the bit intervals 1≤m≤32, and be off otherwise. For SW 2, the opposite occurs.

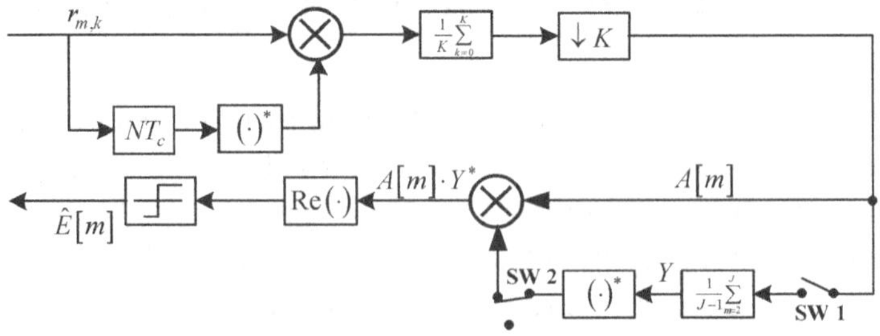

Figure 1: The optimal receiver structure.

We observe immediately from Equation (9) that the arctan operation cannot be avoided in the postcompensation process of the optimal receiver. This requires a computational complexity that renders its theoretical approach impractical in WSNs, where low power and low cost are of paramount importance. Therefore, the creative approximation, thus avoiding intractable complexity in this process, is imperative for IEEE 802.15.4 WSNs where computation of the optimal detector is too difficult and uneconomical.

SUMMARY AND ANALYSIS OF SOME SIMPLIFIED ESTIMATION SCHEMES

Simplified Estimation—Estimation A

Assuming that small CFO and high chip SNR remain $\left|\frac{Im(Y)}{Re(Y)}\right|$ sufficiently small, we can use the simple approximation that $\tan^{-1}(x) \approx x$, for small x, and Equation (9) can be directly approximated as

$$\widehat{\varphi} \approx \begin{cases} \frac{Im(Y)}{Re(Y)}, & \text{if } Re(Y) > 0, \\ \frac{\pi}{2}, & \text{if } Re(Y) = 0, Im(Y) \geq 0, \\ -\pi + \frac{Im(Y)}{Re(Y)}, & \text{if } Re(Y) < 0, \\ -\frac{\pi}{2}, & \text{if } Re(Y) = 0, Im(Y) < 0. \end{cases} \triangleq \widehat{\varphi}_2, \tag{10}$$

The detecor with Equation (10) works only for small CFO and reasonably high SNRs. This is because $\left|\frac{Im(Y)}{Re(Y)}\right| \leq 1$ is guaranteed with very high probability in this condition, and the detection effect of approximate error in $\tan^{-1}(x) \approx x$ will be as small as we expect. Otherwise, large mathematical approximation error will introduce excessive estimation error, and, finally, unacceptable detection penalty cannot be avoided.

Simplified Estimation—Estimation B

In order to obtain a low-complexity estimator that is not only as efficient as Equation (9) but valid for any cases, an equivalent description for Equation (9) was given in [10] by

$$\hat{\varphi} = \begin{cases} \tan^{-1}\left(\frac{Im(Y)}{Re(Y)}\right), & \text{if } \; Re(Y) > 0 \text{ and } |Re(Y)| \geq |Im(Y)|, \\ \frac{\pi}{2} - \tan^{-1}\left(\frac{Re(Y)}{Im(Y)}\right), & \text{if } \; Im(Y) > 0 \text{ and } |Re(Y)| < |Im(Y)|, \\ -\pi + \tan^{-1}\left(\frac{Im(Y)}{Re(Y)}\right), & \text{if } \; Re(Y) < 0 \text{ and } |Re(Y)| \geq |Im(Y)|, \\ -\frac{\pi}{2} - \tan^{-1}\left(\frac{Re(Y)}{Im(Y)}\right), & \text{if } \; Im(Y) < 0 \text{ and } |Re(Y)| < |Im(Y)|. \end{cases} \quad (11)$$

Equations (9) and (11), originally obtained using different intuitive reasonings, are two approaches for computing the principal value of the argument of the measurement Y.

We observe immediately from Equation (11) that the complex observation space is subdivided into four equi-angular sectors illustrated in Figure 2, which are identical to the four regions presented in Table 1 of [8]. Both of them can be distinguished from one another by only simple comparison of the measurement magnitudes and signs of real and imaginary parts of Y. From this viewpoint, Equation (11) differs from Equation (9) in two respects: (1) the observation space is further subdivided into four subspaces, not two; and (2) the subspace locating criteria is redesigned. It is such characteristics that make Equation (11) attractive for devising a simplified estimator as shown in the following.

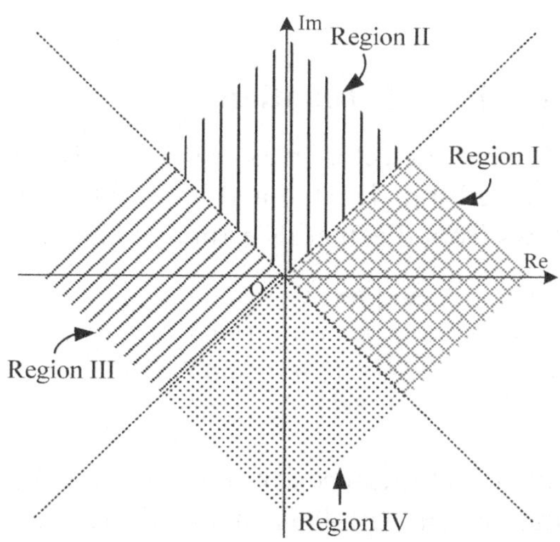

Figure 2: Division of the complex observation space.

Table 1: Summary of the absolute error

Region	Absolute Error |e|		
	Estimator in [8]	Estimator in [4]	Estimator in this work
[−π,−3π/4)	φ+π	(−π+tanφ)−φ	φ−(−π−sinφ)
[−3π/4,−π/2)	−φ−π/2	φ−(−π/2+cotφ)	φ−(−π−sinφ)
[−π/2,−π/4)	φ+π/2	(−π/2−cotφ)−φ	sinφ−φ
[−π/4,0)	−φ	φ−tanφ	sinφ−φ
[0,π/4)	φ	tanφ−φ	φ−sinφ
[π/4,π/2)	−φ+π/2	φ−(π/2−cotφ)	φ−sinφ
[π/2,3π/4)	φ−π/2	(π/2−cotφ)−φ	(π−sinφ)−φ
[3π/4,π)	−φ+π	φ−(π+tanφ)	(π−sinφ)−φ

As indicated in Equation (11), $\left|\frac{Im(Y)}{Re(Y)}\right|$ or $\left|\frac{Re(Y)}{Im(Y)}\right|$ in an observation region is not only never more than 1, but irrelevant to CFOE being estimated or SNR conditions. This appropriately allows us to use the mathematical approximation that $\tan^{-1}(x) \approx x$, for small x, in Equation (11) without undesirable calculation errors. Then, another efficient simplified detection scheme was proposed in our recent work [10], where the estimator is valid for arbitrary CFO and SNR, given by

$$\hat{\varphi} \approx \begin{cases} \frac{Im(Y)}{Re(Y)}, & \text{if} \quad Re(Y) > 0 \text{ and } |Re(Y)| \geq |Im(Y)|, \\ \frac{\pi}{2} - \frac{Re(Y)}{Im(Y)}, & \text{if} \quad Im(Y) > 0 \text{ and } |Re(Y)| < |Im(Y)|, \\ -\pi + \frac{Im(Y)}{Re(Y)}, & \text{if} \quad Re(Y) < 0 \text{ and } |Re(Y)| \geq |Im(Y)|, \\ -\frac{\pi}{2} - \frac{Re(Y)}{Im(Y)}, & \text{if} \quad Im(Y) < 0 \text{ and } |Re(Y)| < |Im(Y)|, \end{cases} \triangleq \hat{\varphi}_3. \tag{12}$$

Simplified Estimation—Estimation C

A complexity efficient detector was also deeply considered in [8]. Actually, Y is quantized into 1, −1, j and −j in this detector, a result later used in [2]. It follows from this that its estimator can be depicted by [10]

$$\hat{\varphi} \approx \begin{cases} 0, & \text{if} \quad Re(Y) > 0 \text{ and } |Re(Y)| \geq |Im(Y)|, \\ \frac{\pi}{2}, & \text{if} \quad Im(Y) > 0 \text{ and } |Re(Y)| < |Im(Y)|, \\ -\pi, & \text{if} \quad Re(Y) < 0 \text{ and } |Re(Y)| \geq |Im(Y)|, \\ -\frac{\pi}{2}, & \text{if} \quad Im(Y) < 0 \text{ and } |Re(Y)| < |Im(Y)|, \end{cases} \triangleq \hat{\varphi}_4. \tag{13}$$

Obviously, Equation (13) can be derived from Equation (11) on condition that the approximation $\tan^{-1}(x) \approx 0$ is involved. This avenue achieves an estimator with perfectly acceptable complexity. The detector with it works only for the CFOE approach to four fixed values, i.e., 0, π/2, −π and −π/2; otherwise, a large error similar to Equation (10) will be introduced. Not surprisingly, this is expected to limit the receiver performance. A quick comparison of Equations (12) and (13) reveals that the estimation error in the latter can be corrected by the former, where two types of adaptive offset terms $\frac{Im(Y)}{Re(Y)}$ and $\frac{Re(Y)}{Im(Y)}$ are additionally provided to improve the estimated accuracy of the time-variant undesired parameter φ.

Remarks

Some comments on the above estimators are as follows:

- These estimators are obtained from approximations made to a structure motivated by DAML estimation of the CPOE.
- The estimator in Equation (10) is convenient and with acceptable complexity as compared with Equation (9). Since it works only for very small frequency-offset values and reasonably high SNRs, it is a CFO and SNR limited estimator. Correspondingly, a severe degradation to the system packet error performance in the detection process will be introduced, which is not suitable for our purposes.
- Equation (12) achieves an estimate with quite reasonable accuracy. It does so by four preconditions, i.e., smart geometric division to the observation space, accurate subspace locating criterion, equivalent avenue for the calculation of ∠Y, and advisable approximation $\tan^{-1}(x) \approx x$. One arctan operation is required for a DAML estimator given in Equation (9). However, this computationally intensive operation is completely avoided in Equation (12). Surprisingly, such a complexity reduction results in almost no frequency-offset invariance degradation in the detection process, as will be inferred in Section 7.
- The canonical approximation $\tan^{-1}(x) \approx x$ involved in Equation (12) is much more accurate than the atypical approximation $\tan^{-1}(x) \approx 0$ involved in Equation (13). This implies that, as such inaccuracy frequently makes the measurement in Equation (13), either a large overestimation or a large underestimation of the CFOE, whereas no such inappropriate estimation is likely to occur in Equation (12). Actually, Equation (12) allows a real-time adjustment with additional one division and one addition operations, which, however, can achieve almost 1.5 dB gains at PER of 1×10^{-3} [10].

In a word, Equation (12) makes a detector with such a reasonable balance between complexity of the postcompensation step and performance of the detection step that it may be more attractive for WSNs. However, a division operation is still required as indicated in Equation (12). Clearly, division is an energy intensive and costly operation [16], and should be avoided as much as possible in WSNs. In the following, we are concerned with further simplication to wipe off this complex nonlinear operation in the first processing, while without unaccepted performance degradation in the second processing.

PROPOSED DIVISOR-FREE CFOE ESTIMATOR

The starting point of our estimation approach is the realization that the modulo-2π reduced phase of the measurement Y can also be expressed in form of $\sin^{-1}(x)$, i.e., Equation (9) can be rewrote as follows:

$$\angle Y = \begin{cases} \sin^{-1}\left(\frac{Im(Y)}{|Y|}\right), & \text{if } \ Re(Y) \geq 0 \\ -\pi - \sin^{-1}\left(\frac{Im(Y)}{|Y|}\right), & \text{if } \ Re(Y) < 0 \end{cases}. \tag{14}$$

Observe from Equation (14) that the absolute value of the magnitude divided argument of the inverse sine function is never more than 1 because length of any right-angle side is never more than that of the hypotenuse. Then, we can immediately simplify Equation (14) as follows:

$$\angle Y \approx \begin{cases} \frac{Im(Y)}{|Y|}, & \text{if } \ Re(Y) \geq 0, \\ -\pi - \frac{Im(Y)}{|Y|}, & \text{if } \ Re(Y) < 0. \end{cases} \tag{15}$$

Note that the mathematical approximation $\sin^{-1}(x) \approx x$, for small x, is involved in Equation (15).

Assuming further that high SNR, known to be useful in estimation problems [26], keeps the magnitude of the non-Gaussian noise term $\eta(1)$ in Equation (3) sufficiently small, we get

$$|Y| = |e^{jK\omega T_c} + \eta(1)| \approx 1. \tag{16}$$

It is worth noting that this simplification is irrespective of the frequency offset being estimated. Substituting Equation (16) into (15), we can finally improve a CFOE estimator as follows:

$$\widehat{\varphi}_{new} \triangleq \begin{cases} Im(Y), & \text{if} \quad Re(Y) \geq 0, \\ -\pi - Im(Y), & \text{if} \quad Re(Y) < 0. \end{cases} \tag{17}$$

Note that Equation (14) can be rewritten in the following form:

$$\angle Y = \begin{cases} \sin^{-1}\left(\frac{Im(Y)}{|Y|}\right), & \text{if} \quad Re(Y) \geq 0, \\ \pi - \sin^{-1}\left(\frac{Im(Y)}{|Y|}\right), & \text{if} \quad Re(Y) < 0. \end{cases} \tag{18}$$

Then, Equation (17) can also be given by:

$$\widehat{\varphi}_{new} \triangleq \begin{cases} Im(Y), & \text{if} \quad Re(Y) \geq 0, \\ \pi - Im(Y), & \text{if} \quad Re(Y) < 0. \end{cases} \tag{19}$$

We provide some interpretations for the receiver whose estimator is expressed by Equation (19).

- Equation (19) are obtained from approximations made to a structure motivated by DAML estimation scheme in Equation (14) or Equation (18), and a simple adaptive offset terms Im(Y) is provided. This enables the receiver to intelligently adjust the estimated CPOE value according to the time-varying channel conditions with respect to each packet transmission.
- The complex observation space is only divided into two subspaces as shown in Figure 3. They can be easily distinguished from one another just with the sign of the real part of *Y*. Thus, the absolute value operation of real and imaginary parts of *Y* in Equation (12) is not required in this case. An useful assumption, namely, the asymptotic or high chip SNR, is considered for simplification of Equation (15). Thus, the nonlinear division operation in Equation (12) is avoided.
- No limitation on the CFOE range is required for the mathematical approximation $\sin^{-1}(x) \approx x$ in our estimator. That is to say, it appears to be a full-range estimator, and the corresponding receiver is a full-range detector, the same feature as the receiver with Equation (12). By the approximation $|Y| \approx 1$ valid at high chip SNR as illustrated in Figure 4, the result in Equation (19) may be shown to approximately hold for arbitrary SNRs, which is also the same as Equation (12). In a word, the CFO and chip SNR conditions do not need a priori knowledge for the mathematical approximation involved in Equation (19). As a result, an excellent detection performance is achieved at the receiver, which can completely

meet the requirements of the IEEE 802.15.4 standard. This will be confirmed by the numerical results shown in the following section.

- Clearly, the estimation error $e \triangleq \hat{\varphi} - \varphi$ is only introduced by the mathematical approximation if the noise is not considered. In this case, the absolute error $|e|$ for three simplified estimators is summarized and compared in Table 1 and Figure 5, respectively. As shown in Figure 5, the absolute error $|e|$ with Equation (12) is no more than $1-\pi/4$ for which the observation space is divided into four subspaces as indicated in Figure 2. It increases to $\pi/4$ for Equation (13). This is exactly half of the radians for the equi-angular sector in Figure 2. This value comes to be $\pi/2-1$ for our improved estimator in Equation (19) where the observation space is only divided into two subspaces. Furthermore, when the CFOE $\phi \in (-\pi, -0.712\pi) \cup (-0.288\pi, 0.288\pi) \cup (0.712\pi, \pi)$, our estimator is more efficient than Equation (12).

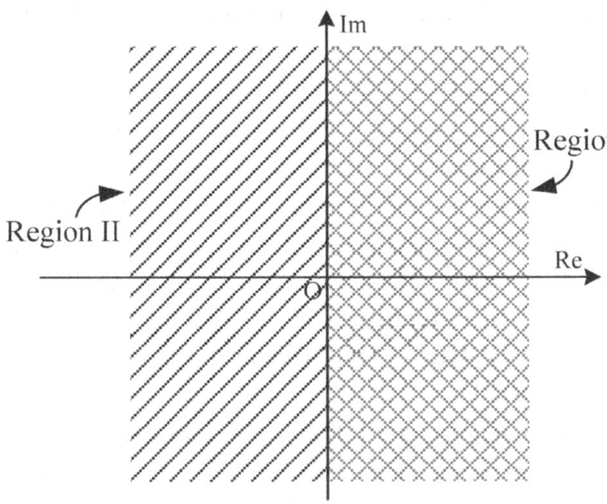

Figure 3: New division of the complex observation space.

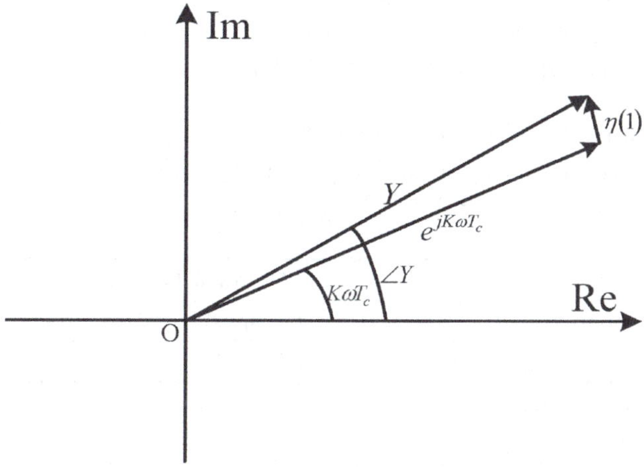

Figure 4: Geomrtirc representation of the measurement Y.

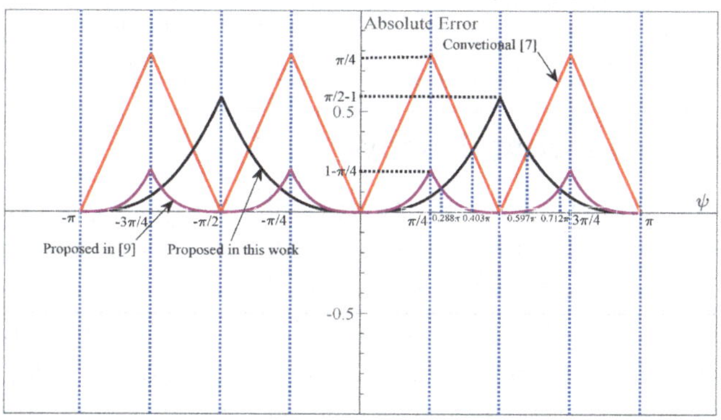

Figure 5: The absolute error |e| of three different estimation methods.

NUMERICAL RESULTS AND DISCUSSION

In this section, we evaluate the BER and PER performances of some BSDCD schemes. In all simulations, the payload length of the physical layer protocol data unit (PPDU) is 20 octets [**1**], and the detection procedure is repeated until enough error packets are collected. The carrier frequency was selected as 924 MHz, i.e., the maximum values required in a 915 MHz band [**1**]. All of these parameters are described in detail in **Table 2**. Note that the bits in the PHR, which indicates the number of octets of the PPDU, are also considered in all simulations. All experiments in this section were developed on a personal computer (3.2 GHz, 8 GB RAM) in a MATLAB (R2017a) (MathWorks, Beijing, China) platform.

Table 2: Parameters used in simulations. AWGN: additive white Gaussian noise; SNR: signal-to-noise ratio; PN: pseudorandom number; PPDU: physical layers protocol data unit; CPO: carrier phase offset

Parameter	**Detailed Description**
Channel condition	Complex AWGN
Power of the complex noise	1/SNR
Detection Scheme	Bit-level noncoherent
Timing synchronisation	Perfect
Generator polynomial of PN code	1+x+x4
Payload length of PPDU (bits)	160
Carrier frequency (MHz)	924
CPO (rads)	Uniform distribution in $(-\pi, \pi)$ or Wiener process

Effect of the Truncation Factors

The truncation factors L_1, L_2 and L_3 regulate the complexity and performance, so we first give evaluation results in Figure 6 for the proposed detector, which provides us with the effect L1 has on detection performance. Here, the CFO was randomly set to range from −80 ppm to 80 ppm with probability density function of symmetric triangular shape [2]. The CPO was considered to be uniform distribution in interval $(-\pi, \pi)$. We observe that performance improves as sample length L_1 increases from 1 to 15, but improvement degrades as sample length L1 increases. For example, Figure 6 shows that for PER of 1×10^{-3}, as sample length L_1 increases from 1 to 3, the SNR gain is about 4.3 dB. As L_1 increases from 5 to 7, the SNR gain reduces to 1.2 dB. As L1 further increases from 13 to 15, the SNR gain becomes 0.5 dB. Similar results were also achieved for L_2 and L_3, which are not given here. In pratice, suitable truncation factors can be selected according to the specific requirements on the performance, and a large degree of freedom in complexity can be achieved. For example, according to [1], the PER should be less than 1×10^{-2} when the SNR is 5 to 6 dB. Referring to Figure 6, five samples are sufficient when the SNR of the input chip signal is 5 dB. In the following, the maximum values are considered for L_1, L_2 and L_3 to ensure the best possible performance.

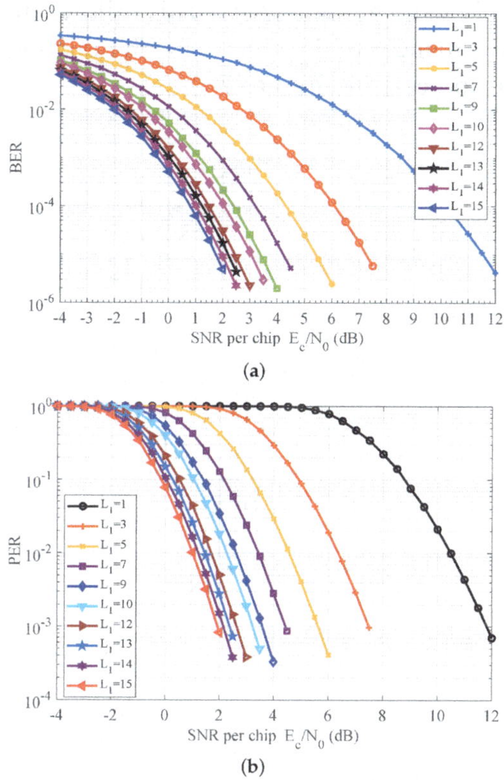

Figure 6: Performance impact of parameter L1. (a) BER impact of L1; (b) PER impact of L1.

Detection Performance of the Receiver

The BER and PER performances of the proposed receiver versus others are shown in Figure 7. The optimal noncoherent receiver with random CPO is considered [6,7]. The ideal coherent (perfect carrier reference phase and no CFO) detection with SD decoding is also given for sake of comparison. Compared to the optimal noncoherent scheme, the CFOE can be compensated by a postcompensation processor to various degrees as shown in Figure 7. At PER of 1×10^{-3}, our detection shceme achieves more than 1.3 dB gains over the method in [8]. The performance gap between the scheme in [10] and our detector is only about 0.2 dB; however, substantial reduction in complexity is achieved. Referring to Figure 7, only 1 dB is enough for our detector to meet specific requirements on the performance in [1]. In addition, our scheme is efficient at all SNR regions, although the approximation $|Y|\approx1$ involved in

Equation (19) is only valid at high SNRs. This follows from the fact that the BER and PER performances of the receiver with Equations (15) and (19) are almost the same as indicated in Figure 7.

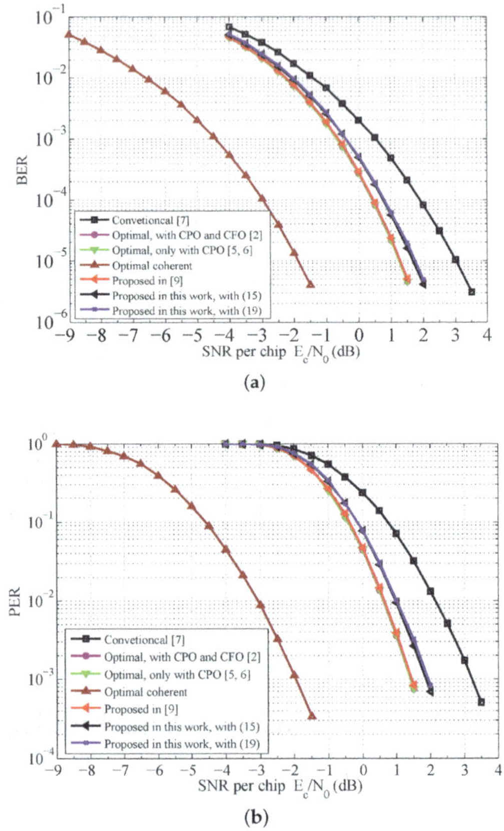

Figure 7: Performance comparison for various receivers. (a) BER performance of different receivers; (b) PER performance of different receivers.

The BER and PER performances of the proposed receiver versus the receiver in [10] when the CFO was randomly set to range from −6.2 ppm (i.e., the CFOE φ is -0.288π) to 6.2 ppm (i.e., the CFOE φ is 0.288π) with triangular symmetric distribution are depicted in Figure 8. As shown in Figure 8, our proposed estimation method in Equation (19) compensates the frequency offset effect more efficiently than that in Equation (12), whereas substantial reduction in complexity is achieved at the same time. This follows from the fact that the absolute estimation error |e| of our proposed estimation method is now smaller than that of Equation (12) as indicated in Figure 5. Similar results were also achieved for the CFOE $\varphi \in (-\pi, -0.712\pi) \cup (0.712\pi, \pi)$, which are not given here.

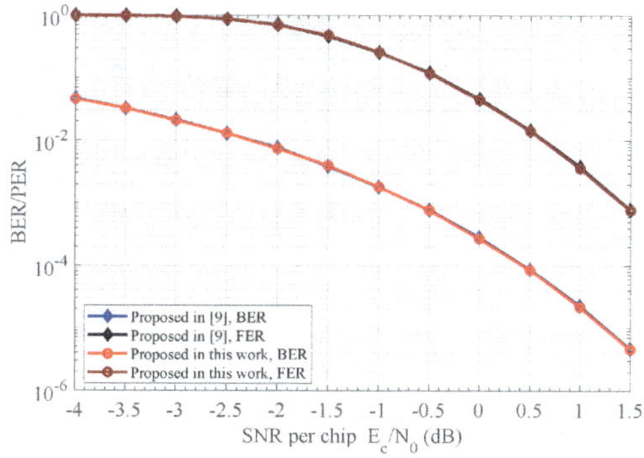

Figure 8: BER and PER comparison for the proposed receiver and the receiver in [10].

Robustness of the Receiver

The performance of the detector in [8] versus CFO is shown in Figure 9. In Figure 9, the horizontal line represents the detection performace when no CFO is considered, which gives us a benchmark for comparison. As indicated in Figure 9, good performance is achieved especially when f equals ±10 ppm for the detector in [8]. This result follows from the number of the gap between actual CFOE and its corresponding quantified value $\pi/2$ being small enough, i.e., only 0.0380π radians. However, when this number greatly increases to 0.2340π radians, i.e., f=±70 ppm, performance is degraded severely by large estimation error (e.g., more than 2 dB at PER of 1×10^{-2} as shown in Figure 10). The analysis result for the simulation in Figure 9 is described in detail in Table 3. The result for the absolute estimation error $|e|$ is also presented in Table 3. Observe from Figure 9 and Table 3 that the smaller the absolute estimation error is, the better the detection performance. When $|e|$ is close to its maximum value 0.25π (i.e., f equals ±5 ppm, ±60 ppm or ±70 ppm), and the receiver suffers from a serious performance penalty. Based on those observations, we see that the detector in [8] works well only for such CFOEs that are close enough to four values, i.e., $0, \pi/2, -\pi$, and $-\pi/2$. Otherwise, a large estimation error will introduce unacceptable performance penalty.

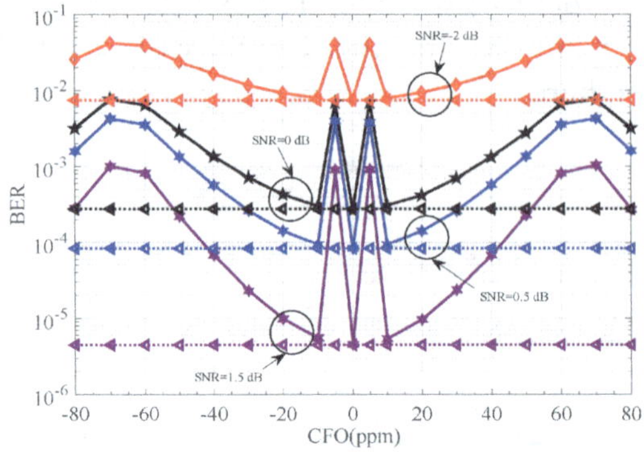

(a)

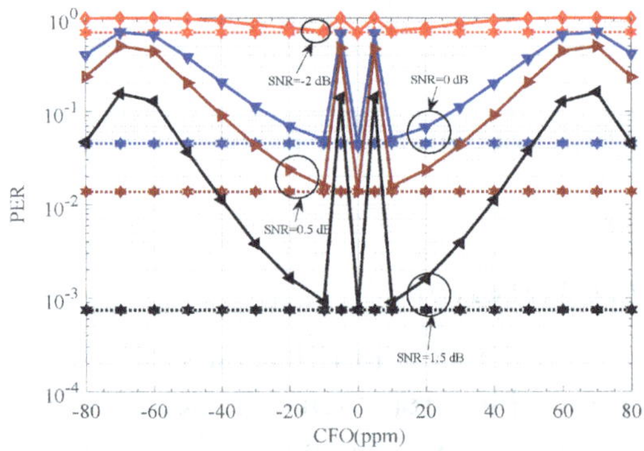

(b)

Figure 9: Detection performance of [8]. (a) BER performance of [8] versus CFO; (b) PER performance of [8] versus CFO. CFO: carrier frequency offset.

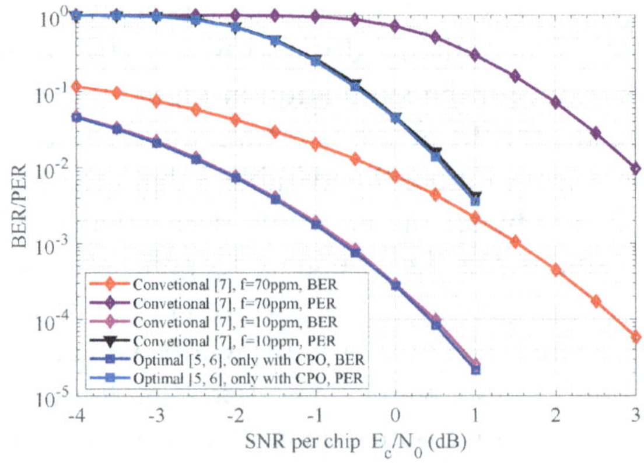

Figure 10: BER and PER performances of [8] versus two particular CFOs. CFO: carrier frequency offset.

Table 3: Robustness analysis for the detector in [8]. CFOE: carrier frequency offset effect.

CFO (ppm)	Performance Rank	Actual CFOE (Rads)	Estimated CFOE (Rads)	|e| **(Rads)**
5	Eighth	0.2310π	0	0.2310π
10	First	0.4620π	π/2	0.0380π
20	Second	0.9240π	−π	0.0768π
30	Third	−0.6275π	−π/2	0.1275π
40	Fourth	−0.1520π	0	0.1520π
50	Fifth	0.3100π	π/2	0.1900π
60	Seventh	0.7720π	−π	0.2280π
70	Ninth	−0.7660π	−π	0.2340π
80	Sixth	−0.3040π	−π/2	0.1960π

The limitation in [8] is avoided to some extent in our detector as depicted in Figure 11. As indicated in Figure 11, good performance is achieved for most considered CFOs. This is because a simple additive adaptive offset term Im(Y) is introduced as shown in Equation (19), and the maximum value for the absolute estimation error |e| is reduced to π/2−1. In addition, only when |φ| is close enough to π/2 (i.e., the CFO *f* equals ±10 ppm), large degradation in performance is observed. This is because a relatively large approximation error in $\sin^{-1}(x) \approx x$ is now achieved as shown in Figure 5. However, note that no limitation on the CFOE range is required for direct application of the mathematical approximation $\sin^{-1}(x) \approx x$, for small *x*, in our estimator.

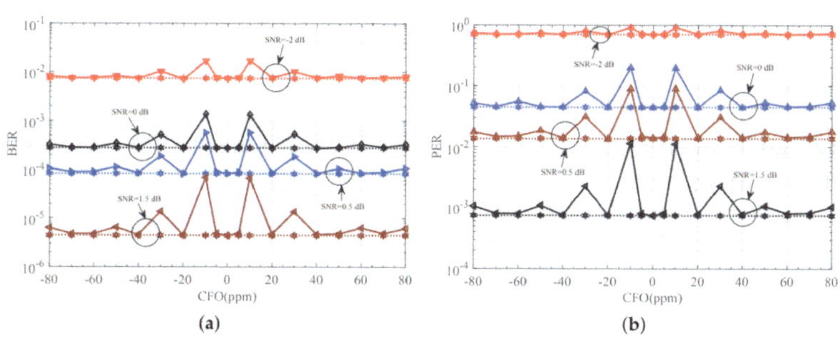

Figure 11: Detection performance of proposed receiver. (a) BER performance of proposed receiver versus CFO; (b) PER performance of proposed receiver versus CFO. CFO: carrier frequency offset.

Performance under Dynamic Channel

Finally, the behavior of the proposed receiver under dynamic channel conditions has been investigated. Our noncoherent receiver is robust to phase jitter as observed from Figure 12 and Figure 13. The phase θ of received chip sequence {rm,k}, 1≤k≤K, is modeled as a Wiener process according to θm+1=θm+Δm, where Δm are zero-mean independent Gaussian random variables with known variance σ2m in each bit interval, and the initial phase θ1 is considered to be uniform distribution. A jitter standard deviation up to four degrees does not degrade significantly the receiver performance. Another feature of the curves in Figure 12 and Figure 13 is that they exhibit an irreducible error floor as SNR increases. Furthermore, the smaller the standard deviation is, the lower the error floor. The explanation is that the random phase increment Δm generates a phasor ejΔm in the autocorrelator output A[m], which, in turn, produces decision errors even in the absence of noise. The smaller the standard deviation, the lesser the effect of the phasor ejΔm on the autocorrelator output A[m].

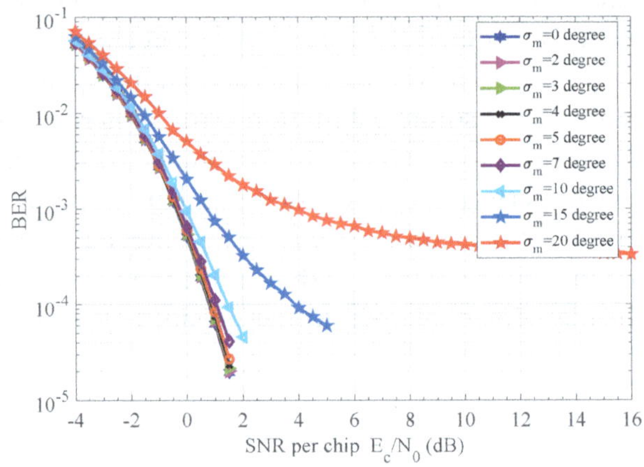

Figure 12: BER performance of proposed receiver versus dynamic CPO. CPO: carrier phase offset.

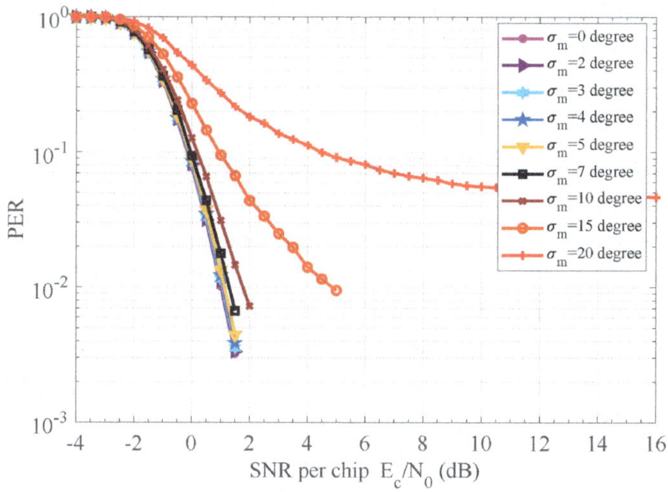

Figure 13: PER performance of proposed receiver versus dynamic CPO. CPO: carrier phase offset.

CONCLUSIONS

We have presented a BSDCD scheme of BPSK signals for IEEE 802.15.4 WSNs. Then, a more meaningful and practical estimate scheme, i.e., an approximate ML estimator with only linear operation, is proposed for extracting a phase reference from preamble signal. Simulations suggest that our lower complexity in implementation does not sacrifice much in terms of detection performance, which can completely meet the requirements of the IEEE 802.15.4 standard. Therefore, it is the most attractive SDCD solution of choice for WSNs, especially when it used in consumer electronics.

It is important to note that our idea in [10] as well as in this work can be used for simplification of Equation (8) in [14], where arctan operation is indispensable to performing an initial CFO estimation process. Loosely speaking, these schemes can be further classified into the hard receiver. It makes use of only the estimates of the nuisance parameters as if they were the true values for the noncoherent detection. In the soft receiver, the postcompensation processor is required to compute posterioris or conditional probability density functions of the nuisance parameters embedded in the chip sequence [12]. Thus, the detection process can incorporate a statistical characterization of the nuisance parameters (instead of the estimated ones). The latter is, of course, nonimplementable in general.

Furthermore, unlike those who were researching ways to use BSDCD to combat the effect of chip-error propagation [9], we add that the theory developed here is easily extended to multiple differential coherent detection (MDCD). That is to say, the extension of the proposed scheme to account for multiple-symbols situation is straightforward. As a consequence, the performance gap between ideal coherent detection and BSDCD, indicated in Figure 7, can be narrowed. Not surprisingly, multiple unknown CFOEs after DD are necessary to be initially removed. Complexity reduction techniques may be developed based on the Viterbi algorithm [27], the fast algorithm in [28] and an algorithm based on subset search [29]. Of course, owing to the reduced PER that this augmented scheme yields, a much more reasonable trade-off can be achieved and is clearly desirable for both the battery-operated transmitter and receiver, which benefit from less energy that will be consumed by retransmissions between sensor nodes [**30**]. Finally, the complexity and energy efficiency can be deeply analyzed with the model given in [**30**]. These subjects will be reported on by the authors in a forthcoming paper.

ACKNOWLEDGMENTS

This work was partially supported by the National Natural Science Foundation of China (Project Nos. 61701172, 61701059, 61701062, 61561045 and 61572114), the Natural Science Foundation of Henan Province (Project No. 162300410097), the Postdoctoral Science Foundation of University of Electronic Science and Technology of China (Project No. Y02006023601721), the Initial Scientific Research Foundation of Doctor in Henan University of Science and Technology (Project No. 13480052), the Joint Funds of the National Natural Science Foundation of China (Project No. U1404615 and U1504619), the Program for Everest Scholar Talents Development in Tibet University, the Program for Science & Technology Innovation Talents in the University of Henan Province (Educational Committee) (Project No. 17HASTIT025), and the Program for International Science & Technology Cooperation Projects of Henan Province (Project No. 172102410072).

AUTHOR CONTRIBUTIONS

Gaoyuan Zhang, Longye Wang and Ping Xie conducted the algorithm design, and Gaoyuan Zhang wrote the paper and performed all the the simulations under the supervision of Hong Wen and Liang Song. Jie Tang and Runfa Liao worked on the image alignment and contributed to preparing and analyzing the simulation data. All authors were involved in modifying the paper, the literature review and the discussion of the results.

REFERENCES

1. IEEE 802.15.4-2011 Standard. *IEEE Standard for Local and Metropolitan area Network-Part 15.4: Low Rate Wireless Personal Area Networks (LR-WPANs)*; IEEE Press: Piscataway, NJ, USA, 2011.
2. Jeon, H.J.; Demeechai, T.; Lee, W.G.; Kim, D.H. IEEE 802.15.4 BPSK receiver architecture based on a new efficient detection scheme. *IEEE Trans. Signal Process.* 2010, *58*, 4711–4719.
3. Tseng, H.W.; Pang, A.C.; Chen, J.H.; Kuo, C.F. An adaptive contention control strategy for IEEE 802.15.4-based wireless sensor networks. *IEEE Trans. Veh. Technol.* 2017, *58*, 5164–5173.
4. Zhang, Y.J.; Zhou, Y.D.; Gao, L.; Qian, Y.W.; Li, J.; Shu, F. A blind adaptive tuning algorithm for reliable and energy-efficient communication in IEEE 802.15.4 networks. *IEEE Trans. Veh. Technol.* 2017, *66*, 8605–8609.
5. Gupta, P.; Kumar, P.R. The capacity of wireless networks. *IEEE Trans. Inf. Theory* 2000, *46*, 388–404.
6. Divsalar, D.; Simon, M.K. Multiple-symbol differential detection of MPSK. *IEEE Trans. Commun.* 1990, *38*, 300–308.
7. Makrakis, D.; Feher, K. Optimal noncoherent detection of PSK signals. *Electron. Lett.* 1990, *26*, 398–400.
8. Lee, S.; Kwon, H.; Jung, Y.; Kim, J.S. Efficient non-coherent demodulation scheme for IEEE 802.15.4 LR-WPAN systems. *Electron. Lett.* 2007, *43*, 879–880.
9. Howlader, M.; Chang, H. Performance of chip-level differential detection for DS-CDMA in multipath fading channels. In Proceedings of the IEEE International Conference on Military Communications, Boston, MA, USA, 13–16 October 2003; pp. 1065–1070.
10. Zhang, G.Y.; Wang, D.; Song, L.; Wu, H.H.; Xie, P.; Ji, B.F.; Wen, H. Simple non-coherent detection scheme for IEEE 802.15.4 BPSK receivers. *Electron. Lett.* 2017, *53*, 628–629.
11. Harry, L.V.T.; Kristine, L.B.; Zhi, T. *Detection, Estimation, and Modulation Theory*, 2nd ed.; Wiley: Hoboken, NJ, USA, 2015.
12. Kam, P.Y.; Ng, S.S.; Ng, T.S. Optimum symbol-by-symbol detection of uncoded digital data over the Gaussian channel with unknown carrier phase. *IEEE Trans. Commun.* 1994, *42*, 2543–2552.

13. Leib, H. Data-aided noncoherent demodulation of DPSK. *IEEE Trans. Commun.* 1995, *43*, 722–725.

14. Liu, J.H.; Stoica, P.; Simon, M.K.; Li, J. Single differential modulation and detection for MPSK in the presence of unknown frequency offset. In Proceedings of the Fortieth Asilomar Conference on Signals, Systems and Computers, Pacific Grove, CA, USA, 29 October–1 November 2006; pp. 1440–1444.

15. Viterbi, A.J.; Viterbi, A.M. Nonlinear estimation of PSK-modulated carrier phase with applications to burst digital transmission. *IEEE Trans. Inf. Theory* 1983, *29*, 543–551.

16. Soderquist, P.; Leeser, M. Area and performance trade-offs in floating-point divide and square root implementations. *ACM Comput. Surv.* 1997, *46*, 518–564.

17. Seimetz, M. *High-Order Modulation for Optical Fiber Transmission*; Springer: Berlin, Germany, 2009.

18. Park, S.; Park, D.; Park, H.; Lee, K. Low-complexity frequency-offset insensitive detection for orthogonal modulation. *Electron. Lett.* 2005, *41*, 1226–1227.

19. Park, D.; Park, C.S.; Lee, K. Simple Design of Detector in the Presence of Frequency Offset for IEEE 802.15.4 LR-WPANs. *IEEE Trans. Circuits Syst. II Express Briefs* 2010, *56*, 330–334.

20. Buetefuer, J.L.; Cowley, W.G. Frequency offset insensitive multiple symbol detection of MPSK. In Proceedings of the IEEE International Conference on Acoustics, Speech, and Signal Processing, Istanbul, Turkey, 5–9 June 2000; pp. 2669–2672.

21. Leib, H.; Pasupathy, S. The phase of a vector perturbed by Gaussian noise and differentially coherent receivers. *IEEE Trans. Inf. Theory* 1988, *34*, 1491–1501.

22. Ferrari, G.; Colavolpe, G.; Raheli, R. *Detection Algorithms for Wireless Communications with Applications to Wired and Storage Systems*; John Wiley & Sons Ltd.: Chichester, UK, 2004.

23. Proakis, J. *Digital Communications*; McGraw-Hill: New York, NY, USA, 1989.

24. Madhow, U. *Fundamentals of Digital Communication*; Cambridge University Press: New York, NY, USA, 2008.

25. Oh, J.Y.; Kwak, B.J.; Kim, J.Y. Carrier frequency recovery for in non-coherent demodulation for IEEE802.15.4 system. In Proceedings of

the IEEE Vehicular Technology Conference, Montreal, QC, Canada, 25–28 September 2006; pp. 1–5.

26. Kay, S.M. *Fundamentals of Statistical Signal Processing: Detection Theory*; Prentice-Hall PTR: Upper Saddle River, NJ, USA, 1998.
27. Adachi, F. Viterbi-decoding differential detection of DPSK. *Electron. Lett.* 1992, *28*, 2196–2198.
28. Mackenthun, K.M. A fast algorithm for multiple-symbol differential detection of MPSK communications. *IEEE Trans. Commun.* 1994, *42*, 1471–1474.
29. Li, B. A new reduced-complexity algorithm for multiple-symbol differential detection. *IEEE Commun. Lett.* 2003, *7*, 269–271.
30. Maunder, R.G.; Weddell, A.S.; Merrett, G.V.; Al-Hashimi, B.M.; Hanzo, L. Iterative decoding for redistributing energy consumption in wireless sensor networks. In Proceedings of the 17th International Conference on Computer Communications and Networks, St. Thomas, US Virgin Islands, USA, 3–7 August 2008; pp. 1–6.

SECTION 4:
SIGNALING OVER BAND-LIMITED CHANNELS

10

Water-filling for Full-duplex Multiple-hop MIMO Relay System

Pham Thanh Hiep[1,2] and Ryuji Kohno[1]

[1]Yokohama National University, 79-5 Tokiwadai, Hodogaya, Yokohama 240-8501, Japan
[2]Le Quy Don Technical University, 236 Hoang Quoc Viet, Nghia Do, Cau Giay, Ha Noi 100000, Vietnam

ABSTRACT

A water-filling (WF) method is one of the techniques to improve a performance of multiple-input multiple-output (MIMO) systems. However, the application of the WF method to multiple-hop MIMO relay systems (MMRSs) has not been discussed, especially in amplify-and-forward (AF) schemes. In this paper, the WF method for MMRSs with AF scheme is proposed and evaluated in both perfect and imperfect channel state information (CSI). The effect of imperfect CSI on optimization of transmit power at all transmitters and on co-channel interference is taken into account.

Citation: Thanh Hiep and Kohno: Water-filling for full-duplex multiple-hop MIMO relay system. EURASIP Journal on Wireless Communications and Networking 2014 2014:174. https://doi.org/10.1186/1687-1499-2014-174

Compared to the average transmit power, the end-to-end channel capacity of the WF method is higher in the perfect CSI; however, it is subject to the effect of imperfect CSI. Additionally, half-duplex and full-duplex transmission methods are also discussed in this paper. The end-to-end channel capacity of full duplex is higher and more robust than that of the half duplex due to a double of allocation time and a half of delay time.

INTRODUCTION

Multiple-input multiple-output (MIMO) relay systems have been discussed in several literatures [1–4]. According to transmission of MIMO schemes, a channel capacity of system was improved. Moreover, the allocation of transmit power for each antenna based on channel model, meaning water-filling (WF), also has been applied to improve the channel capacity of MIMO systems [5–8].

Additionally, in order to reduce the transmit power and/or improve the performance of MIMO system, a multiple-hop MIMO relay system (MMRS) has been analyzed [9–11]. In MMRSs, when the distance between the base station (Tx) and the final receiver (Rx) is fixed, the distance between the Tx to a relay (RS), a RS to a RS, a RS to the Rx called the distance between each transceiver, is shorten. Consequently, the signal-to-noise ratio (SNR) and the channel capacity are increased. However, according to the number of relays, the location, and the transmit power of each relay, the channel capacity of each hop is changed. In addition, the end-to-end channel capacity is restricted by the bottleneck node. Therefore, to obtain the upper bound of end-to-end channel capacity, the location of each relay node, meaning, the distance between each transceiver and the transmit power of each relay, should be optimized. We have analyzed the performance of MMRS with amplify-and-forward (AF) scheme [12, 13] and decode-and-forward (DF) scheme [14]. The distance between each transceiver as well as the transmit power for each relay are optimized to obtain the upper bound of end-to-end channel capacity. However, the transmit power of each relay was divided equally to all antennas in the relay.

Since the DF relay decodes the received data before forwarding, it is known as a regenerative strategy. Furthermore, since each hop is independent of the other hops in the DF scheme MIMO relay system, the original WF method of a MIMO system can be directly applied. On the other hand, the

AF relay amplifies the received data only; therefore it is known as a non-regenerative strategy. The computational simplicity in AF relay makes it a highly attractive and a strong candidate for the real-time application. Therefore, in this paper, we focus on the AF scheme MIMO relay system.

There are many literatures that focus on the AF scheme MIMO relay network [9, 10]. In addition, the mutual information and the total mean square error criteria were selected as objective functions [15, 16]. The diversity multiplexing tradeoff for MIMO relays also was investigated [17]. There are some works on the beamforming design for the special type of half-duplex AF scheme MIMO relay system [18]. However, in these literatures, the WF method was not applied.

In short, the WF method was proposed for the original MIMO system and the relay system which has only one relay. The WF method can be directly applied to DF scheme MMRSs. Moreover, the perfect channel state information (CSI) and the half duplex were assumed. In this paper, the WF method for AF scheme MMRSs is proposed and compared to the original WF of MIMO system. The half-duplex as well as the full-duplex transmission schemes are analyzed based on the end-to-end channel capacity. Moreover, the perfect and the imperfect CSI are taken into account. Three schemes of channel model are proposed, and the end-to-end channel capacity is analyzed based on the proposed scheme by the theoretical calculation method.

The rest of the paper is organized as follows. We introduce the concept of MMRSs in Section 2. The WF method for AF scheme MMRSs is proposed in Section 3, and the numerical evaluation is described in Section 4. Finally, Section 5 concludes the paper.

AF SCHEME MMRSS

Description of MMRSs

As shown in Figure 1, the MMRS that consists of a single transmitter, single receiver, and m relays is considered. Let the Tx, the Rx, and $RS_i (i=1,\cdots,m)$ denote the transmitter, the receiver, and i th relay, respectively. M, N, and $K_i (i=1,\cdots,m)$ are the number of antennas at the Tx, the Rx, and each relay. However, in this paper, the number of antennas of all relays is assumed to be the same as the number of antennas of the transmitter and the final receiver, $M=N=K_i$ $(i=1,\cdots,m)$. Moreover, the total transmit power of all antennas in the i th relay is denoted by W_i, for all $i=0,\cdots,m$. Since the AF method is applied to the channel model, the noise is also amplified and transmitted to the next relay when the signal is amplified at every relay. Mathematical notations used in this paper are as follows. x and X are scalar variables, $\mathbf{x}$ and $\mathbf{X}$ are vector variables or matrix variables, $\mathbf{X}^H$ is conjugate transpose matrix of $\mathbf{X}$.

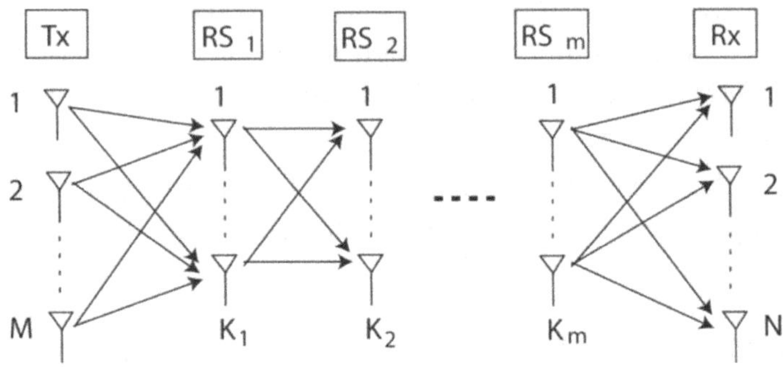

Figure 1: System model of MMRS. The MMRS with m relays is described. m relays are arranged in series from the Tx to the Rx. The number of antennas at all relays, the transmitter, and the receiver is assumed to be the same.

Channel Model

In order to easily describe, the Tx and Rx are also be denoted as the RS_0 and RS_{m+1}, respectively. Since the path loss is taken into consideration, the channel matrix is a composite matrix, and we model as $\sqrt{l_{ii+1}}\mathbf{H}_{ii+1}$,i=0,⋯,m, of which $l_{i\,i+1}$ and $\mathbf{H}_{i\,i+1}$ represent the path loss and the $K_{i+1}\times K_i$ channel matrix between the RS_i and the RS_{i+1}, respectively. $\mathbf{H}_{i\,i+1}$ is a matrix with independent and identical distribution (i.i.d.), zero mean, unit variance, circularly symmetric complex Gaussian entries.

Let $\mathbf{H}_i$ $(i=1,\cdots,m+1)$ denote the channel matrix between the Tx and the RS_i (as expressed in Figure 2). The other parameters of the system, i.e., transmit signal, received signal, noise vector, amplification factor, and total transmit power, are summarized in Table 1.

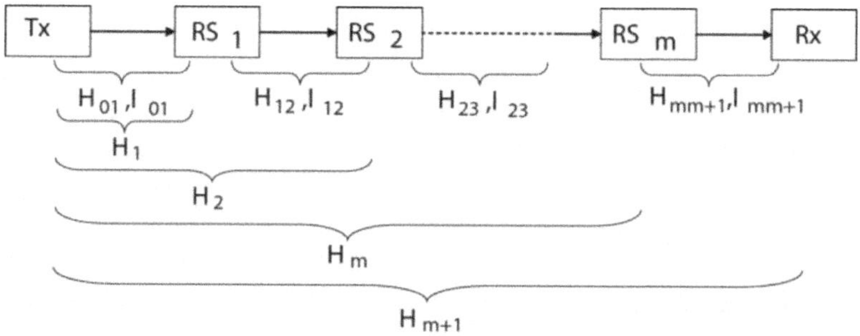

Figure 2: Channel model of MMRS. The channel model of each hop, the channel model from the Tx to each relay, and the path loss between each transceiver are described.

Table 1: System parameters of each transceiver

	Tx	RS_1	...	RS_m	Rx
Transmit signal	$\mathbf{S}_0$	$\mathbf{S}_1$	...	$\mathbf{S}_m$	
Received signal		$\mathbf{R}_1$	...	$\mathbf{R}_m$	$\mathbf{R}_{m+1}$
Noise vector		$\mathbf{n}_1$	...	$\mathbf{n}_m$	$\mathbf{n}_{m+1}$
Amplification factor	$\mathbf{\Gamma}_0$	$\mathbf{\Gamma}_1$	...	$\mathbf{\Gamma}_m$	
Total transmit power	E_0	E_1	...	E_m	

After $M\times 1$ normalized signal **S** is amplified by amplification factor Γ_0, the signal $\mathbf{S}_0=\Gamma_0\mathbf{S}$ is transmitted to the RS_1. The signal $\mathbf{H}_{01}\mathbf{S}_0$ is received by the RS_1, and before being amplified with the amplification factor Γ_1, the received signal is added by the $K_1\times 1$ noise matrix $\mathbf{n}_1$. Therefore, we can separate the received signal and the transmit signal of the RS_1 as

$$\begin{aligned}\mathbf{R}_1 &= \mathbf{H}_{01}\mathbf{S}_0,\\ \mathbf{S}_1 &= \mathbf{\Gamma}_1(\mathbf{R}_1+\mathbf{n}_1).\end{aligned} \quad (1)$$

Similarly, when the signal $\mathbf{S}_{i-1}$ is transmitted from the RS_{i-1}, the received signal and the transmit signal of the RS_i are expressed as

$$\begin{aligned}\mathbf{R}_i &= \mathbf{H}_{i-1i}\mathbf{S}_{i-1},\\ \mathbf{S}_i &= \mathbf{\Gamma}_i(\mathbf{R}_i+\mathbf{n}_i).\end{aligned} \quad (2)$$

The amplification factor of the RS_i is expressed as

$$\mathbf{\Gamma}_0 = \sqrt{\mathbf{P}_0},$$
$$\mathbf{\Gamma}_0 = \sqrt{\frac{\mathbf{P}_i}{|\mathbf{R}_i|^2}}, \quad i = 1, \cdots, m,, \tag{3}$$

where $\mathbf{P}_i = \text{diag}(p_{i1}, p_{i2}, \cdots, p_{iK_i})$ is the transmit power matrix and assumed to be subject to a constraint $\text{Tr}(\mathbf{P}_i) = W_i$. The $\text{Tr}(\cdot)$ and p_{ij} denote the trace and transmit power of the j th antenna of the RS_i, respectively. Finally, the received signal at the Rx can be expressed as

$$\mathbf{S}_{m+1} = \sqrt{\frac{l_{mm+1}\mathbf{P}_m \cdots l_{01}\mathbf{P}_0}{|\mathbf{R}_m|^2 \cdots |\mathbf{R}_1|^2}} \mathbf{H}_{mm+1} \cdots \mathbf{H}_{12}\mathbf{H}_{01}\mathbf{S}_0 + \mathbf{n}. \tag{4}$$

Here, the **n** denotes the noise vector of the system. Therefore, the system channel matrix is written by

$$\mathbf{H}_{m+1} = \mathbf{H}_{mm+1} \cdots \mathbf{H}_{12}\mathbf{H}_{01}. \tag{5}$$

According to using the system channel matrix $\mathbf{H}_{m+1}$, the MMRS can be analyzed as same as a conventional MIMO system.

End-to-end Channel Capacity

The noise vector of the system when m relays intervened becomes

$$\begin{aligned}\mathbf{n} = & \sqrt{\frac{l_{mm+1}\mathbf{P}_m \cdots l_{12}\mathbf{P}_1}{|\mathbf{R}_m|^2 \cdots |\mathbf{R}_1|^2}} \mathbf{H}_{mm+1} \cdots \mathbf{H}_{12}\mathbf{n}_1 \\ & + \sqrt{\frac{l_{mm+1}\mathbf{P}_m \cdots l_{23}\mathbf{P}_2}{|\mathbf{R}_m|^2 \cdots |\mathbf{R}_1|^2}} \sqrt{|\mathbf{R}_1|^2} \mathbf{H}_{mm+1} \cdots \mathbf{H}_{23}\mathbf{n}_2 \\ & + \cdots \\ & + \sqrt{\frac{l_{mm+1}\mathbf{P}_m}{|\mathbf{R}_m|^2 \cdots |\mathbf{R}_1|^2}} \sqrt{|\mathbf{R}_{m-1}|^2 \cdots |\mathbf{R}_1|^2} \mathbf{H}_{mm+1}\mathbf{n}_m \\ & + \mathbf{n}_{m+1}.\end{aligned} \tag{6}$$

However, at first, the CSI is assumed to be perfect; the imperfect CSI is considered in the following section. The channel matrix $\mathbf{H}_{01}$ can be expressed as follows in a singular-value decomposition (SVD) [19, 20].

$$\mathbf{H}_{01} = \mathbf{E}_{01_r}\mathbf{D}_{01}\mathbf{E}_{01_t}^{\mathrm{H}} = \sum_{i=1}^{M} \sqrt{\lambda_i} e_{01_{r,i}} e_{01_{t,i}}^{\mathrm{H}}; \tag{7}$$

here,

$$\begin{aligned} \mathbf{D}_{01} &= \operatorname{diag}\left[\sqrt{\lambda_{01_1}}\ \sqrt{\lambda_{01_2}}\ \cdots\ \sqrt{\lambda_{01_M}}\right], \\ \mathbf{E}_{01_r} &= \left[\mathbf{e}_{01_{r,1}}\ \mathbf{e}_{01_{r,2}}\ \cdots\ \mathbf{e}_{01_{r,M}}\right], \\ \mathbf{E}_{01_t} &= \left[\mathbf{e}_{01_{t,1}}\ \mathbf{e}_{01_{t,2}}\ \cdots\ \mathbf{e}_{01_{t,M}}\right]. \end{aligned} \tag{8}$$

λ_i $(i=1,2,\cdots,M)$ is the i th eigenvalue of the correlation matrix $\mathbf{H}_{01}\mathbf{H}_{01}^{\mathrm{H}}$ (or $\mathbf{H}_{01}^{\mathrm{H}}\mathbf{H}_{01}$), $\mathbf{e}_{01_{r,i}}$ is the eigenvector belonging to the eigenvalue λ_i of correlation matrix $\mathbf{H}_{01}\mathbf{H}_{01}^{\mathrm{H}}$, and $\mathbf{e}_{01_{t,i}}$ is the eigenvector belonging to the eigenvalue λ_i of correlation matrix $\mathbf{H}_{01}^{\mathrm{H}}\mathbf{H}_{01}$. The SVD is applied similarly to the other channel matrices.

The channel matrix and the noise vector are assumed to be uncorrelated, and the covariance of noise vector in all relays is denoted by σ^2. Therefore, the correlation of noise vector of system is represented as follows.

$$\mathbf{n}\mathbf{n}^{\mathrm{H}} = \frac{\sum_{i=0}^{m}\prod_{j\neq i}^{m} l_{jj+1}\mathbf{P}_j\mathbf{D}_{jj+1}^2\sigma^2 + \sigma^4 f\left(l_{ii+1}\mathbf{P}_i\mathbf{D}_{ii+1}^2, \sigma^2\right)}{|\mathbf{R}_m|^2\cdots|\mathbf{R}_1|^2}; \tag{9}$$

here, $f(l_{ii+1}\mathbf{P}_i\mathbf{D}_{ii+1}^2, \sigma^2)$ denotes a polynomial of $l_{ii+1}\mathbf{P}_i\mathbf{D}_{ii+1}^2$ and σ^2. Since $\sigma^2 \leq 1$ and the covariance of noise vector is smaller than the received power, the component containing σ^4 can be ignored. Thus, (9) can be changed as

$$\mathbf{n}\mathbf{n}^{\mathrm{H}} = \frac{\sum_{i=0}^{m}\prod_{j\neq i}^{m} l_{jj+1}\mathbf{P}_j\mathbf{D}_{jj+1}^2\sigma^2}{|\mathbf{R}_m|^2\cdots|\mathbf{R}_1|^2}. \tag{10}$$

Consequently, the end-to-end channel capacity of AF scheme MMRS is expressed as

$$\begin{aligned} C &= \log_2\left(\det\left(\frac{\mathbf{S}_{m+1}\mathbf{S}_{m+1}^{\mathrm{H}}}{\mathbf{n}_{m+1}\mathbf{n}_{m+1}^{\mathrm{H}}}\right)\right), \\ &= \log_2\left(\det\left(\mathbf{I}_M + \frac{\prod_{i=0}^{m} l_{jj+1}\mathbf{P}_j\mathbf{D}_{jj+1}^2}{\sigma^2\sum_{i=0}^{m}\prod_{j\neq i}^{m} l_{jj+1}\mathbf{P}_j\mathbf{D}_{jj+1}^2}\right)\right) \\ &= \log_2\left(\det\left(\mathbf{I}_M + \frac{1}{\sigma^2\sum_{i=0}^{m}\frac{1}{l_{ii+1}\mathbf{P}_i\mathbf{D}_{ii+1}^2}}\right)\right); \end{aligned} \tag{11}$$

here, $\mathbf{I}_M$ denotes the $M\times M$ unit matrix.

Let

$$f(\lambda) = \sigma^2 \sum_{i=0}^{m} \frac{1}{l_{ii+1}\mathbf{P}_i\mathbf{D}_{ii+1}^2}. \quad (12)$$

Thus, the end-to-end channel capacity of system is inversely proportional to $f(\lambda)$. Therefore, $f(\lambda)$ can be considered instead of the end-to-end channel capacity. In order to maximize the end-to-end channel capacity, the $f(\lambda)$ has to be minimized. The $f(\lambda)$ is a function of transmit power and path loss, meaning, the distance between each transceiver, the variance of noise vector, and the matrix of eigenvalue of channel model. The optimization of distance and transmit power was described in [12, 13]; however, the transmit power of each antenna in the same relay is assumed to be equal. In order to improve the performance of a system, the WF method for the AF scheme MMRS is proposed. The $f(\lambda)$ can be changed as

$$f(\lambda) = \sum_{i=1}^{m+1} \frac{1}{\mathbf{SNR}_{i-1i}\mathbf{D}_{i-1i}^2}; \quad (13)$$

here, $\mathbf{SNR}_{i-1i} = \frac{l_{i-1i}\mathbf{P}_{i-1}}{\sigma^2} = \mathrm{diag}\left(\gamma_{i-1i_1}, \gamma_{i-1i_2}, \cdots, \gamma_{i-1i_M}\right)$ represents the SNR of all paths of hop between the RS_{i-1} and the RS_i regardless the other relays. Since the transmit power and the distance are fixed after optimization as optimization method in [12, 13] and notice that $\mathrm{Tr}(\mathbf{P}_i)=W_i$, trace of $\mathbf{SNR}_{i-1i}$ is fixed as $\frac{l_{i-1i}W_{i-1}}{\sigma^2} = \mathrm{snr}_i$. The snr_i is the ratio of received signal power to noise power where the signal is transmitted and received by one antenna.

WF METHOD FOR AF SCHEME MMRS

WF Method for AF Scheme MMRS with Perfect CSI

The WF method can be applied to the AF scheme MMRS by three schemes. Scheme 1: the WF method is applied to each hop in MMRS. Scheme 2: the WF method is applied to each process. Scheme 3: the WF method is applied to all channel models of MMRS. The detail of each scheme is explained as follows.

Scheme 1: the WF method for each hop of MMRS: In this scheme, the channel model of each hop is used to optimize the transmit power of transmitter. The hop between the RS_{i-1} and the RS_i is considered and regardless of the other relays. As described in [7, 8], in the system with only one transmitter and one receiver, the channel capacity is described as follows.

$$C_{i-1i} = \log_2 \prod_{k=1}^{M} \left(1 + \lambda_{i-1i_k}\gamma_{i-1i_k}\right), \qquad ($$

$$= \sum_{k=1}^{M} \log_2 \lambda_{i-1i_k} + \sum_{k=1}^{M} \log_2 \left(\frac{1}{\lambda_{i-1i_k}} + \gamma_{i-1i_k}\right) \tag{14}$$

The channel capacity maximization problem is now changed to the choice of the maximal number of real, non-negative values γ_{i-1i_k} subject to the power constraint $\sum_{k=1}^{M} \gamma_{i-1i_k} = \mathrm{snr}_i$.

The well-known approach to solving constrained optimization problems is the method of Lagrange multipliers. Accordingly, we need to maximize

$$g(\gamma) = \sum_{k=1}^{M} \log_2 \left(\frac{1}{\lambda_{i-1i_k}} + \gamma_{i-1i_k}\right) + \alpha \left(\sum_{k=1}^{M} \gamma_{i-1i_k} - \mathrm{snr}_i\right), \tag{15}$$

where α is the Lagrange multiplier. The partial differentiation with respect to γ_{i-1i_k} leads to

$$\frac{\partial g(\gamma)}{\partial \gamma_{i-1i_k}} = \frac{\lambda_{i-1i_k}}{1 + \lambda_{i-1i_k}\gamma_{i-1i_k}} + \alpha. \tag{16}$$

The optimized γi−1ikγi-1ik can be obtained when $\frac{\partial g(\gamma)}{\partial \gamma_{i-1i_k}} = 0$. Hence,

$$\gamma_{i-1i_k} = -\frac{1}{\alpha} - \frac{1}{\lambda_{i-1i_k}}. \tag{17}$$

The Lagrange multiplier is relabeled as $\beta = -\frac{1}{\alpha}$ to simplify the notation in later formulas. This problem really has two constraints: the power constraint and being non-negative γ_{i-1i_k}. The latter constraint can be described as

$$\begin{aligned}\gamma_{i-1i_k} &= \max\left\{0, \beta - \frac{1}{\lambda_{i-1i_k}}\right\}, \\ &\equiv \left[\beta - \frac{1}{\lambda_{i-1i_k}}\right]^+. \end{aligned} \tag{18}$$

The former constraint can be described as

$$\begin{aligned}\mathrm{snr}_i &= \sum_{k=1}^{M}\left[\beta - \frac{1}{\lambda_{i-1i_k}}\right]^+, \\ &= \sum_{k=1}^{q}\left[\beta - \frac{1}{\lambda_{i-1i_k}}\right]^+, \\ &= q\beta - \sum_{k=1}^{q}\frac{1}{\lambda_{i-1i_k}}. \end{aligned} \tag{19}$$

Thus, the Lagrange multiplier and the optimal SNR, meaning, the optimal transmit power of each path responding to eigenvalues, are obtained.

$$\begin{aligned}\beta &= \frac{1}{q}\left(\sum_{k=1}^{q}\frac{1}{\lambda_{i-1i_k}} + \mathrm{snr}_i\right), \\ \gamma_{i-1i_k} &= \frac{1}{q}\left(\sum_{k=1}^{q}\frac{1}{\lambda_{i-1i_k}} + \mathrm{snr}_i\right) - \frac{1}{\lambda_{i-1i_k}}. \end{aligned} \tag{20}$$

A computational procedure for determining the optimal value of q is to compute for $q=M,M\text{-}1,\cdots$ the value of γ_{i-1i_k} until this quantity is greater than zero for all k from 1 to q. By substituting the optimal γ_{i-1i_k} to (13) and (11), the end-to-end channel capacity can be obtained.

Similar to scheme 1, in schemes 2 and 3, the transmit power of transmitter also can be optimized based on multiplication of channel models. In scheme 2: the WF method for each process, the transmit power of the transmitter is optimized based on multiplication of channel models from the Tx to each receiver. It means that the WF of hop between the RS_{i-1} and the RS_i is analyzed based on the channel model $\mathbf{H}_i=\mathbf{H}_{i\text{-}1i}\cdots\mathbf{H}_{01}$ (refer to Figure 2). Let $\mathbf{D}_i$ denote the eigenvalue matrix of $\mathbf{H}_i$ in SVD; thus,

$$\mathbf{D}_i = \mathbf{D}_{i-1i} \cdots \mathbf{D}_{01}, \tag{21}$$

and the $\mathbf{SNR}_{i\text{-}1i}$ is optimized based on $\mathbf{D}_i$ as mentioned above.

In scheme 3: the WF method for all channel models, the transmit power is optimized based on channel matrix of system, $\mathbf{H}_{m+1}$ in (5). The eigenvalue matrix of $\mathbf{H}_{m+1}$ is denoted by $\mathbf{D}_{m+1}$ that is represented as follows.

$$\mathbf{D}_{m+1} = \mathbf{D}_{mm+1} \cdots \mathbf{D}_{01}. \tag{22}$$

The WF method of all hops is dependent on $\mathbf{D}_{m+1}$; it means that the WF method of all hops is the same.

Up to now, the perfect CSI at both the transmitter and the receiver has been assumed, and the WF method for MMRS has been analyzed based on the proposed scheme. However, in actuality, the perfect CSI assumption is not always practical due to channel estimation errors, feedback delay, and noise. Compared to channel estimation errors, the CSI imperfection introduced by feedback channel delay is sometimes more significant and inevitable.

WF Method for AF Scheme MMRS with Imperfect CSI

Similar to the previous section, the channel model between the $RS_{i\text{-}1}$ and the RS_i is considered. The channel time variation is described by the first-order Markov process as follows.

$$\mathbf{H}_{i-1i}(t) = \rho\mathbf{H}_{i-1i}(t-\tau) + \delta\bar{\mathbf{H}}_{i-1i}(t), \tag{23}$$

where τ denotes the total delay caused by signal processing, feedback, and other system delays; $\rho = J_0(2\varpi f_D \tau)$ is the time correlation coefficient; $J_0(\cdot)$ is the zero-th order Bessel function of the first kind; and f_D denotes the maximum Doppler frequency shift. δ denotes $\sqrt{1-\rho^2}$. The innovation term $\bar{\mathbf{H}}_{i-1i}$ also has i.i.d entries, zero mean, unit variance, and circularly symmetric complex Gaussian entries. For notation brevity, we drop off the time index

$$\mathbf{H}_{i-1i} = \rho\hat{\mathbf{H}}_{i-1i} + \delta\bar{\mathbf{H}}_{i-1i}, \tag{24}$$

where $\hat{\mathbf{H}}_{i-1i}$ denotes the outdated channel matrix while $\hat{\mathbf{H}}_{i-1i}$ is the true one. Notice that there are two ways the outdated CSI affects the system. One is the effect on the separation of subchannel between each transceiver; the other one is the effect on optimization of transmit power of all transmitters. The latter one depends on the scheme of the WF method. The former one depends on the CSI between each transceiver regardless of other CSIs, and it is the same as the CSI of scheme 1 of the latter one. The CSI of the latter one is explained as follows. Notice that the full duplex is assumed, all terminals transmit and receive in the same time. The delay of half duplex is a double of full duplex.

For scheme 1: we assume that transmitters add the CSI symbol in data packets and transmit to receivers; hence, the receiver knows both $\bar{\mathbf{H}}_i$ and $\hat{\mathbf{H}}_i$. After receiving the CSI symbol, the receiver feedbacks the CSI to the transmitter immediately. However, the transmitter receives the CSI after a delay. Therefore, the CSI that is used at the transmitter is the outdated CSI. It means that the transmitter only knows the outdated channel model, $\hat{\mathbf{H}}_i$, while the receiver knows both $\bar{\mathbf{H}}_i$ and $\hat{\mathbf{H}}_i$. The delay at each hop is assumed to be the same as τ. Therefore, in scheme 1, the delay of CSI at each transmitter is the same as τ.

For scheme 2: the CSI symbol that was added at the Tx is assumed to be forwarded at all relays; therefore, all relays and the Rx can know the CSI from the Tx to itself. However, when a relay receives the CSI symbol, the CSI of the previous hop becomes the outdated. Notice that the delay of forward link (from the Tx to the Rx) and backward link (from the Rx to the Tx) was assumed to be the same as τ. The RS_{i-1} starts to optimize the transmit power after receiving the feedback CSI from the RS_i; therefore, although the RS_{i-1} receives the CSI of hop between the Tx and the RS_1 after the delay of $(i-2)\tau$, however, it should wait to receive the feedback from the RS_i; hence the RS_{i-1} starts to optimize the transmit power when the delay is $i\,\tau$. Similarly, the delay of channel model between the RS_{j-1} and the RS_j becomes $(i-j+1)\tau$. As a result, the delay of CSI of $\mathbf{H}_{i-1i},\cdots,\mathbf{H}_{01}$ is $1\tau,\cdots,i\,\tau$, respectively.

For scheme 3: we assumed that a relay (RS_i) not only feedbacks the CSI to the previous relay (RS_{i-1}) but also forwards the CSI received from the next relay (RS_{i+1}) to the previous relay (RS_{i-1}). The process is repeated until the Tx receives the CSI from the Rx. All transmitters optimize their transmit power every time the CSI is updated; however, as analyzed in the previous section, at all relays, the transmit power is optimized based on the same multiplication of channel models. Therefore, the delay of CSI is calculated based on the same channel models. When the Tx transmits a signal, the delay of CSI of $\mathbf{H}_{01},\cdots,\mathbf{H}_{m\,m+1}$ is $\tau,\cdots,(m+1)\tau$, respectively.

Compared to the transmission of the Tx, the RS_i transmits a signal after the delay of $i\tau$; therefore, the delay of CSI of $\mathbf{H}_{01},\cdots,\mathbf{H}_{m\,m+1}$ is $(i+1)\tau,\cdots,(i+m+1)\tau$, respectively.

The influence of outdated CSI on the separation of subchannel of MIMO scheme is explained as follows. In a SVD-based MIMO system, the RS_{i-1} steers the modulated signal vector, $\mathbf{S}_{i-1}$, within the eigenspace spanned by the right singular vectors contained in $\hat{\mathbf{E}}_{i-1i_t}$.

$$\mathbf{S}_{i-1_t} = \hat{\mathbf{E}}_{i-1i_t}\mathbf{S}_{i-1}. \tag{25}$$

The RS_i receives the signal with noise vector as the same in Section 2.

$$\mathbf{R}_{i_r} = \mathbf{H}_{i-1i}\mathbf{S}_{i-1_t} + \mathbf{n}_i, \tag{26}$$

and the preprocesses $\mathbf{R}_{i_r}$ using $\hat{\mathbf{E}}_{i-1i_r}$

$$\begin{aligned}
\mathbf{R}_i &= \left(\hat{\mathbf{E}}_{i-1i_r}\right)^{\mathrm{H}}\mathbf{R}_{i_r}, \\
&= \left(\hat{\mathbf{E}}_{i-1i_r}\right)^{\mathrm{H}}\mathbf{H}_{i-1i}\hat{\mathbf{E}}_{i-1i_t}\mathbf{S}_{i-1} + \left(\hat{\mathbf{E}}_{i-1i_r}\right)^{\mathrm{H}}\mathbf{n}_i, \\
&= \rho\hat{\mathbf{D}}_{i-1i}\mathbf{S}_{i-1} + \delta\left(\hat{\mathbf{E}}_{i-1i_r}\right)^{\mathrm{H}}\bar{\mathbf{H}}_{i-1i}\hat{\mathbf{E}}_{i-1i_t}\mathbf{S}_{i-1} \\
&\quad + \left(\hat{\mathbf{E}}_{i-1i_r}\right)^{\mathrm{H}}\mathbf{n}_{i+1}.
\end{aligned} \tag{27}$$

The p th component of vector $\mathbf{x}$ and the p th row of matrix $\mathbf{X}$ are denoted by $\mathbf{x}[p]$ and $\mathbf{X}[p]$, respectively. The component-wise form is described as follows.

$$\begin{aligned}
\mathbf{R}_i[p] &= \left(\rho\hat{\mathbf{D}}_{i-1i}[p] + \delta\left(\hat{\mathbf{E}}_{i-1i_r}\right)^{\mathrm{H}}\bar{\mathbf{H}}_{i-1i}\hat{\mathbf{E}}_{i-1i_t}[p]\right)\mathbf{S}_{i-1} \\
&+ \delta\sum_{t\neq p}\left(\hat{\mathbf{E}}_{i-1i_r}\right)^{\mathrm{H}}\bar{\mathbf{H}}_{i-1i}\hat{\mathbf{E}}_{i-1i_t}[t]\,\mathbf{S}_{i-1} \\
&+ \left(\hat{\mathbf{E}}_{i-1i_r}\right)^{\mathrm{H}}\mathbf{n}_{i+1}[p],
\end{aligned} \tag{28}$$

where the received signal consists of three components: the information carrying term, the interference term, and the noise term. For each component-wise, there are M-1 interference components that are multiplied by δ. The correlation matrix is approximate to the unit matrix. Therefore, (13) is changed as

$$
\begin{aligned}
&f(\lambda) \\
&= \frac{\sigma^2 \sum_{i=0}^{m} \prod_{j\neq i}^{m} l_{jj+1}\mathbf{P}_j\mathbf{D}_{jj+1}^2 + \delta(M-1)\prod_{i=0}^{m} l_{ii+1}\mathbf{P}_i\mathbf{D}_{ii+1}^2}{\prod_{i=0}^{m} l_{ii+1}\mathbf{P}_i\mathbf{D}_{ii+1}^2}, \\
&= \sum_{i=1}^{m+1} \frac{1}{\mathbf{SNR}_{i-1i}\mathbf{D}_{i-1i}^2} + \delta(M-1).
\end{aligned} \tag{29}
$$

Compared to the system has the perfect CSI at both the transmitter and the receiver (13), in the function $f(\lambda)$ of the system that has the outdated CSI at the transmitter (29), the co-channel interference by the outdated CSI is added.

NUMERICAL EVALUATION

Effects of WF Method

The transmit power and the distance have been optimized based on different numbers of relays in [12, 13]; thus, in this paper, the number of relays is fixed as 4, and the SNR of all relays is assumed to be the same. The number of antennas of all relays, the Tx and the Rx is set as 6. Since the performance of the system should be discussed in both the low-SNR region and the high-SNR region, the $snr_i, i=1,\cdots,m+1$ is changed from 0 to 30 dB. Figure 3 shows the end-to-end channel capacity of the proposed WF method. In order to evaluate the proposed WF method, the result of the average transmit power, meaning, the transmit power of all antennas in one relay is equal, also is represented. The end-to-end channel capacity of average transmit power and three WF methods is compared. The end-to-end channel capacity of scheme 3 is the highest and that of scheme 2 is higher than that of scheme 1. It can be explained that the transmit power of all transmitters in scheme 3 is the most suitable to channel model, and the transmit power of all transmitters in scheme 2 is more suitable to channel model than that of scheme 1. Similarly, the system with WF methods has more diversity than the system with average transmit power; therefore, the end-to-end channel capacity of WF methods is higher than that of the average transmit power. Although the number of antennas in the full-duplex system is a half of the number of antennas in the half-duplex system, the allocation time of the full-duplex system is a double of allocation time of the half-duplex system. Hence, compared to the half duplex, the end-to-end channel capacity of full duplex is higher.

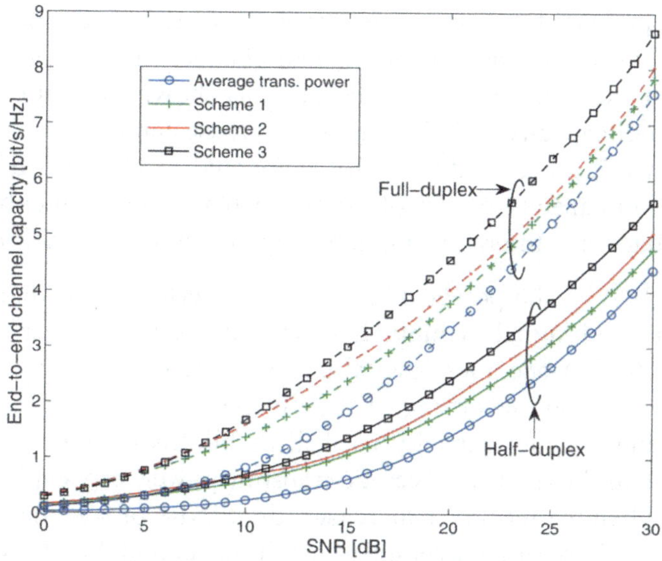

Figure 3: The end-to-end channel capacity of half-duplex and full-duplex AF scheme MMRS with proposed WF method. The end-to-end channel capacity of the average transmit power, and the three WF methods are compared. The end-to-end channel capacity of scheme 3 is the highest, and that of scheme 1 is the lowest. However, the end-to-end channel capacity of the WF method is higher than that of average transmit power. Compared to the half duplex, the end-to-end channel capacity of full duplex is much higher.

System with Outdated CSI

As explained in Section 3.2, there are two ways the outdated CSI affects the system. One is the effect on optimizing the transmit power of all transmitters; the other one is the effect on co-channel interference. At first, the end-to-end channel capacity is calculated when the term $f_D\tau$ is changed; the snr_i, i=1,⋯,m+1 is fixed as 5 dB (in the low-SNR region) and 20 dB (in the high-SNR region). The full-duplex and the half-duplex systems are also compared. Figures 4 and 5 show the end-to-end channel capacity of full duplex and half duplex when the snr_i is 5 and 20 dB, respectively. The end-to-end channel capacity decreases when the term $f_D\tau$ increases; however, the end-to-end channel capacity of WF methods decreases more quickly and be close to that of the average transmit power (Figure 4). The reason is that since the effect of the outdated CSI on co-channel interference is the same in both the WF method and the average transmit power, the reduction of end-to-end channel capacity should be the same. However, the optimization of transmit power of WF methods also be affected by the outdated CSI. The gap between the transmit power and channel models increases when the term $f_D\tau$ increases. On the other hand, as shown in Figure 5, the end-to-end channel capacity of WF methods decreases and drops below that of average transmit power. Additionally, compared to snr $_i$ of 5 dB, the end-to-end channel capacity of snr $_i$ of 20 dB decreases more quickly. It can be explained that since the ratio of desired signal and co-channel interference is fixed as $\frac{1}{\delta(M-1)}$, the power of co-channel interference is high when the power of desired signal is high. It means the signal-to-interference plus noise ratio (SINR) becomes low when the SNR is high. As explained in the previous section, according to the diversity gain of the WF methods, the end-to-end channel capacity as well as the SNR of the WF methods are higher than those of the average transmit power, especially scheme 3. Therefore, when the term $f_D\tau$ increases, the SINRs of schemes 1, 2, and 3 are lower than those of the average transmit power and the SINR of scheme 3 becomes the lowest. It means that the average transmit power is the most robust, and scheme 3 tends to be subjected to the influence of the outdated CSI. Compared to the half duplex, the end-to-end channel capacity of the full duplex in both the low- and high-SNR region is more robust due to the half of delay time.

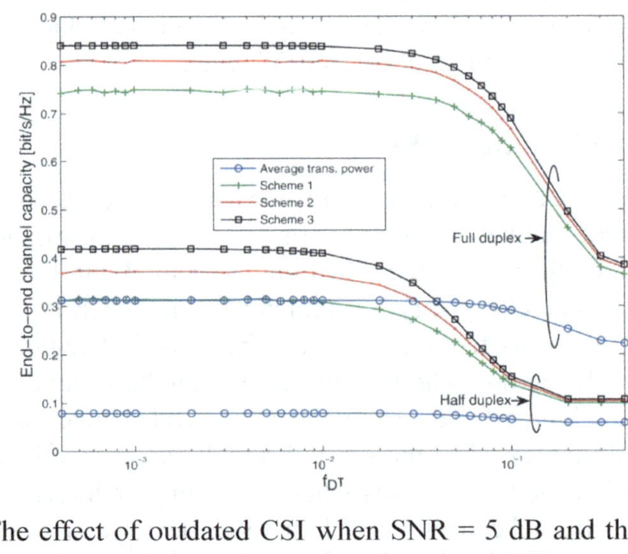

Figure 4: The effect of outdated CSI when SNR = 5 dB and the term $f_D\tau$ is changed. The end-to-end channel capacity of outdated CSI system is described. It considerably decreases when the term $f_D\tau$ is increased. Compared to that of the half duplex, the end-to-end channel capacity of the full duplex decreases more slowly.

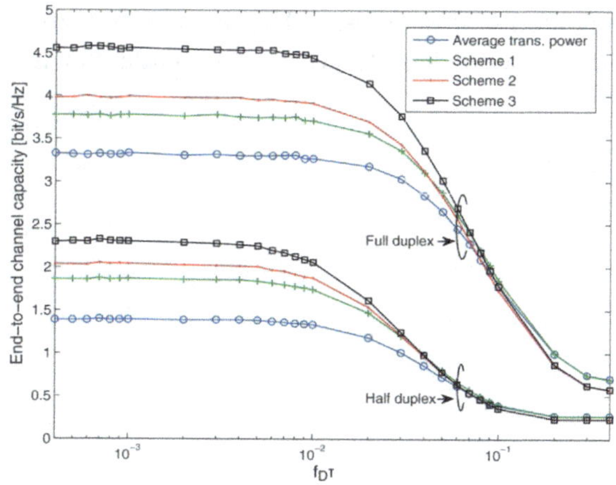

Figure 5: The effect of outdated CSI when SNR = 20 dB and the term $f_D\tau$ is changed. The end-to-end channel capacity of the outdated CSI system is described in the high-SNR region. The end-to-end channel capacity in the high-SNR region decreases more quickly than that in the low-SNR region.

The result as explained above can be recognized from Figures 6 and 7 that show the end-to-end channel capacity of half duplex and full duplex, respectively, when the term $f_D\tau$ is fixed as 0.001 and 0.05. When the term $f_D\tau$ is small (0.001), the effect of the outdated CSI is not so high; therefore, the end-to-end channel capacity increases when the snr_i increases. However, when the term $f_D\tau$ is large (0.05), the effect of the outdated CSI becomes higher; therefore, the end-to-end channel capacity increases slowly, especially in the half-duplex system.

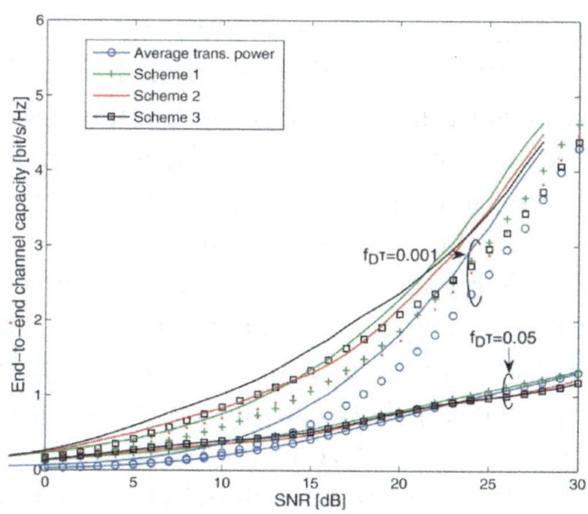

Figure 6: The end-to-end channel capacity of a half-duplex system when snr_i is changed. The higher the SNR is, the more rapidly the end-to-end channel

capacity of the outdated CSI decreases. The average transmit power system is the most robust, and scheme 3 is affected by the outdated CSI much more than the others.

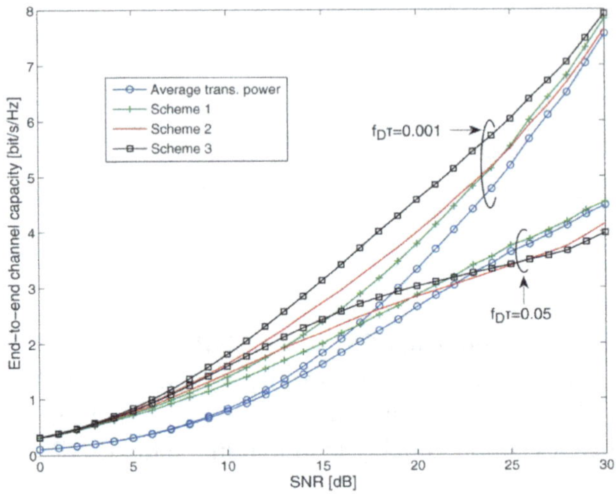

Figure 7: The end-to-end channel capacity of a full-duplex system when snr_i is changed. Compared to the half duplex, the end-to-end channel capacity of the full-duplex system is more robust due to the half of delay time.

CONCLUSIONS

In this paper, we have analyzed the end-to-end channel capacity of MMRSs according to the proposed WF method. The WF method is divided into three schemes. In scheme 1, the WF method is applied to each hop. In scheme 2, the transmit power is optimized based on the multiplication of channel models from the Tx to each relay, and in scheme 3, transmit power of all transmitters is the same and optimized based on multiplication of channel models from the Tx to the Rx. The end-to-end channel capacity of the proposed WF method is higher than that of the average transmit power, and the end-to-end channel capacity of scheme 3 is the highest in the perfect CSI. However, when the imperfect CSI is taken into account, the outdated CSI not only affects the gap between the optimized transmit power and channel models but also is effective in the co-channel interference. In the outdated CSI system, the end-to-end channel capacity decreases when the term $f_D\tau$ increases, especially in the high-SNR region due to the high power of co-channel interference. Additionally, the end-to-end channel capacity of the WF methods decreases close to or drops below that of average transmit power. The average transmit power system is the most robust, and scheme 3 is the most sensitive. The half duplex and the full duplex also have been discussed in this paper. Since the full duplex has a double of allocation time and a half of delay time, the full-duplex system is more robust than the half-duplex system, and the end-to-end channel capacity of the full-duplex system is higher than that of the half-duplex system. In this paper, since the full duplex has been taken into consideration, at least four antennas are necessary. The full duplex has more advantages when the number of antennas is greater than or equal to four, whereas the half duplex has more advantages when the number of antennas is less than four. Additionally, the greater the number of antennas is, the higher the end-to-end channel capacity achieves. Moreover, the signal processing of full duplex is considered to be more complicated than that of the half duplex.

Since the distance and the transmit power have been optimized in other literatures, in this paper, the same SNR at all transmitters was assumed. However, the optimization of distance and transmit power based on each WF method should be considered to improve the performance of MMRS. Additionally, the transmission of all transmitters in both the full duplex and the half duplex was assumed to be controlled on MAC layer. However, the detail of control should be discussed. We leave them to the future works.

REFERENCES

1. Wang B, Zhang J, Host-Madsen A: On the capacity of MIMO relay channel. *IEEE Trans. Inf. Theory* 2005, 51: 29-43.
2. Shiu DS, Foschini GJ, Gans MJ, Kahn JM: Fading correlation and its effect on the capacity of multi-element antenna systems. *IEEE Trans. Commun* 2000, 48: 502-513. 10.1109/26.837052
3. Gesbert D, Bolcskei H, Gore DA, Paulraj AJ: MIMO wireless channel: capacity and performance prediction. In *Proceedings of GLOBECOM*. San Francisco, CA, USA; 2000:1083-1088.
4. Liang Y, Veeravalli VV: Gaussian orthogonal relay channels: optimal resource allocation and capacity. *IEEE Trans. Inf. Theory* 2005, 51: 3284-3289. 10.1109/TIT.2005.853305
5. Makki B, Eriksson T: On the capacity of Rayleigh-fading correlated spectrum sharing networks. *EURASIP J. Wireless Commun. Netw* 2011. doi:10.1186/1687-1499-2011-83
6. Kim K, Ko K, Lee J: Adaptive selection of antenna grouping and beamforming for MIMO systems. *EURASIP J. Wireless Commun. Netw* 2011. doi:10.1186/1687-1499-2011-154
7. Shen Z, Heath RW Jr, Andrews JG, Evans BL: Space-time water-filling for composite MIMO fading channels. *EURASIP J. Wireless Commun. Netw* 2006. doi:10.1155/WCN/2006/16281
8. Lu Y, Zhang W: Water-filling capacity analysis in large MIMO systems. In *Proceedings of Computing, Communications and IT Applications Conference*. ComComAp, Hong Kong; 2013.
9. Lee K-J, Kim J-S, Caire G, Lee I: Asymptotic ergodic capacity analysis for MIMO amplify-and-forward relay networks. *IEEE Trans. Commun* 2010, 9: 2712-2717.
10. Jin S, McKay MR, Zhong C, Wong K-K: Ergodic capacity analysis of amplify-and-forward MIMO dual-hop systems. *IEEE Trans. Inf. Theory* 2010, 56: 2204-2224.
11. Gastpar M, Vetterli M: On the capacity of large Gaussian relay networks. *IEEE Trans. Inf. Theory* 2005, 51: 765-779. 10.1109/TIT.2004.842566
12. Hiep PT, Kohno R: Optimizing position of repeaters in distributed MIMO repeater system for large capacity. *IEICE Trans. Commun* 2010, E93-B: 3616-3623. 10.1587/transcom.E93.B.3616

13. Hiep PT, Fumie O, Ryuji K: Optimizing distance, transmit power and allocation time for reliable multi-hop relay system. *EURASIP J. Wireless Commun. Netw* 2012, 2012: 153. 10.1186/1687-1499-2012-153
14. Hiep PT, Hoang NH, Chika S, Ryuji K: End-to-end channel capacity of MAC-PHY cross-layer multiple-hop MIMO relay system with outdated CSI. *EURASIP J. Wireless Commun. Netw* 2013, 2013: 144. 10.1186/1687-1499-2013-144
15. Tang X, Hua Y: Optimal design of non-regenerative MIMO wireless relays. *IEEE Trans. Wireless Commun* 2007, 6: 1398-1407.
16. Hammerstrom I, Wittneben A: Power allocation schemes for amplify-and-forward MIMO-OFDM relay links. *IEEE Trans. Wireless Commun* 2007, 6: 2798-2802.
17. Yang S, Belfiore JC: Towards the optimal amplify-and-forward cooperative diversity scheme. *IEEE Trans. Inform. Theory* 2007, 53: 3114-3126.
18. Khoshnevis B, Yu W, Adve R: Grassmannian beamforming for MIMO amplify-and-forward relaying. *IEEE J. Sel. Areas Commun* 2008, 26: 1397-1407.
19. Andersen JB: Array gain and capacity for known random channels with multiple element arrays at both ends. *IEEE J. Sel. Areas Commun* 2000, 18: 2172-2178.
20. Karasawa Y: MIMO propagation channel modeling. *IEICE Trans. Commun* 2005, E88-B: 1829-1842. 10.1093/ietcom/e88-b.5.1829

11

Relevance Vector Machines for Enhanced BER Probability in DMT-Based Systems

Ashraf A. Tahat[1] and Nikolaos P. Galatsanos[2]

[1]School of Electrical Engineering, Princess Sumaya University for Technology, Amman 11941, Jordan
[2]Electrical and Computer Engineering Department, University of Patras, Rio 26500, Greece

ABSTRACT

A new channel estimation method for discrete multitone (DMT) communication system based on sparse Bayesian learning relevance vector machine (RVM) method is presented. The Bayesian frame work is used to obtain sparse solutions for regression tasks with linear models. By exploiting a probabilistic Bayesian learning framework, sparse Bayesian learning provides accurate models for estimation and consequently equalization. We consider frequency domain equalization (FEQ) using the

---

Citation: Ashraf A. Tahat and Nikolaos P. Galatsanos. "Relevance Vector Machines for Enhanced BER Probability in DMT-Based Systems" Journal of Electrical and Computer Engineering Volume 2010, Article ID 191808, 8 pages https://doi.org/10.1155/2010/191808

proposed channel estimate at both the transmitter (*preequalization*) and receiver (*postequalization*) and compare the resulting bit error rate (BER) performance curves for both approaches and various channel estimation techniques. Simulation results show that the proposed RVM-based method is superior to the traditional least squares technique.

INTRODUCTION

One of the research goals during the last decades was to provide broadband communication capabilities to and from the customer premises [1, 2]. To cope with the time dispersive transmission character of wireline and wireless communications, multicarrier modulation (MC) offers a viable solution. Since the early seventies a widespread interest was created due to an all digital implementation based on the cost-effective fast Fourier transform (FFT) algorithm [3]. Nowadays, MC systems are included in digital audio or video broadcasting (DAB/DVB) [3, 4] in wireless local area networks such as IEEE802.11a/g [5] and HIPERLAN2 [6], wireless wide area networks like WiMAX and LTE systems [7], and in wireline communication over twisted pairs such as in ADSL, ADSL2+,VDSL [8, 9] and also in power-line local area networks [10].

At the present time, the transmission format as presented in [3] is more commonly known as Discrete Multi-tone (DMT) modulation or Orthogonal Frequency Division Multiplexing (OFDM). An inherent feature of DMT over traditional OFDM is that DMT systems make use of the channel state information (CSI) at the transmitter [11, 12].

DMT divides the available bandwidth into several parallel frequency bins or tones, where each tone is quadrature amplitude modulated (QAM). Modulation and demodulation are efficiently carried out by means of an inverse Fourier transform (IFFT) and a fast Fourier transform (FFT), respectively. Contrary to analog MC systems, FFT-based implementation permits considerable overlap between sub-channels and therefore provides a high bandwidth effiency.

As mentioned, the various flavors of Digital Subscriber Line technologies (xDSL) represent a good example of DMT practical applications over traditional twisted pair phone lines to connect between customer premises and the Central Office (CO). We note that the bit rates highly depend on the length of the twisted pair: higher bit rates are only offered on relatively short loops due to signal attenuation caused by the telephone line. Since line attenuation increases with length and frequency, short loops are able to exploit a broader frequency spectrum and hence enable higher data rates. Tones with higher Signal-to-Noise Ratio (SNR) are able to carry more bits for a predetermined Bit Error Rate (BER) than those with low SNR. The measurement of SNR is usually performed during the initialization phase of the modem during which the frequency response of the channel is also estimated.

To deliver high bit rates towards the customer, DMT modems depend on advanced digital signal processing to mitigate several loop (channel) impairments such as time dispersion, noise, echo, and radio frequency interference (RFI) [13, 14]. In this paper, we formulate a Relevance Vector Machine (RVM) method [15] for a DMT-baseband system in a regression mode to improve channel estimation when the SNR is low to moderate, which in turn, will improve the probability of error performance when compared with the traditional Least Squares (LS) approach [16].

In Section 2 of this paper, we describe the models used for analysis and design. In Section 3 we present the RVM channel estimation methodology. In Section 4 we present the channel equalization approaches used. In Section 5 we present simulation results, and we conclude in Section 6.

SYSTEM MODEL

Consider the system block diagram shown in Figure 1, which represents a baseband Discrete Multi-Tone (DMT) link that includes a DMT modulator, the channel model, DMT demodulator, and frequency domain equalizer (FEQ).

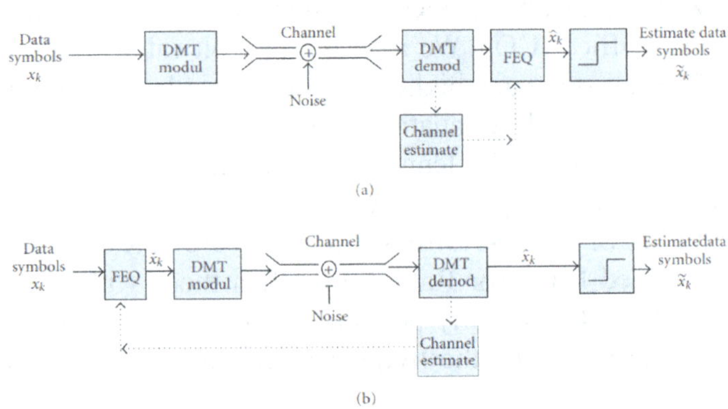

Figure 1: (a) DMT system block diagram with FEQ at receiver. (b) DMT system block diagram with FEQ at transmitter.

Figure 1(a) depicts the DMT system with FEQ implemented at the receiver (*postequalization*), while in Figure 1(b) we show the configuration of FEQ performed at the transmitter (*preequalization*) [11].

The DMT Modulator

DMT transmission [3] splits a high-rate data stream, R, into N lower rate streams transmitted simultaneously over subcarriers. The symbol rate for any of the N complex data streams is R/N symbols/second, which are sent to a $2N$-point inverse fast Fourier transform (IFFT) block after zero padding, forming a complex conjugate symmetry around the center of the IFFT and converting the frequency-domain data symbols into $2N$ time-domain real-valued samples as shown in Figure 2.

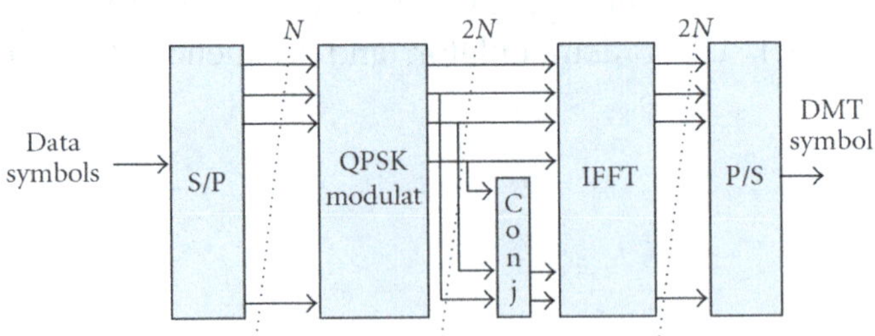

Figure 2: Block diagram of the baseband DMT modulator.

A cyclic prefix (CP) of length v is prepended to the $2N$-point time-domain samples to form the cyclically extended DMT symbol.

This is similar to the configuration used for DMT-based Asymmetric Digital Subscriber Line (ADSL), but with fixed modulation on each subchannel (i.e., QPSK). The cyclic prefix length, v, is chosen to encompass the maximum *delay spread* of the channel to prevent intersymbol interference (ISI) and making the DMT symbol appear periodic over the time span of interest.

The Channel Model

When the channel impulse response (CIR) is smaller than or equal in length to the CP length, v, the $2N$-DMT-sampled time points (Baud) and the channel are circularly convolved. This enables straight forward frequency-domain channel equalization at the receiver. However, when the order of the CIR is larger than the cyclic prefix, more complex equalization techniques are required [17]. In this paper, the channel that we will consider has a CIR that does not exceed the CP. Hence, inserting of the CP at the transmitter and then discarding the first v samples at the receiver give the received real-valued sampled-time domain samples over one DMT data block (baud) as follows:

$$\mathbf{r} = \mathbf{C}\mathbf{x} + \mathbf{n}, \qquad (1)$$

where the vector $r = [r_0, r_1, r_2, \ldots, r_{2N-1}]^T$ contains the received time-domain samples, the vector $x = [x_0, x_1, x_2, \ldots, x_{2N-1}]^T$ contains the transmitted time-domain samples, and the vector $n = [n_0, n_1, n_2, \ldots, n_{2N-1}]^T$ contains the samples of the band-limited additive noise with average power (variance) σ_n^2. The matrix multiplication operator in (1) represents circular convolution without intersymbol interference (ISI). The circulant matrix $\mathbf{C} \in \Re^{2N \times 2N}$

$$\mathbf{C} = \begin{bmatrix} h_0 & 0 & \ldots & 0 & h_L & \cdots & h_1 \\ h_1 & h_0 & & & & & \vdots \\ \vdots & & \ddots & & & & h_L \\ h_L & & & & & & 0 \\ 0 & h_L & & & & & \\ \vdots & \ddots & \ddots & & \ddots & \ddots & \vdots \\ 0 & \cdots & 0 & h_L & \cdots & h_1 & h_0 \end{bmatrix}, \qquad (2)$$

can be diagonalized by pre- and post-multiplication with the $2N$-point discrete Fourier transform (DFT) matrix, $\mathbf{W}_{2N}$, and inverse DFT (IDFT) matrix $(\mathbf{W}_{2N})^{-1}$ [18] that is,

$$\mathbf{W}_{2N}\mathbf{C}\,\mathbf{W}_{2N}^{-1} = \mathrm{diag}\{[H_0 \ldots H_{2N-1}]\}, \quad (3)$$

where $[H_0, H_1, H_2, \ldots, H_{2N-1}]^T$ is the frequency response of the channel.

When the transmitted symbols are generated in the frequency domain (i.e., $\mathbf{x} = \mathbf{W}_{2N}^{-1}[X_0, X_1, X_2, \ldots, X_{2N-1}]^T$), and the received samples are converted to the frequency domain (i.e., $[R_0, R_1, R_2, \ldots, R_{2N-1}]^T = \mathbf{W}_{2N}\mathbf{r}$), we obtain the following very simple input-output relationship:

$$\begin{bmatrix} R_0 \\ \vdots \\ R_{2N-1} \end{bmatrix} = \begin{bmatrix} H_0 & \cdots & 0 \\ \vdots & \ddots & \vdots \\ 0 & \cdots & H_{2N-1} \end{bmatrix} \begin{bmatrix} X_0 \\ \vdots \\ X_{2N-1} \end{bmatrix} + \begin{bmatrix} N_0 \\ \vdots \\ N_{2N-1} \end{bmatrix}, \quad (4)$$

or, alternatively as follows:

$$\begin{bmatrix} R_0 \\ \vdots \\ R_{2N-1} \end{bmatrix} = \begin{bmatrix} X_0 & \cdots & 0 \\ \vdots & \ddots & \vdots \\ 0 & \cdots & X_{2N-1} \end{bmatrix} \begin{bmatrix} H_0 \\ \vdots \\ H_{2N-1} \end{bmatrix} + \begin{bmatrix} N_0 \\ \vdots \\ N_{2N-1} \end{bmatrix}, \quad (5)$$

where $[N_0, N_1, N_2, \ldots, N_{2N-1}]^T = \mathbf{W}_{2N}\,\mathbf{n}$. From (4) and (5), we can see that each received frequency-domain symbol on each of the subchannels is simply a scaled version of the transmitted frequency-domain symbol plus white Gaussian noise. Moreover, every sub-channel can be processed independently of the other subchannels. In other words, block transmission with CP has converted a time dispersive channel into $2N$ parallel, narrowband flat sub-channels, or tones each having a channel gain H_k and additive Gaussian noise, hence the term discrete multi-tone transmission (DMT).

CHANNEL ESTIMATION

An important aspect of DMT is the fact that it uses an optimized frequency division allocation of energy and bits to maximize the achievable data rates that can be transmitted over band-limited channels. During startup of the DMT modem, the receiver measures the quality of the signals received (SNR) on each tone and reports this information to the transmitter [9], this is repeated periodically for dynamic channels. To estimate the channel during initialization, a pseudorandom frequency-domain training (pilot) sequence, $X_b(k)$, is employed. Where X_b is some constant modulus frequency-domain training sequence (i.e., $|X_b(k)|$ = constant for k = 0, 1, ..., 2N −1). Here, without loss of generality, we normalize X_b so that $|X_b(k)| = 1$. Now, the received pilot symbols can be expressed in vector form as follows:

$$\mathbf{R}_b = \mathbf{X}_b\mathbf{H} + \mathbf{N}. \tag{6}$$

In (6), R_b is the $2N \times 1$ received signal vector, the diagonal matrix X_b is the transmitted (pilot) signal, N is a vector containing the complex noise of the 2N subcarriers, and H is a vector that contains the overall complex channel gains between transmitter and receiver.

The problem at hand is to estimate the frequency-domain channel vector, H_k, $(k = 0, 1, 2, 3,, N\ 1)$ as the FFT of L unknown sample-spaced time-domain tap gains, where L is chosen to encompass the maximum expected delay spread and does not exceed the CP length, v. The frequency-domain model of the channel for each of the $2N$ tones is given by

$$H_k = \sum_{n=0}^{L-1} h_n e^{-j2\pi kn/2N}, \tag{7}$$

where h_n is channel tap at discrete time n.

To perform channel estimation from (6), we start by multiplying each of the frequency-domain symbols by the conjugate of the training symbols to produce the vector T_b as follows:

$$\mathbf{T}_b = \mathbf{X}_b^H\mathbf{R}_b = \mathbf{X}_b^H\mathbf{X}_b\mathbf{H} + \mathbf{X}_b^H\mathbf{N}, \tag{8}$$

where the superscript H denotes conjugate-transpose. Because of the property of unitary magnitude of the training symbols, (8) can be expressed as

$$\mathbf{T}_b = \mathbf{X}_b^H\mathbf{R}_b = \mathbf{H} + \mathbf{N}'. \tag{9}$$

In the noise-free case, Tb in (9) will contain the perfect frequency-domain channel gains. Due to channel noise, Tb contains the channel estimates that are corrupted by the additive noise, N . This is equivalent to least squares (LS) estimation [16] in which no assumption about the channel impulse response length has been incorporated (i.e., hn can have values larger than zero for n = 0, 1, ..., 2N-1).

The LS estimator for the impulse response, H, minimizes the criterion $(R_b\text{-}X_b\ H)^H\ (R_b\text{-}X_b\ H)$ and is given by

$$\mathbf{H}_{\mathrm{LS}} = \left(\mathbf{X}_b^H \mathbf{X}_b\right)^{-1} \mathbf{X}_b^H \mathbf{R}_b, \tag{10}$$

where due to the unitary magnitude of the training symbols we have

$$\mathbf{H}_{\mathrm{LS}} = \mathbf{X}_b^H \mathbf{R}_b = \mathbf{T}_b. \tag{11}$$

When the values of the channel impulse response are forced to be zero for sample numbers larger than the cyclic prefix duration, v, then this will be equivalent to "modified LS" estimation [16].

Our objective is to improve the performance of the channel estimation at low to moderate SNR after truncation of the channel impulse response to v [19]. Specifically, we will use a regression model similar to the one used for the relevance vector machine method (RVM) [15, 20]. We apply this model to baseband-DMT channel estimation by first taking the inverse Fourier transform of (9) to fit a regression model as follows:

$$\mathbf{t}_b = \mathbf{h} + \mathbf{n}', \tag{12}$$

where $t_b = [t_0, t_1, t_2, \ldots, t_{2N-1}]^T$ is the target vector time domain samples, $h = [h_0, h_1, h_2, \ldots, h_v, 0, \ldots, 0]^T$ is the $2N \times 1$ channel vector, and the vector $n' = [n_0, n_1, n_2, \ldots, n_{2N-1}]^T$ contains samples of the band-limited additive Gaussian noise with variance σ'^2.

In order to estimate h from the observations of t_b in (12) and ameliorate the effects of the noise, we will use sparse Bayesian regression. This approach has the property of setting the appropriate regression weights to zero automatically, avoiding the fitting of noise in the signal t_b.

Assume that the channel can be approximated using the function y(n) which is the linear combination of kernel functions as follows:

$$y(n) = \sum_{i=0}^{v} w_i \phi(n-i). \tag{13}$$

This is a convolution that can be written in matrix vector as

$$\mathbf{y} = \mathbf{\Phi w}. \tag{14}$$

where $w = [w_0, w_1, w_2, \ldots, w_i, \ldots, w_v]^T$ are the model weights, and Φ is the $v \times (v+1)$ convolution (design) matrix that is created from the kernel we select. One commonly used kernel is the Gaussian-shaped given by

$$\varphi(n-i) = \exp\left(\frac{-\|n-i\|^2}{2\sigma^2}\right). \tag{15}$$

However, there are many other choices (e.g., spline, polynomial, etc.) [20]. Now, our model becomes

$$\mathbf{t}_b = \mathbf{y} + \mathbf{e} = \mathbf{\Phi w} + \mathbf{e}, \tag{16}$$

where $y = [y_0, y_1, y_2, \ldots, y_v]^T$ is an approximation function and $e = [e_0, e_1, e_2, \ldots, e_v]^T$ is the vector with the regression model error.

The Bayesian model used assumes that the errors are modeled as independent identically distributed zero-mean Gaussian random variables, with variance σ^2, so

$$p(\mathbf{e}) = \prod_{i=1}^{v} \mathcal{N}(e_i \mid 0, \sigma^2). \tag{17}$$

The parameter σ^2 can be set in advance if known, but can be also estimated from the data. This error model implies a multivariate Gaussian likelihood for the target vector

$$p(\mathbf{t}_b \mid \mathbf{w}, \sigma^2) = (2\pi\sigma^2)^{-2N/2} \exp\left(\frac{-\|\mathbf{t}_b - \mathbf{\Phi w}\|^2}{2\sigma^2}\right). \tag{18}$$

Maximum-likelihood estimation of **w** from (18) is equivalent to the Least Squares solution given by

$$\mathbf{w}_{\mathrm{LS}} = \left(\mathbf{\Phi}^T\mathbf{\Phi}\right)^{-1}\mathbf{\Phi}^T\mathbf{t}_b, \tag{19}$$

and leads to severe overfitting of the data [20].

In order to overcome this problem, a flexible Gaussian prior over the weights **w** and Bayesian inference will be used [20]. More specifically,

$$p(\mathbf{w}, \boldsymbol{\alpha}) = \prod_{i=1}^{v} \mathcal{N}\left(w_i \mid 0, \alpha_i^{-1}\right), \tag{20}$$

with α is a vector of v hyperparameters [15, 20]. The flexibility of this prior is based on the fact that it uses a separate hyperparameter for every weight w_i.

The weights **w** are marginalized in according to

$$p(\mathbf{t}_b; \mathbf{a}, \sigma^2) = \int p\left(\frac{\mathbf{t}_b}{\mathbf{w}}; \sigma^2\right) p(\mathbf{w}; \mathbf{a}) d\mathbf{w}. \tag{21}$$

This integral can be found in closed form and is given by

$$p(\mathbf{t}_b \mid \boldsymbol{\alpha}, \sigma^2) = N\left(\mathbf{t}_b \mid \mathbf{0}, \mathbf{B}^{-1} + \boldsymbol{\Phi}\mathbf{A}^{-1}\boldsymbol{\Phi}^{\mathrm{T}}\right), \tag{22}$$

where we have defined $A = \mathrm{diag}(\alpha_0, \alpha_1, \alpha_2, \ldots, \alpha_v)$, $B = \sigma^2 I_v$, and I_v is the v × v identity matrix. The pdf $p(t_b \mid \alpha, \sigma^2)$ is called the marginal likelihood (or evidence) for the hyperparameters α and σ^2 [21]. Because the involved pdfs are Gaussian, calculation of the posterior in closed form is also possible and is given by

$$p(\mathbf{w} \mid \mathbf{t}_b, \boldsymbol{\alpha}, \sigma^2) = N\left(\mathbf{w} \mid \boldsymbol{\mu}_{w/t}, \boldsymbol{\Sigma}_{w/t}\right), \tag{23}$$

with

$$\Sigma_{w/t} = \left(\Phi^T \mathbf{B} \Phi + \mathbf{A}\right)^{-1}, \qquad \mu_{w/t} = \Sigma_{w/t} \Phi^T \mathbf{B} \mathbf{t}_b. \tag{24}$$

The regression estimates obtained by Bayesian Inference are given by $\Phi\mu_{w/t}$ where the hyperparameters α and σ^2 used in $\mu_{w/t}$ are the Maximum Likelihood (ML) estimates given the observations t_b and are found by maximizing $p(\alpha, \sigma^2 \mid t_b)$ [15]. However, maximization of (22) with respect to α and σ^2, which is termed as the Type II ML [20, 21], is equivalent to finding the maximum of $p(\alpha, \sigma^2 \mid t_b)$ assuming a uniform hyperprior [15, 20].

The values of α and σ^2 which maximize (22) cannot be obtained in closed form. However, they can be found by stationary point analysis of (22). More specifically, they are found by the following iterative update equations [22]:

$$\alpha_i^{\text{new}} = \frac{\gamma_i}{\mu_i^2}, \qquad (\sigma^2)^{\text{new}} = \frac{\left\|\mathbf{t}_b - \Phi\mu_{w/t}\right\|^2}{v - \sum_i \gamma_i}, \tag{25}$$

where μ_i is the ith component of the posterior mean $\mu_{w/t}$ from (24) and the quantities γ_i are defined by $\gamma_i = 1 - \alpha_i\Sigma_{ii}$ with Σ_{ii} being the ith diagonal element of the posterior weight covariance and $\Sigma_{w/t}$ from (24) computed with the current α and σ^2 values.

The learning algorithm thus proceeds by iterating between (24) and (25) until some convergence criteria have been satisfied [15, 22]. It can be shown that this iteration is equivalent to an expectation-maximization (EM) algorithm [20]. The E-step is (24), and the M-step is (25). This relation guarantees convergence of the proposed algorithm [20].

An important property of this model is that the optimal value of most hyperparameters α_i is typically infinite [15, 22]. From (24) this implies that the corresponding μi is zero. Thus, the regressor $y(n) = \sum_{i=0}^{V} \mu_i \varphi(n-i)$ is sparse since many of its weights μ_i are zero. In matrix form, the RVMchannel estimate is expressed as

$$h_{\mathrm{RVM}} = \Phi\mu_{w/t}. \tag{26}$$

FREQUENCY-DOMAIN CHANNEL EQUALIZATION

If the channel is known for the noise-free case, we can perform the equalization by inverting (1) assuming **C** is non-singular. This is the zero-forcing (ZF) equalizer [23]. The eigenvalues of the $2N \times 2N$ circulant matrix are equal to the DFT coefficients of the first column [24]. Since the first column has the channel coefficients in ascending order, from (6) these eigenvalues are

$$H_k = \sum_{n=0}^{L-1} h_n e^{-j2\pi kn/2N} = \sum_{n=0}^{L-1} h_n W^{-kn}, \tag{27}$$

where $W = e^{j2\varpi/2N}$, and H_k represents the channel frequency response. Note that $h_n = 0$ for $L < n < 2N$.

The circulant matrix **C** can be diagonalized with the DFT matrix [24] as mentioned in (3) as follows:

$$\mathbf{W}_{2N}\mathbf{C}\mathbf{W}_{2N}^{-1} = \mathrm{diag}\{[H_0 \cdots H_{2N-1}]\} = \Lambda_C. \tag{28}$$

Because the DFT (and thus its matrix $\mathbf{W}_{2N}$) is invertible, we deduce that the circulant matrix, **C**, is invertible if and only if Λ_C is invertible. And, the diagonal matrix will be invertible if it has no zero on its diagonal (i.e., the channel transfer function has no zero on the DFT grid). The diagonal elements of $[\Lambda_C]^{-1}$ are $1/H_k$, which can be regarded as a set of DFT-domain equalizers. The symbols are equalized at the receiver as in Figure 1(a) in the frequency domain before detection as follows:

$$\hat{\mathbf{X}} = \Lambda_C^{-1}\mathbf{R}, \tag{29}$$

where $\hat{\mathbf{X}}$ is a $2N \times 1$ vector, and R is a $2N \times 1$ received signal vector. Of course, in practice channel noise can be amplified by $[\Lambda_C]^{-1}$ if the coefficients, $1/H_k$, are large. This is the result of an optimization based on the ZF criterion which aims at equalizing the symbols regardless of noise level. Where on each subcarrier, the received symbol is multiplied by a complex multiplier to off set the effect of the channel, which is manifested in the form of attenuation and phase rotation. We can express the equalized received symbol on each subcarrier in equation form as

$$\hat{X}_k = X_k + \frac{1}{H_k}N_k. \tag{30}$$

In this version (*postequalization*), the receiver has all the complexity.

If the channel is known at the transmitter (e.g., transmitter has access to CSI through a feedback channel [25]), as it is the case in DMT [11, 12]. We can move $[\Lambda_C]^{-1}$ to the transmitter. Equalization in this configuration is implemented at the transmitter (*preequalization*) as shown in Figure 1(b). This results in superior performance [11] as will be demonstrated in the simulation section. The symbols are equalized in the frequency domain before modulation and transmission as follows:

$$\breve{\mathbf{X}} = \Lambda_C^{-1}\mathbf{X}, \tag{31}$$

where $\breve{\mathbf{X}}$ is a $2N \times 1$ symbols vector. And the received symbol on each subcarrier after demodulation is expressed as

$$\hat{X}_k = X_k + N_k. \tag{32}$$

We can see that channel noise will not be amplified for this second version. In addition, this yields a useful configuration for the cases where the receiver is simpler.

SIMULATION RESULTS

To evaluate the performance of the DMT/RVM system, it was simulated using MATLAB. The RVM-channel estimation scheme was evaluated and compared to other methods in addition to the ideal perfect case. The following are the common simulation parameters: $2N = 512$ is the DMT FFT size; $v = 32$ is the Cyclic Prefix length. The additive channel noise is band-limited white Gaussian with power σ_n^2. The actual channel length is $L = 32$. The used subchannels consisted of tones 6 to 255. Also, each of the subcarriers uses a QPSK modulation scheme.

In Figure 3, we present an example target vector time domain samples of (12), which in the noise-free case should be the actual impulse response of the channel and are superimposed on the same plot.

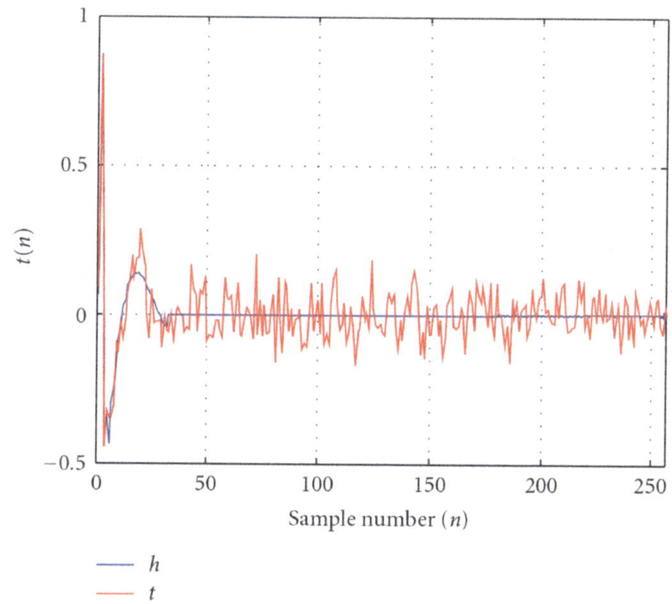

Figure 3: An example target vector.

The channels were assumed to remain constant after each sounding using the pilot sequence. Figure 4 depicts the actual sampled-time domain channel in the cyclic prefix range where it is equal to zero for values of time larger than the cyclic prefix. The target vector is also plotted in the same time range, which is equivalent to the Least Squares (LS) estimate of the channel. The proposed Bayesian sparse regression algorithm is used to process the target vector samples to filter out the noise at an SNR of 5 dB. The locations of the non-zero regression weights are identified on the plot with circles. Notice that we use *only 7 nonzero weights* from a total of 32. The remaining 25 are set to zero by the sparse model.

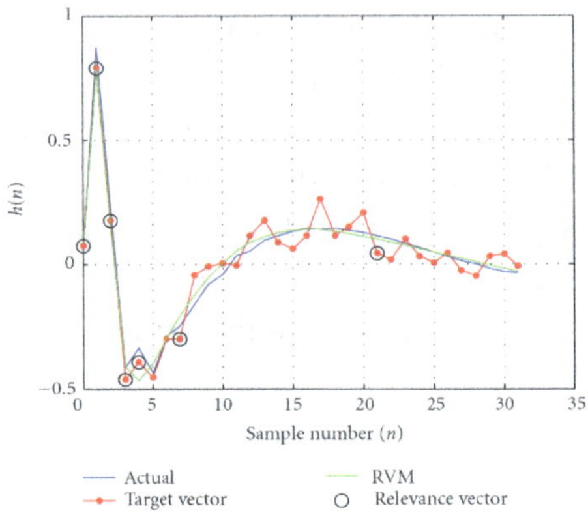

Figure 4: Channel impulse response and its estimate obtained using the RVM method.

To demonstrate the effects of estimating the channel impulse response on the frequency attenuation of each subcarrier (and hence equalization), in Figure 5 we plotted the frequency response of the estimated channel, the actual channel, and the original noise corrupted channel.

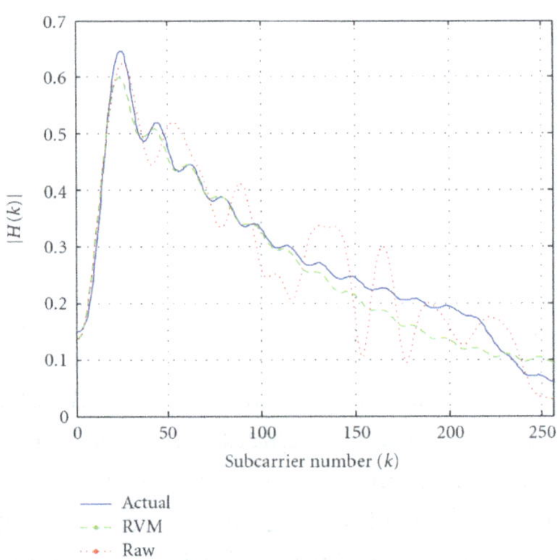

Figure 5: Frequency response of actual and estimated channel.

From Figure 5 we observe that in the frequency domain the estimated channel by the sparse model has smoother variation between adjacent tones and is closer to the desired frequency response.

When FEQ is implemented at the transmitter using the estimated channel at the receiver, BER performance curves (labeled TX-) demonstrate in Figure 6 the superior performance attained as a result of the more accurate sparse regression channel (RVM) estimate over the LS one. Also, the BER performance curve based on the actual channel (perfect estimate) is superimposed on the same plot of Figure 6 as a performance bench mark.

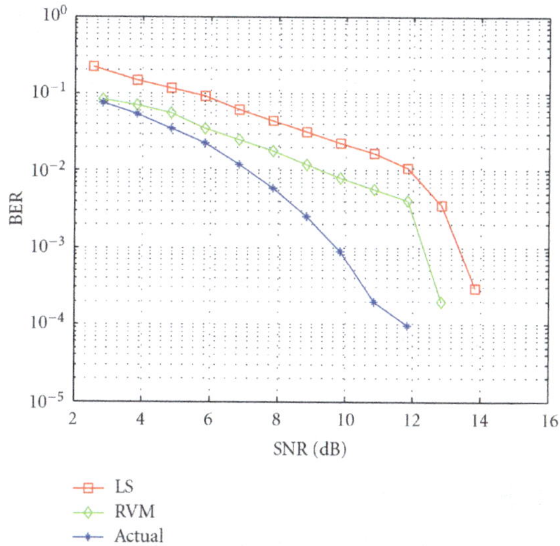

Figure 6: BER curves versus SNR with equalization at Transmitter.

We also compare in Figure 7 the accuracy in channel estimation when FEQ is implemented at the receiver as compared to that at the transmitter. We can make two observations First, even though it is intuitive that FEQ at the transmitter (*preequalization*) will achieve better performance, we demonstrate that it is far superior to that at the receiver (*postequalization*) for the same channel estimate. Second, the improved accuracy in channel estimation has limited effect on BER performance curves when FEQ is implemented at the receiver (labeled RX-), and BER curves are virtually identical for the actual channel, sparse regression (RVM) channel estimate, and the LS channel estimate. Low SNR on attenuated subcarriers renders the improved channel accuracy not important in this case, since equalization will also enhance noise simultaneously.

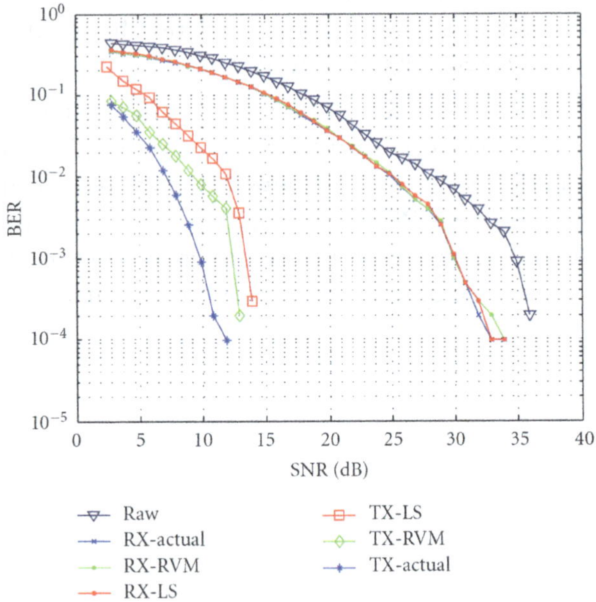

Figure 7: Comparisons of BER curves with FEQ at receiver to that at transmitter.

CONCLUSIONS

Proper equalization is commonly used to improve utilization of available bandwidth in a DMT-data transmission system. In this paper, an improved channel estimation method in a DMT communication system based on sparse Bayesian learning and the RVM method was presented. Deploying the RVM method, more accurate channel estimates in a low to medium SNR environments were attained. Although, frequency-domain equalization at the transmitter (*preequalization*) will intuitively yield better BER performance, we demonstrated the performance enhancement achieved when using the improved channel estimate via the RVM method. We also show through simulations that preequalization is more sensitive to channel estimation accuracy than frequency equalization at the receiver side (*postequalization*). Moreover, when compared to the traditional least squares technique, the proposed enhanced channel estimation scheme had superior performance and exhibited promising characteristics, particularly when applied to an ADSL system.

REFERENCES

1. Broadband Forum, "Next generation broadband access white paper," Marketing Report 185, August 2009.
2. E. Dahlman, S. Parkvall, and P. Beming, *3G Evolution: HSPA and LTE for Mobile Broadband*, Academic Press, Oxford, UK, 2007.
3. S. B. Weinstein and P. M. Ebert, "Data transmission by frequency-division multiplexing using the discrete Fourier transform," *IEEE Transactions on Communications*, vol. 19, no. 5, part 1, pp. 628–634, 1971.
4. The European Telecommunications Standards Institute, "Digital video broadcasting (DVB); framing structure, channel coding and modulation for terrestrial digital terrestrial television," Tech. Rep. EN 300 744 V1.4.1, ETSI, 2001.
5. The Institute of Electrical and Electronics Engineers, "Wireless LAN medium access control (MAC) and physical layer (PHY) specification," IEEE standard 802.11a, 1999.
6. The European Telecommunications Institute, "Broadband radio access networks (BRAN); high performance radio local area networks (HIPERLAN) type 2; system overview," Tech. Rep. 101 683 V1.1.1, ETSI, 2000.
7. S. Glisic, *Advanced Wireless Communications: 4G Technologies*, John Wiley & Sons, Chechester, England, 2004.
8. P. J. Kyees, R. C. McConnell, and K. Sistanizadeh, "ADSL. A new twisted-pair access to the information highway," *IEEE Communications Magazine*, vol. 33, no. 4, pp. 52–59, 1995.View at: Publisher Site | Google Scholar
9. H. Dedieu, *Fundamentals of DSL Technology*, Auerbach, Boca Raton, Fla, USA, 2006.
10. S. Baig and N. D. Gohar, "A discrete multitone transceiver at the heart of the PHY layer of an in-home power line communication local area network," *IEEE Communications Magazine*, vol. 41, no. 4, pp. 48–53, 2003.
11. Z. Wang and G. B. Giannakis, "Wireless multicarrier communications: where Fourier meets Shannon," *IEEE Signal Processing Magazine*, vol. 17, no. 3, pp. 29–48, 2000.

12. F. J. Simois and J. I. Acha, "Linear IIR precoding for non-redundant transmission over minimum-phase channels in wireline multitone systems," *IEEE Transactions on Communications*, vol. 57, no. 5, pp. 1496–1504, 2009.
13. L. de Clercq, M. Peeters, S. Schelstraete, and T. Pollet, "Mitigation of radio interference in xDSL transmission," *IEEE Communications Magazine*, vol. 38, no. 3, pp. 168–173, 2000.
14. C.-C. Li and Y.-P. Lin, "Receiver window designs for radio frequency interference suppression in DMT systems," *IET Signal Processing*, vol. 3, no. 1, pp. 33–39, 2009.
15. M. E. Tipping, "Sparse Bayesian learning and the relevance vector machine," *Journal of Machine Learning Research*, vol. 1, no. 3, pp. 211–244, 2001.
16. J. van de Beek, O. Edfors, M. Sandell, S. K. Wilson, and P. O. Borjesson, "On channel estimation in OFDM systems," in *Proceedings of the IEEE 45th Vehicular Technology Conference*, pp. 815–819, July 1995.
17. T. Pollet, M. Peeters, M. Moonen, and L. Vandendorpe, "Equalization for DMT-based broadband modems," *IEEE Communications Magazine*, vol. 38, no. 5, pp. 106–113, 2000.
18. G. Golub and C. Van Loan, *Matrix Computations*, The John Hopkins University Press, Baltimore, Md, USA, 3rd edition, 1996.
19. A. A. Tahat and N. P. Galatsanos, "Relevance vector machines for DMT based systems," in *Proceedings of the 20th International Conference on Electronics Communications and Computers (CONIELECOMP '10)*, pp. 31–36, February 2010.
20. C. Bishop, *Pattern Recognition and Machine Learning*, Springer, Berlin, Germany, 2006.
21. J. Berger, *Statistical Decision Theory and Bayesian Analysis*, Springer, New York, NY, USA, 1985.
22. A. Faul and M. Tipping, "Analysis of sparse Bayesian learning," in *Proceedings of the Neural Information Processing Systems: Natural and Synthetic (NIPS '01)*, pp. 383–389, December 2001.
23. J. Proakis, *Digital Communications*, McGraw-Hill, New York, NY, USA, 3rd edition, 1995.
24. A. Papoulis, *Signal Analysis*, McGraw Hill, New York, NY, USA, 1977.

25. D. J. Love, R. W. Heath Jr., V. K. N. Lau, D. Gesbert, B. D. Rao, and M. Andrews, "An overview of limited feedback in wireless communication systems," *IEEE Journal on Selected Areas in Communications*, vol. 26, no. 8, pp. 1341–1365, 2008.

SECTION 5:
SIGNALING OVER FADING CHANNELS

12

Bit Error Rate Closed-Form Expressions for LoRa Systems under Nakagami and Rice Fading Channels

Claudio Ferreira Dias [1], Eduardo Rodrigues de Lima [1] and Gustavo Fraidenraich [2]

[1] Department of Digital Hardware Design, Instituto Eldorado, Eldorado, Alan Turing - 275, Campinas 13083-898, Brazil

[2] Department of Communications, University of Campinas, Albert Einstein - 400, Campinas 13083-852, Brazil

ABSTRACT

We derive exact closed-form expressions for Long Range (LoRa) bit error probability and diversity order for channels subject to Nakagami-*m*, Rayleigh and Rician fading. Analytical expressions are compared with numerical results, showing the accuracy of our proposed exact expressions. In the limiting case of the Nakagami and Rice parameters, our bit error probability expressions specialize into the non-fading case.

Citation: Ferreira Dias, C.; Rodrigues de Lima, E.; Fraidenraich, G. "Bit Error Rate Closed-Form Expressions for LoRa Systems under Nakagami and Rice Fading Channels". Sensors 2019, 19, 4412. https://doi.org/10.3390/s19204412

Keywords: LPWAN, Internet of Things, MIMO, Nakagami, Rayleigh, Rice

INTRODUCTION

Low Power Wide Area Networks (LPWAN) are the new trend in the development of wireless communication technologies in the era of the Internet of Things (IoT) [1]. Long Range (LoRa) [2] technology has awakened the interest of the research community due to its growing usage for LPWAN deployments [2,3,4,5].

As clearly stated in [6,7,8,9], LoRa technology is mainly based on the features of the chirp spread spectrum (CSS) modulation. In [10], it establishes a more solid mathematical representation of the modulation/demodulation process addressing the bit error rate (BER) performance for AWGN and frequency selective channels. In [11], the BER was derived assuming CSS modulation and the use of moment matching to obtain analytical BER approximations for AWGN and Rayleigh scenarios. The BER performance under the perspective of orthogonal signaling described in [12] can be used to assess the performance of chirp spread spectrum modulation, as pointed out in [10].

The primary motivation of this investigation is to find new expressions for the bit error rate of chirp spread spectrum (CSS) modulation in the essential fading scenarios. For instance, Rayleigh statistics models well the signal power in a multipath environment [13]. On the other hand, Rice appropriately models multipath in situations with a dominant line-of-sight signal. In complex dense urban environments, we expect all kinds of small-scale fading, i.e., Ricean fading for strong, static links, and a smooth transition from Rayleigh fading down to Double-Rayleigh fading for moving links. A mathematically convenient method to approximate these distributions is by using the Nakagami distribution [14]. The use of Nakagami mathematical method is broad and can describe the amplitude of a received signal after maximum ratio diversity combining, since the *k*-branch MRC with Rayleigh-fading signals results in a Nakagami with m=k. In this way, it is possible to use the Nakagami-*m* distribution to investigate the use of multiple antennas (MIMO) in LoRa gateway (or LoRa base station). The Nakagami distribution is also the best match for many fading measured data, as related in [15,16,17].

The main contribution of our work is the derivation of exact closed-form expressions to calculate the CSS BER performance in diverse scenarios such as Rayleigh, Rician, and Nakagami fading channels. We believe that our main results extend the knowledge in this topic by including two new scenarios: the performance analysis under the Nakagami-*m* case and the Rice case. As it is well known in the literature, as the Nakagami parameter m=1, or the Rice parameter κ=0, our results specialize into the Rayleigh case. Also, the authors in [11] provided an approximate expression for the Rayleigh case, whereas we have provided an exact and elegant result.

We organize the remainder of this paper as follows: Section 2 refers to the underlying system model; Section 3 refers to the exact bit error probability performance derivation. Section 4 confirms the accuracy of the proposed exact expressions. Finally, Section 5 points out the main aspects of our achievements.

SYSTEM MODEL

Figure 1 illustrates the system model in which the designer can choose one out of three options of essential fading environments. LoRa devices use shift chirp modulation scheme for communication. The modulation consists of a signal which chirps within a frequency range in a given period. It is also known as spread spectrum modulation, and different symbols can be composed by varying bandwidth and spread spectrum parameters. The spreading factor (SF) determinates the number of samples within a symbol duration such that $T_s = \frac{2^{SF}}{B}$, where *B* is the signal bandwidth and SF∈{6,7,...,12} which are the values available for typical commercial devices [18].

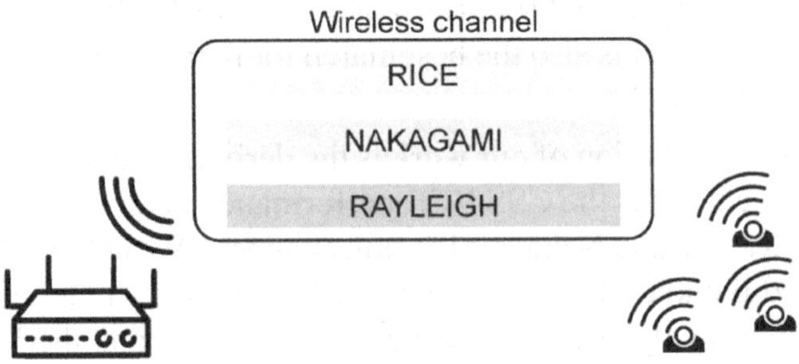

Figure 1: System Model.

The encoder maps every SF bits to a symbol s_k, such that $k \in \{0,1,\ldots,2^{SF}-1\}$. We then understand that k is related to a $f_k = B\frac{k}{2^{SF}}$ offset frequency. Each chirp signal has a specific starting frequency, and it wraps around bandwidth B as long as frequency keeps increasing. The chirping rate is given by $\frac{B^2}{2^{SF}}$ and signal decodes according to the position offset of f_k. The transmitted waveform can be written as [11]

$$s_k(nT) = \sqrt{E_s}\omega_k(nT)$$
$$= \sqrt{\frac{E_s}{2^{SF}}}e^{j2\pi\left[(k+n) \mod 2^{SF}\right]\frac{n}{2^{SF}}}, \tag{1}$$

where $T = \frac{1}{B}$ is the sampling period, $n=0,1,2,\ldots,(2^{SF}-1)$ is the sample index at time nT, E_s is the signal energy, and $\omega_k(nT)$ are the 2^{SF} orthonormal basis functions. The receiver demodulates the signal using the outputs of the correlator as [11]

$$\sum_{n=0}^{2^{SF}-1} r_k(nT) \cdot \omega_i^*(nT) = \begin{cases} \sqrt{E_s} + \phi_i, & i = k \\ \phi_i, & i \neq k \end{cases}, \tag{2}$$

where $r_k(\cdot)$ is the received signal, ϕ_i depicts a complex Gaussian noise process, and $\omega_i^*(nT)$ is the complex conjugate of the i-th basis function.

As shown by (2), the chirp signal demodulation is based on orthogonality properties. Thus, the detection of the symbols happens with the index selection of the basis waveform that has the highest correlation magnitude with respect to the received signal. Therefore, for a received waveform of $r_k(nT)$, the detected index symbol is computed as [11]

$$\tilde{k} = \{i \mid \arg_i\max\left(|\delta_{k,i}\sqrt{E_s} + \phi_i|\right)\}, \tag{3}$$

where $|\cdot|$ is the absolute value operator, $\delta_{i,k}=1$ for i=k and $\delta_{i,k}=0$ otherwise. We also must define γ as the signal-to-noise ratio (SNR) which is expressed as [11]

$$\gamma = \frac{E_s}{T_s}\frac{1}{N_0 B} = \frac{E_s}{N_0 \cdot 2^{SF}}. \tag{4}$$

BIT ERROR PROBABILITY

Error Probability for AWGN Channels

From (2), we can derive a random variable such that $\rho_i=|\phi_i|_{i\neq k}$. Since ϕ_i is a complex zero-mean Gaussian noise process, then ρ_i will be a Rayleigh random distributed variable such that [12]

$$F_{\rho_i}(x) = 1 - \exp\left[-\frac{x^2}{2\sigma^2}\right], \tag{5}$$

where $\sigma^2 = \frac{N_0}{2}$ and N_0 is the noise spectral density. The symbol error probability is given such that

$$P_{e|k} = \Pr\left[\max_{i,i\neq k}(\rho_i) > \beta_k\right], \tag{6}$$

where $\beta_k = |\sqrt{E_s} + \phi_k|$ follows a Rician distribution with shape parameter $K = \frac{E_s}{2\sigma^2} = \frac{E_s}{N_0}$. In this way, the probability density function of β_k can be written as

$$f_{\beta_k}(y) = \frac{y}{\sigma^2}\exp\left[-\frac{(y^2+E_s)}{2\sigma^2}\right] I_0\left(\frac{y\sqrt{E_s}}{\sigma^2}\right). \tag{7}$$

The random variable $\rho=\max(\rho_i)|_{i\neq k}$ is defined as the maximum of $(2^{SF}-1)$ i.i.d. Rayleigh random variables. Since the $(2^{SF}-1)$ variables are independent, the cumulative distribution function for ρ can be given as [12]

$$F_\rho(x) = \left[1 - \exp\left[-\frac{x^2}{2\sigma^2}\right]\right]^{2^{SF}-1}. \tag{8}$$

Using (6)–(8), and equally probable symbols, we can express the average bit error probability P_b as

$$P_{e|k} = \int_0^\infty \left[1 - \left[1 - \exp\left[-\frac{y^2}{2\sigma^2}\right]\right]^{2^{SF}-1}\right] f_{\beta_k}(y)dy. \quad (9)$$

Expression (9) is the average bit error rate probability for AWGN channels. We can further simplify (9) using the Newton's binomial identity

$$1-(1-e^x)^N = \sum_{q=1}^{N}\binom{N}{q}(-1)^{q+1}e^{xq}, \quad (10)$$

then (9) can be written as

$$P_{e|k} = \sum_{q=1}^{2^{SF}-1}(-1)^{q+1}\binom{2^{SF}-1}{q}\int_0^\infty e^{-\frac{qy^2}{2\sigma^2}} f_{\beta_k}(y)dy. \quad (11)$$

Now, if we replace (7) into (11), we have

$$P_{e|k} = \sum_{q=1}^{2^{SF}-1}(-1)^{q+1}\binom{2^{SF}-1}{q}e^{-\frac{E_S}{2\sigma^2}\frac{q}{q+1}} \times \int_0^\infty \frac{y}{\sigma^2} I_0\left(\frac{\sqrt{E_S}y}{\sigma^2}\right)e^{-\frac{(q+1)y^2+\frac{E_S}{q+1}}{2\sigma^2}}dy. \quad (12)$$

Fortunately, the integral in (12) can be computed as

$$\int_0^\infty \frac{y}{\sigma^2} I_0\left(\frac{\sqrt{E_S}y}{\sigma^2}\right)e^{-\frac{(q+1)y^2+\frac{E_S}{q+1}}{2\sigma^2}}dy = \frac{1}{1+q}, \quad (13)$$

Then, using (4), we find a closed-form expression for the error probability given the symbol *k* [12]

$$P_{e|k} = \sum_{q=1}^{2^{SF}-1}\frac{(-1)^{q+1}}{q+1}\binom{2^{SF}-1}{q}\exp\left[-\frac{q}{q+1}\gamma\cdot 2^{SF}\right]. \quad (14)$$

Then, according to [12], we finally find the average bit error rate such that

$$P_b = \frac{2^{SF-1}}{2^{SF}-1}\sum_{q=1}^{2^{SF}-1}\frac{(-1)^{q+1}}{q+1}\binom{2^{SF}-1}{q}\exp\left[-\frac{q}{q+1}\gamma\cdot 2^{SF}\right]. \quad (15)$$

Error Probability for Nakagami-*m* Channels

We first consider that a random variable $\alpha \sim \text{Gamma}(k,\theta)$. Then, we relate the Nakagami-m parameter to this distribution such that $k=m$ and $\theta=\Omega/m$. In the case of Nakagami-*m* channels, the correlation output at the demodulator is given as

$$\sum_{n=0}^{2^{SF}-1} r_k(nT) \cdot \omega_i^*(nT) = \begin{cases} \sqrt{\alpha E_s} + \phi_i, & i = k, \\ \phi_i, & i \neq k \end{cases} \tag{16}$$

where $\sqrt{\alpha}$ follows a Nakagami-m distribution. Accordingly, (6) can be modified to include the Nakagami-*m* random variable as

$$P_{e|\alpha} = Pr\left[\rho > |\sqrt{\alpha E_s} + \phi_i|\right]. \tag{17}$$

Following similar rationale as in (9), the bit error probability for Nakagami-*m* can now be expressed as

$$P_{e|\alpha} = \int_0^\infty \int_0^\infty \left[1 - \left[1 - e^{-\frac{y^2}{2\sigma^2}}\right]^{2^{SF}-1}\right] \Psi_m(y,\alpha) dy d\alpha, \tag{18}$$

where

$$\Psi_m(y,\alpha) = f_{\beta_k|\alpha}(y) \frac{m^m}{\Gamma(m)} \alpha^{m-1} e^{-m\alpha}, \tag{19}$$

and

$$f_{\beta_k|\alpha}(y) = \frac{y}{\sigma^2} \exp\left[-\frac{(y^2 + \alpha E_s)}{2\sigma^2}\right] I_0\left(\frac{y\sqrt{\alpha E_s}}{\sigma^2}\right). \tag{20}$$

Note that $\Omega=1$ in the term $\frac{m^m}{\Gamma(m)} x^{m-1} e^{-mx}$ is the normalized channel power for Nakagami-*m* distribution. The expression in (18) is the average bit error rate probability for Nakagami-*m* channels in the integral form. A closed-form expression of (18) can be obtained following the same rationale in the steps of (10)–(13), so that

$$P_{e|\alpha} = \sum_{q=1}^{2^{SF}-1} \int_0^\infty \int_0^\infty \binom{2^{SF}-1}{q} (-1)^{q+1} e^{-\frac{qy^2}{2\sigma^2}} \times \Psi_m(y,\alpha) dy d\alpha, \tag{21}$$

In the sequel, we must apply a change of variable of the form $z=y^2$ and perform the integration in order to get

$$P_{e|\alpha} = \sum_{q=1}^{2^{SF}-1} \int_0^\infty \frac{\binom{2^{SF}-1}{q}(-1)^{q+1}e^{-\frac{\gamma q 2^{SF}}{q+1}}}{(q+1)} \frac{m^m \alpha^{m-1} e^{-m\alpha}}{\Gamma(m)} d\alpha \qquad (22)$$

finally, integrating with respect to the *x* variable, we can get the following expression

$$P_{e|\alpha} = \sum_{q=1}^{2^{SF}-1} \frac{(-1)^{q+1}}{q+1} \binom{2^{SF}-1}{q} \left[1 + \frac{q}{m(q+1)}\gamma \cdot 2^{SF}\right]^{-m} \qquad (23)$$

where γ was defined in (4). Note that

$$\lim_{m\to\infty} \left[1 + \frac{q}{m(q+1)}\gamma \cdot 2^{SF}\right]^{-m} = \exp\left(-\frac{q}{q+1}\gamma 2^{SF}\right) \qquad (24)$$

and for this case (23) coincides exactly with (14), that is, the Gaussian case is the limit when the Nakagami-*m* parameter tends to infinity.

Again, according to [12], we finally find the average bit error rate for Nakagami-m such that

$$P_b = \frac{2^{SF-1}}{2^{SF}-1} \sum_{q=1}^{2^{SF}-1} \frac{(-1)^{q+1}}{q+1} \binom{2^{SF}-1}{q} \left[1 + \frac{q}{m(q+1)}\gamma \cdot 2^{SF}\right]^{-m} \qquad (25)$$

Error Probability for Rayleigh Channels

Our new expression (23) is general and valid for any positive *m* parameter. The symbol error probability for Rayleigh channels can be obtained for the particular case of m=1 in (23), thus

$$P_{e|\alpha} = \sum_{q=1}^{2^{SF}-1} (-1)^{q+1} \binom{2^{SF}-1}{q} \frac{\left[1 + \frac{q}{q+1}\gamma \cdot 2^{SF}\right]^{-1}}{q+1} \qquad (26)$$

The Gaussian hypergeometrical function is defined as [19]

$$_2F_1\left[\begin{matrix} -m & b \\ & c \end{matrix}; z\right] = \sum_{n=0}^{m} (-1)^n \binom{m}{n} \frac{(b)_n}{(c)_n} z^n, \qquad (27)$$

where $(\zeta)_n$ is the Pochhammer symbol defined as

$$(\zeta)_n = \begin{cases} 1 & , n = 0 \\ \zeta(\zeta+1)...(\zeta+n-1) & , n > 0 \end{cases} \qquad (28)$$

We can rewrite (26) as

$$\begin{aligned} P_{e|\alpha} &= 1 + \left(-1 - \sum_{q=1}^{2^{SF}-1} (-1)^{q+1} \binom{2^{SF}-1}{q} \frac{\left[1 + \frac{q}{q+1}\gamma \cdot 2^{SF}\right]^{-1}}{q+1}\right) \\ &= 1 - \sum_{q=0}^{2^{SF}-1} (-1)^q \binom{2^{SF}-1}{q} \frac{1}{1+q+q2^{SF}\gamma}. \end{aligned} \qquad (29)$$

Notice that

$$\frac{\left(\frac{1}{1+2^{SF}\gamma}\right)_q}{\left(\frac{1}{1+2^{SF}\gamma}+1\right)_q}=\frac{\left(\frac{1}{1+2^{SF}\gamma}\right)\left(\frac{1}{1+2^{SF}\gamma}+1\right)\left(\frac{1}{1+2^{SF}\gamma}+2\right)\cdots\left(\frac{1}{1+2^{SF}\gamma}+q-1\right)}{\left(\frac{1}{1+2^{SF}\gamma}+1\right)\left(\frac{1}{1+2^{SF}\gamma}+2\right)\cdots\left(\frac{1}{1+2^{SF}\gamma}+q-1\right)\left(\frac{1}{1+2^{SF}\gamma}+q\right)}. \tag{30}$$

Except for the first term in the numerator, and the last term in the denominator, all the terms in the numerator and denominator are the same and therefore can be canceled. We can finally, write (30) as

$$\frac{\left(\frac{1}{1+2^{SF}\gamma}\right)_q}{\left(\frac{1}{1+2^{SF}\gamma}+1\right)_q}=\frac{\left(\frac{1}{1+2^{SF}\gamma}\right)}{\left(\frac{1}{1+2^{SF}\gamma}+q\right)}=\frac{1}{1+q+q2^{SF}\gamma}. \tag{31}$$

If we consider that z=1 from (27) and use (31), we can write

$$\begin{aligned} P_{e|\alpha} &= 1-\sum_{q=0}^{2^{SF}-1}(-1)^q\binom{2^{SF}-1}{q}\frac{1}{1+q+q2^{SF}\gamma} \\ &= 1-{}_2F_1\left[\begin{matrix}1-2^{SF} & \frac{1}{1+2^{SF}\gamma} \\ \frac{2+2^{SF}\gamma}{1+2^{SF}\gamma} & \end{matrix};1\right]=1-\frac{\Gamma(2^{SF})\Gamma\left(\frac{2+\gamma\cdot 2^{SF}}{1+\gamma\cdot 2^{SF}}\right)}{\Gamma\left(\frac{1+2^{SF}+\gamma\cdot 2^{2SF}}{1+(\gamma\cdot 2^{SF})}\right)}. \end{aligned} \tag{32}$$

The average error bit probability for Rayleigh is derived such that [12]

$$P_b=\frac{2^{SF-1}}{2^{SF}-1}\left[1-\frac{\Gamma(2^{SF})\Gamma\left(\frac{2+\gamma\cdot 2^{SF}}{1+\gamma\cdot 2^{SF}}\right)}{\Gamma\left(\frac{1+2^{SF}+\gamma\cdot 2^{2SF}}{1+(\gamma\cdot 2^{SF})}\right)}\right]. \tag{33}$$

It is important to emphasize that our closed-form solution given in (33) is exact, whereas (33) in [11], although very accurate, provides an approximation.

The new expression in (33), obtained here, allows getting some intuition with respect the equivalent signal to noise ratio γ. Note that when $\gamma \to 0$, that is for very low values of the equivalent signal to ratio, the ratio given by $\frac{\Gamma(2^{SF})\Gamma\left(\frac{2+\gamma\cdot 2^{SF}}{1+\gamma\cdot 2^{SF}}\right)}{\Gamma\left(\frac{1+2^{SF}+\gamma\cdot 2^{2SF}}{1+(\gamma\cdot 2^{SF})}\right)}$ tends to 0 and therefore $P_b \to \frac{1}{2}$ (for SF>6), as expected. On the other hand, for high values of γ, the same ratio tends to 1 and therefore Pb$\to$0.

Rician Fading

In the case of Rician fading channels, the correlation output at the demodulator is given as

$$\sum_{n=0}^{2^{SF}-1} r_k(nT)\cdot\omega_i^*(nT) = \begin{cases}\sqrt{(\alpha+v)E_s}+\phi_i, & i=k,\\ \phi_i, & i\neq k\end{cases} \tag{34}$$

where v is related to the direct path component. Following similar rationale as in (9), the bit error probability for Rician can now be expressed as

$$P_{e|v} = \int_0^\infty\int_0^\infty \left[1-\left[1-e^{-\frac{y^2}{2\sigma^2}}\right]^{2^{SF}-1}\right]\Psi_r(y,\alpha)dy d\alpha, \tag{35}$$

where

$$\Psi_r(y,\alpha) = f_{\beta_k|\alpha}(y)\frac{\alpha}{\sigma_r^2}e^{-\frac{-\alpha^2+v^2}{2\sigma_r^2}} I_0\left(\frac{v\alpha}{\sigma_r^2}\right). \tag{36}$$

where $f_{\beta k|\alpha}(\cdot)$ is given in (20), v^2 is the direct component power, σ_r^2 is the variance of the Rice distribution. We introduce the variable κ as the shape parameter which is the ratio of the power contributions by line-of-sight path to the remaining multi-paths, and Ω as the scale parameter related with the total power received in all paths. Furthermore, we can establish the following relations

$$v^2 = \frac{\kappa}{\kappa+1}\Omega, \qquad \sigma_r^2 = \frac{\Omega}{2(1+\kappa)}. \tag{37}$$

Expression (35) is the average bit error rate probability for Rician channels in the integral form. A closed-form expression of (35) can be obtained following the same rationale in the steps of (10)–(13), and substituting (37) so that

$$P_{e|v} = \sum_{q=1}^{2^{SF}-1}\frac{(-1)^{q+1}}{1+q+\frac{2^{SF}q\gamma\Omega}{1+\kappa}}\binom{2^{SF}-1}{q}\times\exp\left[-\frac{\gamma 2^{SF}q\kappa\Omega}{1+\kappa+q(1+\kappa+\gamma 2^{SF}\Omega)}\right] \tag{38}$$

Note, when Ω=1, we have that

$$\lim_{\kappa\to\infty}\left[\exp\left(-\frac{\gamma 2^{SF}q\kappa\Omega}{1+\kappa+q(1+\kappa+\gamma 2^{SF}\Omega)}\right)\right] = \exp\left(-\frac{q}{q+1}\gamma\cdot 2^{SF}\right), \tag{39}$$

and

$$\lim_{\kappa\to\infty}\left[\frac{(-1)^{q+1}}{1+q+\frac{2^{SF}q\gamma\Omega}{1+\kappa}}\right] = \frac{(-1)^{q+1}}{q+1} \tag{40}$$

which agrees with the limit case given in (14) for AWGN scenario when the Rician shape parameter tends to infinity. On the other hand, we have that

$$\lim_{\kappa\to 0}\left[\exp\left(-\frac{\gamma 2^{SF} q\kappa\Omega}{1+\kappa+q\left(1+\kappa+\gamma 2^{SF}\Omega\right)}\right)\right]=1, \tag{41}$$

and

$$\lim_{\kappa\to 0}\left[\frac{(-1)^{q+1}}{1+q+\frac{2^{SF}q\gamma\Omega}{1+\kappa}}\right]=\frac{(-1)^{q+1}}{1+q+2^{SF}q\gamma} \tag{42}$$

which agree with the limit case in the (26) of Rayleigh scenario when the Rician shape parameter tends to zero.

The average error bit probability for Rice case is derived such that [12]

$$P_b=\frac{2^{SF-1}}{2^{SF}-1}\sum_{q=1}^{2^{SF}-1}\frac{(-1)^{q+1}}{1+q+\frac{2^{SF}q\gamma\Omega}{1+\kappa}}\binom{2^{SF}-1}{q}\times\exp\left[-\frac{\gamma 2^{SF}q\kappa\Omega}{1+\kappa+q\left(1+\kappa+\gamma 2^{SF}\Omega\right)}\right]. \tag{43}$$

NUMERICAL RESULTS

In this section, we will evaluate the numerical expressions and compare the results. Figure 2a presents comparisons of the derived Nakagami BER given in (23) and (33) versus the theoretical BER performance solved numerically using the integrals in (18) and (26), respectively. We choose SF∈{6,12} as the lowest and highest spread factor values to produce two sets of curves and each group with m∈{1,2,10}. As shown in Figure 2, the expressions in (23) and (33) are indistinguishable from (18) and (26). Also, note that as *m* increases, the bit error rate curve gets very close to the AWGN curve, as expected. With SF =12 and m=10, there is a gain of almost 30 dB for a BER of 10^{-4} in a multipath environment.

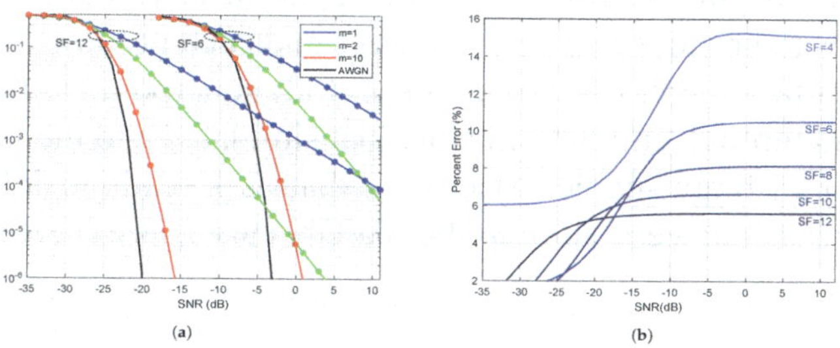

Figure 2: (a) Bit Error Rate for Nakagami-*m* and Rayleigh cases. The continuous blue curve is given by (26) (Rayleigh Integral) and the blue stars are computed

using (33) (Rayleigh Exact). The green and red curves are the BER for the Nakagami-*m* case, the continuous and star lines are given by (18) (Nakagami-*m* Integral) and (23) (Nakagami-*m* Exact), respectively. Analytic AWGN curve is provided for reference. (b) Percentage error between the Rayleigh BER approximation in [11] and our *exact* analytical solution given in (33).

Figure 2b shows the percentage error of the difference between the Rayleigh BER approximation in [11] and (33). The difference can vary from 6% (when SF=12) to 15% (when SF=4).

Next, we focus on the accuracy of the results given by numerical methods considering integrals and a finite series. We calculated the percentage error between on (18) and (23). The data is summarized in the Table 1. The data was generated such that parameters assume SNR(dB)∈{−30,−10,10}, SF∈{12,10,8,6,4}, and m∈{1,3,5}. The numbers in the table are tiny, and therefore, it is necessary to multiply by a factor of 10−6 to get the percentage error.

Table 1: Table with percentage error values between (18) and (23). The value of each cell corresponds to the number times 10−6 %

SNR = −30 dB					**SNR = −10 dB**					**SNR = 10 dB**				
	m	**1**	**3**	**5**		m	**1**	**3**	**5**		m	**1**	**3**	**5**
SF					SF					SF				
12		0.7	2	0.0	**12**		0.5	5.0	4.0	**12**		2.1	11.0	4.0
10		1.5	8.0	0.1	**10**		2.2	1.0	1.2	**10**		0.2	7.8	80.0
8		65.6	1.9	1.5	**8**		0.1	0.2	2.6	**8**		23.4	0.6	13.3
6		25.4	0.7	0.2	**6**		2.7	2.3	1.9	**6**		0.9	4.6	3.3
4		21.7	0.4	0.7	**4**		0.8	13.0	4.2	**4**		2.4	1.2	3.5

Figure 3 presents the derived BER in (38) versus simulation for Rician environments. Again, we choose SF∈{6,12} as the lowest and highest spread factor values to produce two sets of curves and each group with κ∈{0,5,∞}. As shown in the figure, the expression in (38) superposes with the discrete values found through simulation. Note that simulated and theoretical curves are almost indistinguishable. Also, note that as κ tends to infinity, the bit error rate curve gets very close to the AWGN curve, as expected.

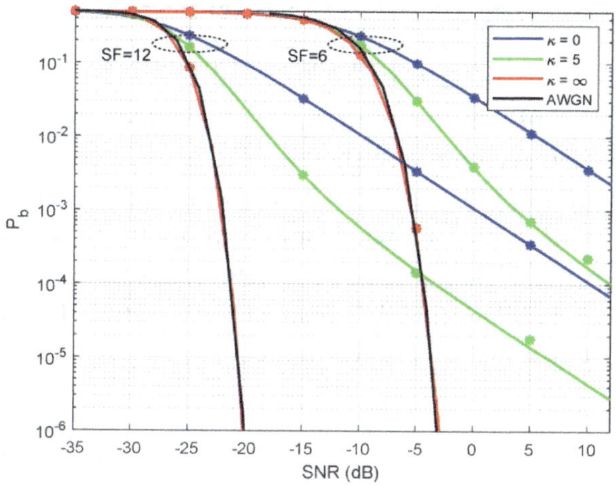

Figure 3: Bit error rate for the Rician case. The continuous lines is given by (38) and stars are given by the Matlab simulations. Analytic AWGN curve is provided for reference.

APPLICATION CASE

Since the signal to noise ratio γ in (23) is related to the maximum link distance, we can compute this distance in a system using the LoRa technology where the communication link is under the Nakagami-*m* distribution.

We choose the device SX1272 [18] as reference. For instance, the datasheet report that the receiver sensitivity is P_{RX}=−137 dBm for B=125 kHz and SF=12. Using (23), we can set a target bit error probability of P_b=10−4 and analyze what is the impact of the Nakagami-*m* factor in the maximum link distance.

We choose Okumura-Hata [20] as our path-loss model with urban area parameters. The heights of the base station and mobile stations are 40 meters and 1 m, respectively. The system operates in the 900 MHz frequency, which agree with the limits presented in Table 2.

Table 2: Regions and Operating Frequencies for LoRa devices [18]

Region	Frequency (MHz)
EU	863–870
US	902–928

AU	915–928
CN	779–787 and 470–510

Figure 4 presents the maximum link distance for different values of SF and *m*. When we compare these curve against the AWGN curve (no fading case) given in (14), we notice that the distance decreases from 5 km to 200 m for SF=12, which indicates a loss of about 96% of link range. For scenarios where the Nakagami-*m* parameter ranges from 2 to 6, the link range loss reduces to 80% and 60%, respectively. For m=10, the performance becomes closer to the AWGN performance.

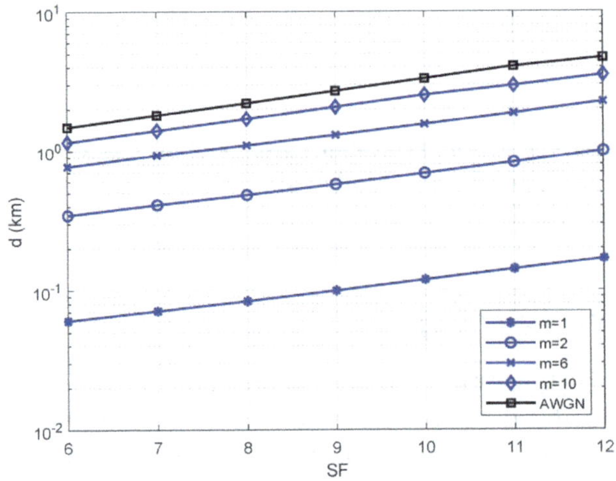

Figure 4: Maximum distance using the Okumura-Hata path-loss model for an urban area and a bit error probability of Pb=10−4.

CONCLUSIONS

In this paper, we derived new expressions to assess the BER performance of CSS in the essential fading scenarios, i.e., Rayleigh, Rician, and Nakagami-*m* fading channels. All the results are in terms of LoRa parameters such as spreading factor SF, bandwidth *B*, and symbol to noise ratio E_s/N_0. The derived bit error probability expressions will be handy for any IoT provider that aims to predict the network coverage for different scenarios where the fading may vary. Also, we have addressed the limiting cases where $m \to \infty$ (Nakagami parameter) and $\kappa \to \infty$ (Rice parameter), showing that they converge to the non-fading case or the case under the influence only of the additive white Gaussian noise. All the expressions have been validated by simulation and when applicable, compared with previous results.

ACKNOWLEDGMENTS

The authors would like to thank the anonymous reviewers for their constructive comments, which certainly helped them to improve the manuscript.

AUTHOR CONTRIBUTIONS

C.F.D. participated in the tasks: conceptualization, formal analysis, research, software, visualization and writing—original draft. E.R.d.L. participated in the tasks: conceptualization, funding acquisition, project administration and writing—review & editing. G.F. participated in the tasks: visualization, project administration and writing—review & editing.

REFERENCES

1. Centenaro M., Vangelista L., Zanella A., Zorzi M. Long-range communications in unlicensed bands: The rising stars in the IoT and smart city scenarios. IEEE Wirel. Commun. 2016;23:60–67. doi: 10.1109/MWC.2016.7721743.
2. LoRa Alliance. [(accessed on 27 September 2019)]; Available online: https://lora-alliance.org/
3. Mikhaylov K., Petaejaejaervi J., Haenninen T. Analysis of capacity and scalability of the LoRa low power wide area network technology; Proceedings of the European Wireless 2016 22th European Wireless Conference; Oulu, Finland. 18–20 May 2016; pp. 1–6.
4. Aref M., Sikora A. Free space range measurements with Semtech LoRa™ technology; Proceedings of the 2014 2nd International Symposium on Wireless Systems within the Conferences on Intelligent Data Acquisition and Advanced Computing Systems; Offenburg, Germany. 11–12 September 2014; pp. 19–23.
5. Augustin A., Yi J., Clausen T., Townsley W. A study of LoRa: Long range & low power networks for the internet of things. Sensors. 2016;16:1466.
6. Sforza F. Communication System. 8406275 B2. U.S. Patent. 2013 Mar 26;
7. Goursaud C., Gorce J.M. Dedicated networks for IoT: PHY/MAC state of the art and challenges. EAI Endorsed Trans. Internet Things. 2015 doi: 10.4108/eai.26-10-2015.150597.
8. Reynders B., Meert W., Pollin S. Range and coexistence analysis of long range unlicensed communication; Proceedings of the 2016 23rd International Conference on Telecommunications (ICT); Thessaloniki, Greece. 16–18 May 2016; pp. 1–6.
9. Reynders B., Pollin S. Chirp spread spectrum as a modulation technique for long range communication; Proceedings of the 2016 Symposium on Communications and Vehicular Technologies (SCVT); Mons, Belgium. 22 November 2016; pp. 1–5.
10. Vangelista L. Frequency shift chirp modulation: The LoRa modulation. IEEE Signal Process. Lett. 2017;24:1818–1821. doi: 10.1109/LSP.2017.2762960.
11. Elshabrawy T., Robert J. Closed-form approximation of LoRa modulation BER performance. IEEE Commun. Lett. 2018;22:1778–1781. doi: 10.1109/LCOMM.2018.2849718.

12. Proakis J.G., Salehi M. Digital Communications. Volume 4 McGraw-Hill; New York, NY, USA: 2001.
13. Mendes J.R., Yacoub M.D., Fraidenraich G. Closed-form generalized power correlation coefficient of the Hoyt fading signal. IEEE Commun. Lett. 2006;10:94–96. doi: 10.1109/LCOMM.2006.02009.
14. Nakagami M. Statistical Methods in Radio Wave Propagation. Elsevier; Los Angeles, CA, USA: 1960. The m-distribution—A general formula of intensity distribution of rapid fading; pp. 3–36.
15. Suzuki H. A statistical model for urban radio propogation. IEEE Trans. Commun. 1977;25:673–680. doi: 10.1109/TCOM.1977.1093888.
16. Dias U.S., Yacoub M.D., Santos Filho J.C.S., Fraidenraich G., da Costa D.B. On the Nakagami-m autocorrelation and power spectrum functions: Field trials and validation; Proceedings of the 2006 International Telecommunications Symposium; Fortaleza, Brazil. 3–6 September 2006; pp. 253–256.
17. Porto I.B., Yacoub M.D., Santos Filho J.C.S., Cotton S.L., Scanlon W.G. Nakagami-m phase model: Further results and validation. IEEE Wirel. Commun. Lett. 2013;2:523–526. doi: 10.1109/WCL.2013.070113.130327.
18. SX1272/73—860 MHz to 1020 MHz Low Power Long Range Transceiver. [(accessed on 27 September 2019)]; Available online: https://www.semtech.com/uploads/documents/sx1272.pdf.
19. Abramowitz M., Stegun I.A. Handbook of Mathematical Functions with Formulas, Graphs, and Mathematical Tables. Dover; New York, NY, USA: 1964. Ninth Dover Printing, Tenth gpo Printing ed.
20. Yacoub M.D. Foundations of Mobile Radio Engineering. Routledge; Abingdon, UK: 2019.

13

Ergodic Rate for Fading Interference Channels with Proper and Improper Gaussian Signaling

Mohammad Soleymani [1], Ignacio Santamaria [2], Christian Lameiro [1] and Peter J. Schreier [1]

[1]Signal and System Theory Group, Universität Paderborn, 33098 Paderborn, Germany
[2]Department of Communications Engineering, University of Cantabria, 39005 Santander, Cantabria, Spain

ABSTRACT

This paper studies the performance of improper Gaussian signaling (IGS) over a 2-user Rayleigh single-input single-output (SISO) interference channel, treating interference as noise. We assume that the receivers have perfect channel state information (CSI), while the transmitters have access to only statistical CSI. Under these assumptions, we consider a signaling scheme, which we refer to as proper/improper Gaussian signaling or PGS/IGS, where at most one user may employ IGS. For the Rayleigh fading

Citation: Mohammad Soleymani, Ignacio Santamaria, Christian Lameiro and Peter J. Schreier. "Ergodic Rate for Fading Interference Channels with Proper and Improper Gaussian Signaling" Entropy 2019, 21(10), 922; https://doi.org/10.3390/e21100922

channel model, we characterize the statistical distribution of the signal-to-interference-plus-noise ratio at each receiver and derive closed-form expressions for the ergodic rates. By adapting the powers, we characterize the Pareto boundary of the ergodic rate region for the 2-user fading IC. The ergodic transmission rates can be attained using fixed-rate codebooks and no optimization is involved. Our results show that, in the moderate and strong interference regimes, the proposed PGS/IGS scheme improves the performance with respect to the PGS scheme. Additionally, we numerically compute the ergodic rate region of the full IGS scheme when both users can employ IGS and their transmission parameters are optimized by an exhaustive search. Our results suggest that most of the Pareto optimal points for the 2-user fading IC channel are attained when either both users transmit PGS or when one transmits PGS and the other transmits maximally improper Gaussian signals and time sharing is allowed.

Keywords: achievable ergodic rate region, fading channels, improper Gaussian signaling, 2-user interference channel

INTRODUCTION

Interference management techniques play a key role in modern wireless communication systems, which are mostly interference-limited [1]. The capacity-achieving signaling in point-to-point communication systems is proper Gaussian signaling (PGS) [2]. Nevertheless, this is not generally the case in interference-limited systems, where it has been shown that improper (or asymmetric) Gaussian signaling (IGS) can improve the performance of various interference-limited systems [3,4,5,6,7,8,9,10,11,12,13,14,15,16,17,18,19,20,21]. In contrast to proper signals, the real and imaginary parts of an improper signal are correlated and/or have unequal powers [22,23].

Related Work

The Gaussian interference channel (IC) is a fundamental communication scenario where multiple users share the same frequency and time resources for the transmission of the desired messages. It has been vastly studied over the last decades [24,25,26,27,28,29,30,31] but its capacity is still unknown in the general case [24,25,26]. In some particular cases such as the very strong and very weak interference regimes, the capacity-achieving schemes are known [26]. For example, in the very strong interference regime, the capacity of the 2-user IC is achieved by successive interference cancellation (SIC), in which the interfering signals are firstly decoded and then canceled from the received signal [26,27,28]. In other words, in the strong interference regime, the interference has to be decoded in order to achieve capacity. Obviously, SIC is not a linear operation and has a high computational cost, specially in a multiuser setting. On the other hand, when the interference level is very low, which is also referred as the very weak interference regime, treating interference as noise (TIN) is the capacity achieving strategy [25,26]. TIN is a simple and practical decoding strategy, which turns out to be optimal from generalized degree of freedom (GDoF) point of view in low interference regimes [31].

In summary and at a coarse level, if the interference is strong, the optimal interference management approach is to decode it along with the desired signal, whereas if the interference is weak, then the interference management approach is to treat interference as noise. If the strength of the interference is comparable to the strength of the desired signal, the degrees-of-freedom (DoF) optimal interference management approach is to apply interference alignment (IA) [32,33,34,35,36]. IA designs transmit precoders that reduce the dimension of the interference subspace, in such a way that the interference can be zero-forced at each receiver thus getting an interference-free subspace for the transmission of the desired signals. However, the feasibility of IA for channels with constant coefficients depends on the number of antennas and hence packing the interference into a low-dimensional subspace is not always possible. Hence, in some practical scenarios, we have to tolerate and manage interference to improve the system performance. As a result, in modern wireless communication systems non-orthogonal multiple access (NOMA) schemes play an essential role [37,38,39].

Another approach to handle interference is to employ IGS, which has been considered as an interference-management technique for the first time in Reference [3]. The authors of Reference [3] showed that IGS can increase the degrees-of-freedom (DoF) of the 3-user single-input single-output (SISO) interference channel (IC) with constant channel coefficients. In Reference [4], this result was extended to the 4-user SISO IC and, in Reference [5], the authors showed that IGS can also increase the DoF of the 2-user multiple-input, multiple-output (MIMO) X channel. In these papers, it is shown that IGS with IA can improve the DoF of different interference-limited systems. However, IGS can also be beneficial from rate and/or power/energy-efficiency point of view when interference is treated as noise [6,7,8,9,10,11,12,13,14,15,16,17,18,19,20,21]. It was shown in Reference

[6], for example, that the secondary user (SU) in an underlay cognitive radio (UCR) system may increase its rate by transmitting IGS codewords while satisfying the rate requirement of the primary user (PU). The superiority of IGS over PGS schemes for the Z-IC was also established in several works [7,8,9] and similar performance improvements have also been shown for the 2-user IC in References [10,11,12]. The benefits of IGS from Energy efficiency perspective were also studied in different scenarios in References [15,16].

All the aforementioned works [3,4,5,6,7,8,9,10,11,12,13,14,15,16,17] assume perfect and instantaneous channel state information at the transmitter side (CSIT) and they additionally involve some optimization of the parameters of the transmitted improper signal (e.g., circularity coefficient and transmit power). However, perfect CSIT is a restrictive assumption in practice.

The number of works studying the performance of IGS in scenarios with either imperfect or statistical CSIT in fading channels is much smaller. For instance, in Reference [18], we proposed a robust IGS for a 2-user IC with imperfect instantaneous CSIT and showed that IGS can enlarge the achievable rate region and outperform PGS. Reference [19] considers an underlay cognitive radio (UCR) network in which the secondary user has only statistical CSIT and optimizes its transmit power and degree of impropriety to maintain a given primary user (PU) outage probability. A more challenging UCR scenario where the PU is assumed to work using full-duplex mode is considered in Reference [20] for Rayleigh fading channels. Nevertheless, References [19,20] only consider block fading models and the probability of outage as the figure of merit to optimize. In Reference [21], the authors studied a full-duplex relay system over Nakagami-*m* fading channels. They derived the outage probability for a given target rate as well as the ergodic rate of the system and showed the benefits of IGS.

Contribution

In this work, we study the achievable ergodic rates of IGS schemes over the 2-user fading IC when interference is treated as noise. To the best of our knowledge, it is the first work that analyzes the ergodic rate of improper signaling in the 2-user IC. For the case where at most one user employs IGS (denoted as PGS/IGS scheme), we derive closed-form expressions for the ergodic rates for both users. These rates can be attained using fixed-rate codebooks and no optimization is involved. For the more general IGS/IGS scheme where both users can employ improper signaling no closed-form expressions for the rates exist and hence we compute numerically the optimal transmission parameters and the boundary of the rate region. In our model, the users have perfect and instantaneous CSI at the receiver side (CSIR) but they only have access to statistical CSIT. In other words, the transmitters know the parameters defining the Rayleigh fading distributions of the direct and interfering links. Note that the perfect instantaneous CSIR can be achieved using, for example, training sequences [18,40]. Under these assumptions, we derive the achievable ergodic rates when the users transmit proper and improper Gaussian codewords.

The main contributions of this paper are as follows:

- We derive closed-form expressions for the achievable ergodic rate of the PGS and IGS users. We also characterize the achievable ergodic rate region of the mixed PGS/IGS scheme by allowing power control.
- Our proposed scheme incurs no additional overhead at the transmitter side when both users transmit at maximum power. Moreover, in order to operate at a specific Pareto-optimal point, the transmitters (or a central entity implementing the power control module) should be informed about the average gain of the direct and interference links, which are scalar values and can be easily sent to the transmitters. Hence, the PGS/IGS scheme is simple to implement in practice.
- Through numerical examples, we show that our proposed PGS/IGS scheme substantially outperforms the PGS scheme in moderate and strong interference regimes. Moreover, we analyze the maximum sum-rate point for the symmetric 2-user IC when both users transmit at maximum power. Under this condition, the sum-rate point is attained either with PGS or with PGS/MIGS (maximally IGS); that is, one user transmits proper signals while the other user employs maximally improper signaling. Then, the maximum sum-rate point does not require any optimization and/or power control.
- We also compare our results with the more general scheme where both users can employ IGS and their transmission parameters are numerically optimized. Our numerical results show that the union

of the PGS and PGS/MIGS schemes with time sharing attains almost the whole Pareto-optimal points in the rate region and little is gained by allowing both users to employ IGS.

Paper Outline

The rest of the paper is organized as follows. In Section 2, we define the system model parameters and formulate the problem. In Section 3 we propose our scheme and derive the ergodic rates of the PGS/IGS scheme. Section 4 presents some numerical examples and the main conclusions are summarized in Section 5.

SIGNAL MODEL AND PROBLEM STATEMENT

Notation and Preliminaries

We denote the probability density function (PDF) of a zero-mean complex Gaussian distribution with variance $p_x = \mathbb{E}\{|X|^2\}$ and complementary variance $q_x = \mathbb{E}\{X^2\}$ as $X \sim \mathcal{CN}(0, p_x, q_x)$. We define the circularity coefficient of X as $\kappa_x = \frac{|q_x|}{p_x}$, where $0 \leq \kappa_x \leq 1$ [22,23]. The variable X is proper if $\kappa_x = 0$. Otherwise, X is improper (a.k.a. asymmetric or non-circular). Moreover, we call X maximally improper if $\kappa_x = 1$.

A random variable X that is Gamma-distributed with shape parameter α and rate parameter β is denoted as $X \sim \Gamma(\alpha, \beta)$. Its PDF is $f_X(x) = \frac{\beta^\alpha x^{\alpha-1}}{\Gamma(\alpha)} e^{-\beta x}$, $x \geq 0$, where $\Gamma(\alpha)$ is the gamma function. If $X \sim \Gamma(1, \beta)$, then X has an exponential distribution.

Signal Model

We consider the 2-user SISO IC depicted in Figure 1. The received signal at receiver $i \in \{1,2\}$ is

$$y_i = h_{1i} x_1 + h_{2i} x_2 + n_i, \tag{1}$$

where x_i, n_i and h_{ji} for $j \in \{1,2\}$ are, respectively, the transmit signal of user i, proper Gaussian noise with variance σ^2 and the channel coefficient for the link between transmitter j and receiver i. We consider Rayleigh fading channels modeled as

$$h_{ii} \sim \mathcal{CN}(0, \sqrt{\mathrm{snr}_i}, 0), \quad (2)$$

$$h_{\bar{i}i} \sim \mathcal{CN}(0, \sqrt{\mathrm{inr}_i}, 0), \quad (3)$$

where snr_i and inr_i for $i, \bar{i} \in \{1,2\}$ and $i \neq \bar{i}$ are, respectively, the average gain of the direct and interference links of receiver *i*. Additionally, the channel gains are Gamma distributed, that is,

$$|h_{ii}|^2 \sim \Gamma(1, 1/\mathrm{snr}_i), \quad (4)$$

$$|h_{\bar{i}i}|^2 \sim \Gamma(1, 1/\mathrm{inr}_i). \quad (5)$$

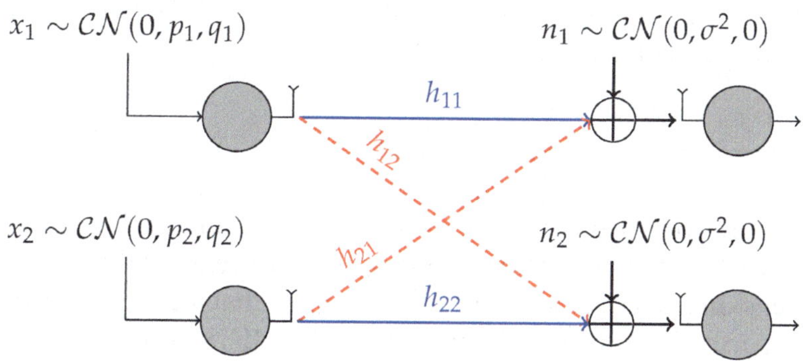

Figure 1: The 2-user interference channel (IC) with proper/improper Gaussian signaling.

We assume that the transmitters know neither the fading states on which they transmit, nor the fading states of the interfering signal. On the other hand, we assume that the receivers have perfect knowledge of the channels.

We treat interference as noise and assume that users may employ IGS. Thus, the instantaneous rate of user *i* is [18]

$$R_i = \frac{1}{2}\log_2\left(\frac{(\sigma^2 + p_1|h_{1i}|^2 + p_2|h_{2i}|^2)^2 - |q_1 h_{1i}^2 + q_2 h_{2i}^2|^2}{(\sigma^2 + p_{\bar{i}}|h_{\bar{i}i}|^2)^2 - |q_{\bar{i}} h_{\bar{i}i}^2|^2}\right), \quad (6)$$

where p_i and q_i are the transmission power and complementary variance of user *i*, respectively. Hereafter, we consider $\sigma^2=1$ without loss of generality. Moreover, the power budget of user *i* is P_i.

We first consider the scenario, where at most one user may employ IGS. For notational simplicity in this section, we consider user 1 as the PGS user without loss of generality. Obviously, there is no difference between the users and either user can be the PGS user. The instantaneous rate of the PGS user is

$$R_1^{\text{PGS}} = \frac{1}{2}\log_2\left(1+\frac{X_1}{1+(1+\kappa)Y_1}\right)+\frac{1}{2}\log_2\left(1+\frac{X_1}{1+(1-\kappa)Y_1}\right), \tag{7}$$

where $X_i = p_i|h_{ii}|^2 \sim \Gamma(1,1/(p_i\text{snr}_i))$ and $Y_i = p_{\bar{\imath}}|h_{\bar{\imath}i}|^2 \sim \Gamma(1,1/(p_{\bar{\imath}}\text{inr}_i))$ are independent Gamma distributed random variables and κ is the circularity coefficient of the IGS user. The instantaneous rate of the IGS user is

$$R_2^{\text{IGS}} = \frac{1}{2}\log_2\left(1+\frac{(1+\kappa)X_2}{1+Y_2}\right)+\frac{1}{2}\log_2\left(1+\frac{(1-\kappa)X_2}{1+Y_2}\right). \tag{8}$$

A special case of this scheme is MIGS, that is, κ=1. In this case, the rate of the PGS user simplifies to

$$R_1^{\text{PGS}} = \underbrace{\frac{1}{2}\log_2(1+X_1)}_{\text{interference free}}+\underbrace{\frac{1}{2}\log_2\left(1+\frac{X_1}{1+2Y_1}\right)}_{\text{with interference}}. \tag{9}$$

Since the MIGS user transmits only on one dimension, there is an interference-free dimension for the PGS user. We can exploit this interference-free dimension even without CSIT due to the fact that the receivers know the channels perfectly. Additionally, the rate of the MIGS user is

$$R_2^{\text{MIGS}} = \frac{1}{2}\log_2\left(1+\frac{2X_2}{1+Y_2}\right). \tag{10}$$

As we observe in Equation (10), the PGS user causes less interference to the MIGS user because both the PGS user and the noise divide their power between the two dimensions. The downside is that the MIGS user loses some rate because only one signal dimension is exploited for transmission. However, the MIGS user can compensate for this rate loss by increasing its transmission power if the power budget has not been fully used and the overall effect can be beneficial in terms of sum-rate. As indicated, the PGS user has an interference-free dimension, which ensures a minimum instantaneous achievable rate of

$$R_1^{\text{PGS}} > \frac{1}{2}\log_2(1+X_1). \tag{11}$$

Problem Statement

In this paper, we derive the achievable ergodic rates for the PGS/IGS scheme under fast fading channels (i.e., when the codewords span a large number of fading states). The achievable ergodic rate is defined as the expected achievable rate $\overline{R} = \mathbb{E}\{R\}$. In order to derive the achievable ergodic rate region, we employ the Pareto boundary, which is defined in Definition 1.

Definition 1

([7]). *The ergodic rate pair* $(\overline{R}_1, \overline{R}_2)$ *is called Pareto-optimal if* $(\overline{R}_1', \overline{R}_2)$ *and* $(\overline{R}_1, \overline{R}_2')$, *with* $\overline{R}_1' > \overline{R}_1$ *and* $\overline{R}_2' > \overline{R}_2$, *are not achievable.*

The whole achievable ergodic rate region can be obtained by varying the powers and complementary variances of the transmit symbols. In References [12,41] it is proven that the Pareto-optimal rates for the 2-user IC when interference is treated as noise are obtained when at least one user transmits at maximum power. This result, which was originally proved for the case where the transmitters have perfect CSIT, is also valid for our ergodic fast fading channel model.

PGS/IGS SCHEME

In this section, our goal is to derive the achievable pair of ergodic rates $\left(\overline{R}_1^{\mathrm{PGS}}, \overline{R}_2^{\mathrm{IGS}}\right)$ obtained by taking the mathematical expectation of R_1^{PGS} in Equation (7) and R_2^{IGS} in Equation (8). We first present Lemma 1 in order to derive the ergodic rate of the users. Then we present the achievable ergodic rates in Theorem 1.

Lemma 1.

Let us define the random variable $Z = \frac{\beta X}{1+\alpha Y}$, *where* $\alpha \geq 0$ *and* $\beta \geq 0$ *are given constants and* $X \sim \Gamma(1, \frac{1}{snr})$ *and* $Y \sim \Gamma(1, \frac{1}{inr})$ *are independent Gamma distributed random variables. The cumulative density function (CDF) of Z is*

$$F_Z(z) = 1 - \frac{e^{-\frac{z}{\beta snr}}}{1 + \frac{\alpha inr}{\beta snr} z}. \tag{12}$$

The expectation of Z is

$$\mathbb{E}\{Z\} = \begin{cases} \beta snr & if \quad \alpha = 0, \\ \left(\frac{\beta snr}{\alpha inr}\right) e^{\frac{1}{\beta snr}} E_1\left(\frac{1}{\beta snr}\right) & if \quad \alpha > 0. \end{cases} \tag{13}$$

Additionally, the expectation of R=log(1+Z) *is*

$$f(\beta\mathrm{snr}, \alpha\mathrm{inr}) \triangleq \overline{R} = \begin{cases} e^{\frac{1}{\beta\mathrm{snr}}} E_1\left(\frac{1}{\beta\mathrm{snr}}\right) & if \quad \alpha = 0, \\ \frac{\beta\mathrm{snr}\left(e^{\frac{1}{\beta\mathrm{snr}}} E_1\left(\frac{1}{\beta\mathrm{snr}}\right) - e^{\frac{1}{\alpha\mathrm{inr}}} E_1\left(\frac{1}{\alpha\mathrm{inr}}\right)\right)}{\beta\mathrm{snr}-\alpha\mathrm{inr}}, & if \quad \beta\mathrm{snr} \neq \alpha\mathrm{inr} > 0 \\ e^{\frac{1}{\beta\mathrm{snr}}} E_2\left(\frac{1}{\beta\mathrm{snr}}\right), & if \quad \beta\mathrm{snr} = \alpha\mathrm{inr} > 0 \end{cases} \tag{14}$$

where the $E_n(x)$ *function is defined by the integral (for* $n \geq 1$ *and* $x>0$*)*

$$E_n(x) = \int_1^{\infty} \frac{e^{-xt}}{t^n} dt. \tag{15}$$

Proof of Lemma 1.

Please refer to **Appendix A.**

Note that when $\beta=0$, the variable Z defined in Lemma 1 is $Z=0$. To include this case, we define $f(0,\alpha inr) \triangleq 0$. Moreover, notice that $E_n(x)$ can be written in terms of the upper incomplete gamma function as $En(x)=x^{n-1}\Gamma(1-n,x)$, where

$$\Gamma(s,x) = \int_x^{\infty} t^{s-1} e^{-t}. \tag{16}$$

$E_1(x)=\Gamma(0,x)$ is the exponential integral. These functions can be computed using standard software packages.

Theorem 1.

The ergodic rates of the PGS and the IGS users are, respectively,

$$\bar{R}_1^{PGS} = f(p_1\mathrm{snr}_1, (1+\kappa)p_2\mathrm{inr}_1) + f(p_1\mathrm{snr}_1, (1-\kappa)p_2\mathrm{inr}_1), \tag{17}$$

$$\bar{R}_2^{IGS} = f((1+\kappa)p_2\mathrm{snr}_2, p_1\mathrm{inr}_2) + f((1-\kappa)p_2\mathrm{snr}_2, p_1\mathrm{inr}_2), \tag{18}$$

where f(x,y) *is given by Equation **(14)**.*

Proof of Theorem 1.

The ergodic rate of users are given by taking expectations of Equations (7) and (8). Hence, by applying the results of Lemma 1 and substituting the corresponding α and β, we can prove the theorem.

Remark 1.

The ergodic rate of the PGS user is strictly increasing in κ, while the ergodic rate of the IGS user is strictly decreasing in κ. This is due to the fact that the instantaneous rate of the PGS (IGS) user is strictly increasing (decreasing) in κ.

We can derive the achievable ergodic rate region by varying the transmission power of the two users. As indicated in Section 2.3, every point on the boundary of the achievable ergodic rate region is obtained when at least one user transmits with maximum power [12,41]. Thus, the whole Pareto optimal region for the PGS/IGS scheme can be derived as

$$\mathcal{S}^{IGS} = \bigcup_{\substack{0\le\kappa_2\le 1\\ 0\le p_2\le P_2}} \left(\overline{R}_1^{\mathrm{PGS}}(P_1,p_2,\kappa_2),\overline{R}_2^{\mathrm{IGS}}(P_1,p_2,\kappa_2)\right) \bigcup_{\substack{0\le\kappa_2\le 1\\ 0\le p_1\le P_1}} \left(\overline{R}_1^{\mathrm{PGS}}(p_1,P_2,\kappa_2),\overline{R}_2^{\mathrm{IGS}}(p_1,P_2,\kappa_2)\right)$$
$$\bigcup_{\substack{0\le\kappa_1\le 1\\ 0\le p_2\le P_2}} \left(\overline{R}_1^{\mathrm{IGS}}(P_1,p_2,\kappa_1),\overline{R}_2^{\mathrm{PGS}}(P_1,p_2,\kappa_1)\right) \bigcup_{\substack{0\le\kappa_1\le 1\\ 0\le p_1\le P_1}} \left(\overline{R}_1^{\mathrm{IGS}}(p_1,P_2,\kappa_1),\overline{R}_2^{\mathrm{PGS}}(p_1,P_2,\kappa_1)\right). \quad (19)$$

Note that the achievable ergodic rate region in Equation (19) is the union of the achievable ergodic rate region of two transmission strategies:

- User 1 employs PGS, while user 2 may employ IGS,
- User 2 employs PGS, while user 1 may employ IGS.

NUMERICAL EXAMPLES

We consider a 2-user fading IC where the channel coefficients of the direct links are drawn from $h_{ii} \sim \mathcal{CN}(0,\sqrt{\mathrm{snr}},0)$ and the channel coefficients of the interfering links are drawn from $h_{i\bar{i}} \sim \mathcal{CN}(0,\sqrt{\mathrm{inr}},0)$. The four channel coefficients are independent random variables. In this way, the performance of the proposed schemes can be analyzed in terms of the average snr and inr, which are the same at both receivers. Moreover, we assume $P_1=P_2=1$ and $\sigma^2=1$. Indeed, we reflect the effect of the transmission power as well as the noise power by the gain of the direct and interference links, that is, snr and inr. Since the fading direct and interference channels are symmetric, the two users send information at the same rate.

When at most one user employs IGS, the instantaneous rates are independent of the phases of the channels. Additionally, it allows us to find closed-form expressions of the ergodic rates. However, the question remains: *How far is the PGS/IGS scheme from the Pareto-optimal design of a scheme that would allow both users to use IGS on the 2-user IC?*

To answer the question, we compare the performance of our scheme with an exhaustive search over all design parameters (κ_1,κ_2,p_1,p_2) when both users may employ IGS. Unfortunately, to the best of our knowledge, there is no closed-form solution for the ergodic rates when both users employ IGS. Thus, we estimate the ergodic rate of the 2-user IC when both users employ IGS by Monte Carlo simulations. That is, we generate a large number (105) of channels and compute the instantaneous rate achieved by the transmission parameters for each channel realization. Finally, we estimate the ergodic rate by averaging the instantaneous rates. Once the ergodic rate pairs have been estimated for many different transmission parameters, the Pareto boundary for the IGS/IGS scheme can readily be estimated.

The considered schemes are denoted as follows.

- **PGS**: The PGS scheme with κ_1=κ_2=0.
- **MIGS**: The PGS/MIGS scheme with κ=1.
- **E-IGS**: The exhaustive search for IGS when both users may employ IGS.
- **O-P/IGS**: The PGS/IGS scheme with the optimal κ.
- **MIGS-U$_i$**: The PGS/MIGS scheme when the MIGS user is user *i*.
- **MIGS-TS**: The PGS/MIGS scheme with time sharing over the points with maximum power for both users.

In Figure 2, we depict the achievable ergodic rate region of our proposed schemes for snr=10 dB. As can be observed, the union of the ergodic rate regions achievable by the PGS and MIGS schemes is very close to the optimal achievable ergodic rate region. If we employ time sharing over the points with maximum power for both users in the union of the PGS and PGS/MIGS schemes (indicated by circles in Figure 2), we attain almost the entire Pareto-optimal region of the 2-user IC when both users are allowed to employ IGS.

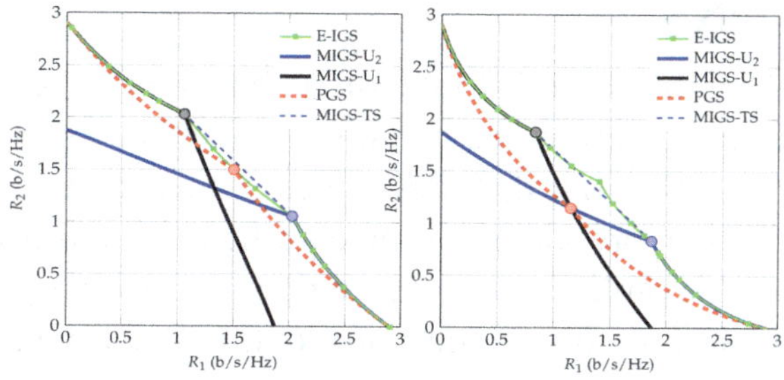

(a) inr = 7 dB. (b) inr = 10 dB.

Figure 2: Proper Gaussian signaling (PGS) and improper Gaussian signaling (IGS) rate regions for a symmetric 2-user IC with snr=10 dB.

In Figure 3, we show the sum-ergodic rate as a function of the inr when both users transmit at maximum power. Note that, in this case, the sum-ergodic rate can be achieved by red points for the PGS scheme and by either black or blue points for the PGS/MIGS scheme in Figure 2. As can be observed in Figure 3, the PGS/MIGS scheme significantly outperforms the PGS scheme when the interference level is high. Interestingly, the sum-rate can be achieved by either κ=0 or κ=1 and no optimization over κ is required for maximizing the sum-rate. Note that, for symmetric channels, both users can achieve the same ergodic rate in the PGS/MIGS scheme by letting each user be the MIGS user 50% of the time. Hence, the MIGS scheme with switching between proper and maximally improper users can substantially improve the performance of the 2-user IC. This scheme requires as little cooperation as possible between users and further it requires no optimization or power control at the transmitter side.

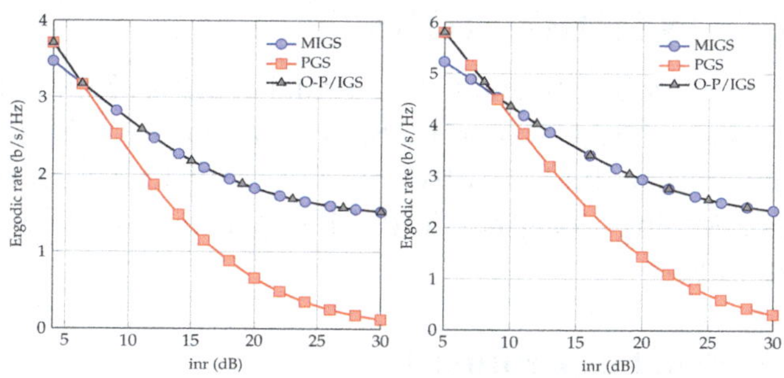

(a) snr = 10 dB. (b) snr = 15 dB.

Figure 3: Sum ergodic rate of the schemes as a function of inr.

CONCLUSIONS

This paper studied the ergodic rate of a PGS/IGS scheme in a 2-user Rayleigh IC treating interference as noise. Under the assumption that the users have perfect CSIR but they have access only to statistical CSIT, we derived closed-form expressions for the ergodic rates and characterized the boundary of the achievable ergodic rate region. We also proposed a practical IGS scheme which does not require any optimization and power control at the transmitter side. In this scheme, both users transmit with maximum power. One user employs MIGS while the other user employs PGS. By numerical examples, we showed that IGS can substantially increase the sum-rate and enlarge the ergodic rate region of the 2-user Rayleigh SISO-IC under strong interference.

Future Studies

This paper, to the best of our knowledge, is the first work on IGS in IC without instantaneous CSIT, which can open a path for further studies on IGS with more practical assumptions when instantaneous CSIT is not available. It has been shown that IGS can improve the performance of multi-user (more than two users) and/or multiple antenna systems in the presence of perfect instantaneous CSIT [3,5,9,17]. However, perfect and instantaneous CSIT might not be available in practical scenarios. Thus, the following question arises, which has to be answered in future works. *Does IGS provide some gains, in multi-user (large scale scenarios) and/or multiple antenna systems when instantaneous CSIT is not available?*

AUTHOR CONTRIBUTIONS

Conceptualization, M.S., I.S. and C.L.; Formal analysis, M.S., I.S. and C.L.; Funding acquisition, I.S., C.L. and P.J.S.; Methodology, M.S., I.S. and C.L.; Project administration, P.J.S.; Software, M.S. and I.S.; Supervision, I.S., C.L. and P.J.S.; Writing—original draft, M.S.; Writing—review and editing, I.S., C.L. and P.J.S.

Appendix A. Proof of Lemma 1

By definition, the CDF of Z is

$$F_Z(z) = Pr(Z \leq z) = 1 - Pr(\beta X \geq z(1+\alpha Y)) = \int_0^\infty \frac{e^{-y/\text{inr}}}{\text{inr}} \int_{\frac{z(1+\alpha y)}{\beta}}^\infty \frac{e^{-x/\text{snr}}}{\text{snr}} dx dy = 1 - \frac{e^{-\frac{z}{\beta \text{snr}}}}{1 + \frac{\alpha \text{inr}}{\beta \text{snr}} z}. \quad \text{(A1)}$$

Furthermore, since Z is a positive random variable, its expectation can be obtained as

$$\mathbb{E}\{Z\} = \int_0^\infty (1 - F_Z(z))\, dz = \begin{cases} \beta\text{snr} & \text{if } \alpha = 0, \\ \left(\frac{\beta\text{snr}}{\alpha\text{inr}}\right) e^{\frac{1}{\beta\text{snr}}} E_1\left(\frac{1}{\beta\text{snr}}\right) & \text{if } \alpha > 0. \end{cases} \tag{A2}$$

It can be easily verified that $\bar{R} = e^{\frac{1}{\beta\text{snr}}} E_1\left(\frac{1}{\beta\text{snr}}\right)$ when $\alpha=0$. Hence, in the following, we consider $\alpha>0$. In this case, the CDF of $R=\log(1+Z)$ is

$$F_R(r) = 1 - e^{\frac{1}{\beta\text{snr}}} e^{-\frac{e^r}{\beta\text{snr}}} \left(\frac{1}{1 - \frac{\alpha\text{inr}}{\beta\text{snr}} + \frac{\alpha\text{inr}}{\beta\text{snr}} e^r}\right), \quad r \geq 0. \tag{A3}$$

Since R is a positive random variable, its expectation may be found from its CDF as

$$\mathbb{E}\{R\} = \int_0^\infty (1 - F_R(r))\, dr. \tag{A4}$$

Making the change of variables $e^r=1+z$ in Equation (A4), $\bar{R}$ is given by the integral

$$\mathbb{E}\{R\} = \int_0^\infty \frac{e^{-\frac{z}{\beta\text{snr}}}}{1 + \left(\frac{\alpha\text{inr}}{\beta\text{snr}} + 1\right) z + \left(\frac{\alpha\text{inr}}{\beta\text{snr}}\right) z^2} dz. \tag{A5}$$

If $\beta\text{snr} \neq \alpha\text{inr}$, the integrand in Equation (A5) has two simple poles at $z=-1$ and $z=-\beta\text{snr}/\alpha\text{inr}$. Decomposing the integrand into simple fractions we arrive, after some change of variables, at

$$\bar{R} = \left(\frac{\beta\text{snr}}{\beta\text{snr} - \alpha\text{inr}}\right) \left[e^{\frac{1}{\beta\text{snr}}} E_1\left(\frac{1}{\beta\text{snr}}\right) - e^{\frac{1}{\alpha\text{inr}}} E_1\left(\frac{1}{\alpha\text{inr}}\right) \right]. \tag{A6}$$

On the other hand, if $\beta\text{snr}=\alpha\text{inr}$, the integrand in Equation (A5) has a double pole at $z=-1$. Then, Equation (A5) becomes

$$\bar{R} = \int_0^\infty \frac{e^{-\frac{z}{\beta\text{snr}}}}{1+z^2} dz = e^{\frac{1}{\beta\text{snr}}} \int_{\frac{1}{\beta\text{snr}}}^\infty \frac{e^{-t}}{t^2} dt = e^{\frac{1}{\beta\text{snr}}} E_2\left(\frac{1}{\beta\text{snr}}\right), \tag{A7}$$

thus proving Lemma 1.

REFERENCES

1. Andrews, J.G.; Buzzi, S.; Choi, W.; Hanly, S.V.; Lozano, A.; Soong, A.C.; Zhang, J.C. What will 5G be? *IEEE J. Sel. Areas Commun.* 2014, *32*, 1065–1082.
2. Cover, T.M.; Thomas, J.A. *Elements of Information Theory*; John Wiley & Sons: Hoboken, NJ, USA, 2012.
3. Cadambe, V.R.; Jafar, S.A.; Wang, C. Interference alignment with asymmetric complex signaling–Settling the Høst-Madsen-Nosratinia conjecture. *IEEE Trans. Inf. Theory* 2010, *56*, 4552–4565.
4. Lameiro, C.; Santamaria, I. Degrees-of-Freedom for the 4-User SISO Interference Channel with Improper Signaling. In Proceedings of the 2013 IEEE International Conference on Communications (ICC), Budapest, Hungary, 9–13 June 2013.
5. Yang, L.; Zhang, W. Interference alignment with asymmetric complex signaling on MIMO X channels. *IEEE Trans. Commun.* 2014, *62*, 3560–3570.
6. Lameiro, C.; Santamaria, I.; Schreier, P.J. Benefits of improper signaling for underlay cognitive radio. *IEEE Wireless Commun. Lett.* 2015, *4*, 22–25.
7. Lameiro, C.; Santamaria, I.; Schreier, P.J. Rate region boundary of the SISO Z-interference channel with improper signaling. *IEEE Trans. Commun.* 2016, *65*, 1022–1034.
8. Kurniawan, E.; Sun, S. Improper Gaussian signaling scheme for the Z-interference channel. *IEEE Trans. Wirel. Commun.* 2015, *14*, 3912–3923.
9. Lagen, S.; Agustin, A.; Vidal, J. On the superiority of improper Gaussian signaling in wireless interference MIMO scenarios. *IEEE Trans. Commun.* 2016, *64*, 3350–3368.
10. Ho, Z.K.; Jorswieck, E. Improper Gaussian signaling on the two-user SISO interference channel. *IEEE Trans. Wirel. Commun.* 2012, *11*, 3194–3203.
11. Zeng, Y.; Yetis, C.M.; Gunawan, E.; Guan, Y.L.; Zhang, R. Transmit optimization with improper Gaussian signaling for interference channels. *IEEE Trans. Signal Process.* 2013, *61*, 2899–2913.
12. Park, H.; Park, S.-H.; Kim, J.-S.; Lee, I. SINR balancing techniques in coordinated multi-cell downlink systems. *IEEE Trans. Wirel. Commun.* 2013, *12*, 626–635.

13. Kariminezhad, A.; Sezgin, A. Heterogeneous multi-tier networks: Improper signaling for joint rate-energy optimization. *IEEE Trans. Wirel. Commun.* 2019, *18*, 680–694.
14. Javed, S.; Amin, O.; Shihada, B.; Alouini, M.-S. Improper Gaussian signaling for hardware impaired multihop full-duplex relaying systems. *IEEE Trans. Commun.* 2019, *67*, 1858–1871.
15. Soleymani, M.; Lameiro, C.; Schreier, P.J.; Santamaria, I. Energy-efficient design for underlay cognitive radio using improper signaling. In Proceedings of the ICASSP 2019—2019 IEEE International Conference on Acoustics, Speech and Signal Processing (ICASSP), Brighton, UK, 12–17 May 2019; pp. 4769–4773.
16. Soleymani, M.; Lameiro, C.; Santamaria, I.; Schreier, P.J. Improper signaling for SISO two-user interference channels with additive asymmetric hardware distortion. *IEEE Trans. Commun.* 2019.
17. Soleymani, M.; Lameiro, C.; Santamaria, I.; Schreier, P.J. Energy-Efficient Improper Signaling for K-User Interference Channels. In Proceedings of the IEEE European Signal Processing Coference (EUSIPCO), A Coruña, Spain, 2–6 September 2019.
18. Soleymani, M.; Lameiro, C.; Santamaria, I.; Schreier, P.J. Robust improper signaling for two-user SISO interference channels. *IEEE Trans. Commun.* 2019, *67*, 4709–4723.
19. Amin, O.; Abediseid, W.; Alouini, M.-S. Underlay cognitive radio systems with improper Gaussian signaling: Outage performance analysis. *IEEE Trans. Wirel. Commun.* 2016, *15*, 4875–4887.
20. Gaafar, M.; Amin, O.; Abediseid, W.; Alouini, M.-S. Underlay spectrum sharing techniques with in-band full-duplex systems using improper Gaussian signaling. *IEEE Trans. Wirel. Commun.* 2017, *16*, 235–249.
21. Gaafar, M.; Khafagy, M.G.; Amin, O.; Schaefer, R.F.; Alouini, M.-S. Full-duplex relaying with improper Gaussian signaling over Nakagami-*m* fading channels. *IEEE Trans. Commun.* 2018, *66*, 64–78.
22. Adali, T.; Schreier, P.J.; Scharf, L.L. Complex-valued signal processing: The proper way to deal with impropriety. *IEEE Trans. Signal Process.* 2011, *59*, 5101–5125.
23. Schreier, P.J.; Scharf, L.L. *Statistical Signal Processing of Complex-Valued Data: The Theory of Improper and Noncircular Signals*; Cambridge University Press: Cambridge, UK, 2010.

24. Dysto, A.; Tuninetti, D.; Devroye, N. Interference as noise: Friend or foe? *IEEE Trans. Inf. Theory* 2016, *62*, 3561–3596.
25. Motahari, A.S.; Khandani, A.K. Capacity bounds for the Gaussian interference channel. *IEEE Trans. Inf. Theory* 2009, *55*, 620–643.
26. Haghi, A.; Mohammadizadeh, N.; Khandani, A.K. Delay in Cooperative Communications: Achieving Higher Multiplexing Gain in Gaussian Interference Channels with Full-Duplex Transmitters. *IEEE Trans. Inf. Theory* 2019.
27. Carleial, A. A case where interference does not reduce capacity. *IEEE Trans. Inf. Theory* 1975, *21*, 569–570.
28. Sato, H. The capacity of the Gaussian interference channel under strong interference. *IEEE Trans. Inf. Theory* 1981, *27*, 786–788.
29. Fasoulakis, M.; Tsiropoulou, E.E.; Papavassiliou, S. Satisfy instead of maximize: Improving operation efficiency in wireless communication networks. *Comput. Netw.* 2019, *159*, 135–146.
30. Fasoulakis, M.; Traganitis, A.; Ephremides, A. The Gaussian interference channel revisited as a non-cooperative game with transmission cost. In Proceedings of the 2017 15th International Symposium on Modeling and Optimization in Mobile, Ad Hoc, and Wireless Networks (WiOpt), Paris, France, 15–19 May 2017.
31. Geng, C.; Naderializadeh, N.; Avestimehr, A.S.; Jafar, S.A. On the optimality of treating interference as noise. *IEEE Trans. Inf. Theory* 2015, *61*, 1753–1767.
32. Tsinos, C.G.; Berberidis, K. Blind Opportunistic Interference Alignment in MIMO Cognitive Radio Systems. *IEEE J. Emerg. Sel. Top. Circuits Syst.* 2013, *3*, 626–639.
33. Cadambe, V.R.; Jafar, S.A. Interference Alignment and Degrees of Freedom of the *K*-User Interference Channel. *IEEE Trans. Inf. Theory* 2008, *54*, 3425–3441.
34. Tsinos, C.G.; Berberidis, K. Blind opportunistic interference alignment in cognitive radio systems. In Proceedings of the 2013 IEEE International Conference on Communications (ICC), Budapest, Hungary, 9–13 June 2013; pp. 4880–4884.
35. Perlaza, S.M.; Fawaz, N.; Lasaulce, S.; Debbah, M. From Spectrum Pooling to Space Pooling: Opportunistic Interference Alignment in MIMO Cognitive Networks. *IEEE Trans. Signal Process.* 2010, *58*, 3728–3741.

36. Maddah-Ali, M.; Motahari, A.; Khandani, A. Communication over MIMO X channels: Interference alignment, decomposition, performance analysis. *IEEE Trans. Inf. Theory* 2008, *54*, 3457–3470.
37. Ding, Z.; Liu, Y.; Choi, J.; Sun, Q.; Elkashlan, M.; Chih-Lin, I.; Poor, V. Application of non-orthogonal multiple access in LTE and 5G networks. *IEEE Commun. Mag.* 2017, *55*, 185–191.
38. Ding, Z.; Lei, X.; Karagiannidis, G.K.; Schober, R.; Yuan, J.; Bhargava, V.K. A survey on non-orthogonal multiple access for 5G networks: Research challenges and future trends. *IEEE J. Sel. Areas Commun.* 2017, *35*, 2181–2195.
39. Liu, Y.; Qin, Z.; Elkashlan, M.; Ding, Z.; Nallanathan, A.; Hanzo, L. Nonorthogonal multiple access for 5G and beyond. *Proc. IEEE* 2017, *105*, 2347–2381.
40. Shen, H.; Wang, J.; Levy, B.C.; Zhao, C. Robust optimization for amplify-and-forward MIMO relaying from a worst-case perspective. *IEEE Trans. Signal Process.* 2013, *61*, 5458–5471.
41. Charafeddine, M.A.; Sezgin, A.; Han, Z.; Paulraj, A. Achievable and crystallized rate regions of the interference channel with interference as noise. *IEEE Trans. Wirel. Commun.* 2012, *11*, 1100–1111.

14

Statistical Modeling, Simulation, and Experimental Verification of Wideband Indoor Mobile Radio Channels

Yuanyuan Ma[1] , Bjørn Olav Hogstad[2] , Matthias Pätzold[3] , and Pedro M. Crespo[4]

[1]College of Information Science and Engineering, Shandong Agricultural University, Shandong 271018, China
[2]Department of Mathematical Science, Norwegian University of Science and Technology, Teknologiveien 22, 2815 Gjøvik, Norway
[3]Faculty of Engineering and Science, University of Agder, 4898 Grimstad, Norway
[4]CEIT and Tecnun, University of Navarra, 20018 San Sebastián, Spain

ABSTRACT

This paper focuses on the modeling, simulation, and experimental verification of wideband single-input single-output (SISO) mobile fading channels for indoor propagation environments. The indoor reference channel model is derived from a geometrical rectangle scattering model, which consists of an infinite number of scatterers. It is assumed that the

Citation: Yuanyuan Ma, Bjørn Olav Hogstad, Matthias Pätzold, and Pedro M. Crespo. "Statistical Modeling, Simulation, and Experimental Verification of Wideband Indoor Mobile Radio Channels" Wireless Communications and Mobile Computing Volume 2018, Article ID 8271765, 13 pages https://doi.org/10.1155/2018/8271765

scatterers are exponentially distributed over the two-dimensional (2D) horizontal plane of a rectangular room. Analytical expressions are derived for the probability density function (PDF) of the angle of arrival (AOA), the PDF of the propagation path length, the power delay profile (PDP), and the frequency correlation function (FCF). An efficient sum-of-cisoids (SOC) channel simulator is derived from the nonrealizable reference model by employing the SOC principle. It is shown that the SOC channel simulator approximates closely the reference model with respect to the FCF. The SOC channel simulator enables the performance evaluation of wideband indoor wireless communication systems with reduced realization expenditure. Moreover, the rationality and usefulness of the derived indoor channel model is confirmed by various measurements at 2.4, 5, and 60 GHz.

INTRODUCTION

Indoor wireless communications has attracted considerable attention in recent years due to a broad set of emerging indoor services offered by personal communication systems [1, 2], wireless local area networks [3, 4], and wireless private branch exchange networks [5]. The interest in indoor wireless communications is growing especially when optical wireless communication techniques [6, 7] are applied to indoor deployment scenarios. Numerous researches on modulation techniques [8], multiple access techniques [9], and propagation modeling [10] have been carried out for indoor optical wireless communications. In order to design an efficient indoor wireless communication system and to predict accurately its system performance, a prerequisite is to develop an appropriate indoor channel model that describes realistically the underlying propagation conditions.

To improve the indoor channel characterization and modeling, numerous measurement campaigns have been conducted in a variety of indoor environments, such as offices, corridors, buildings, and factory halls. Measurement results have been reported in the literature for various frequency bands, including the frequency bands at 900 MHz [11–13], 1.5 GHz [14, 15], 2.4 GHz [16], 4–5.5 GHz [17–19], 17-18 GHz [18, 20], and 60 GHz [21–24]. Based on measurement results, several empirical statistical channel models [13, 24–27] have been developed for the simulation and performance analysis of various indoor communication systems. The advantage of measurement-based channel models is that they are characterizing the fading behavior utterly realistic. By changing the model parameters, the developed statistical channel models can be employed to simulate indoor propagation channels for a broad range of different scenarios. However, to determine appropriate parameters for different propagation scenarios, a large number of measurement data are required, which is time- and material-consuming and leads thus to high development costs. Alternatively, approaches based on ray-tracing techniques [28–31] have been developed to enable the simulation of indoor propagation channels. Ray-tracing channel models [29, 32–34] can efficiently capture the fading behavior of specified indoor environments. Many developed ray-tracing channel models have shown good agreement with measured channels [35–37]. However, ray-tracing models greatly depend on the physical layouts and materials through which electromagnetic waves propagate, for example, walls, floors, and ceilings. In addition, the main drawback of ray-tracing models comes from their computational costs, which are widely determined by the size and complexity of the geographic database as well as the strategy used for the ray search. Therefore, a trade-off has to be found between the prediction accuracy and the simulation efficiency when modeling indoor propagation channels using ray-tracing techniques.

The two main types of indoor channel models, namely, the empirical statistical channel models and the ray-tracing channel models, have their own strengths and limitations. To cope with the drawbacks mentioned above, a geometrical channel model has been proposed in [38] for indoor environments, where it is assumed that the scatterers are randomly distributed inside of a circle with a base station (BS) in its centre. More specifically, this model is based on the assumption that the distance between the BS and the scatterers follows an exponential distribution. Such an assumption imposes severe restrictions on the applicability of the model. In addition, from the physical point of view, it is questionable whether it is reasonable to represent the geometry of indoor scattering areas, such as offices and corridors, by a circle. On the contrary, a rectangle is a more appropriate geometric figure to model typical indoor propagation environments. A geometrical rectangle scattering model has been proposed in [39] to characterize narrowband indoor radio propagation channels. However, to the best of the authors' knowledge, the modeling of wideband indoor radio channels by employing an appropriate geometrical scattering model that takes into consideration the unique indoor propagation conditions is still an open problem.

Motivated by the scarcity of proper wideband geometry-based indoor channel models, such a model was developed in [40] by extending the geometrical rectangle scattering model in [39] with respect to frequency selectivity. In this paper, we improve the statistical models in [39–41] for the characterization of the scatterer distribution. We concentrate on the statistical characterization of a wideband reference channel model, which is based on the assumption that an infinite number of scatterers are randomly distributed over the two-dimensional (2D) horizontal plane of a rectangular room. To be more specific, we assume that the probability for the occurrence of a scatterer decays exponentially with the distance from the walls. This assumption includes the uniform distribution of the scatterers as a specific case. In contrast to [42–44], we apply a different procedure for studying the statistical properties of the resulting reference channel model, which requires no knowledge of the complex channel gain. We derive the analytical expressions for the probability density function (PDF) of the angle of arrival (AOA), the PDF of the propagation path length, the power delay profile (PDP), and the frequency correlation function (FCF). Moreover, we derive a wideband sum-of-cisoids (SOC) channel simulator from the reference model. It is shown that the designed wideband SOC channel simulator matches the underlying reference model accurately with respect to the FCF. The obtained SOC channel simulator enables the simulation of indoor mobile fading channels with reduced realization expenditure. At the end, we demonstrate the closeness to reality of the proposed reference channel model by comparing its mean excess delay and the root mean square (RMS) delay spread with the corresponding empirical quantities obtained from measured channels in laboratories at 2.4 GHz [16] and a conference room at 5 GHz [19] as well as corridors and offices at 60 GHz [23].

The remainder of this paper is structured as follows. In Section 2, we introduce the 2D geometrical indoor scattering model, which serves as a starting point for the derivation of the reference channel model. Section 3 analyzes the statistical characteristics of the wideband reference channel model with emphasis on the PDF of the AOA, the PDF of the propagation path length, the PDP, and the FCF. Section 4 introduces a design procedure for wideband indoor SOC channel simulators. Numerical and experimental results are presented in Section 5 to confirm the correctness of the obtained theoretical results. Section 6 validates the usefulness of the proposed indoor channel model by matching its statistical properties to those of measured (real-world) channels. Finally, Section 7 draws the conclusion.

THE GEOMETRICAL INDOOR SCATTERING MODEL

This section briefly describes the geometrical scattering model, which was first proposed in [40] to characterize indoor propagation scenarios. Although a room has three dimensions, we only consider the 2D horizontal plane in which all local scatterers as well as the BS and the mobile station (MS) are located. As shown in Figure 1, the rectangle represents the 2D horizontal plane of a room. Its length and width are denoted by A and ,B respectively. The BS is considered as the transmitter, leaving the MS to play the role as the receiver. For convenience, it is also assumed that the MS moves along the x-axis. Moreover, we consider single bounce scattering, which means that the plane waves emitted from the BS are only bounced once by scatterers before reaching the MS. The black star in Figure 1 represents a single local scatterer. In real-world environments, the number of scatterers is limited and their locations differ from one propagation scenario to another. To avoid studying the fading characteristics for a specific indoor propagation scenario characterized by a specific realization of a finite number of scatterers, we focus on a general statistical model that results from averaging over all possible propagation scenarios. Therefore, we assume that an infinite number of scatterers are randomly distributed inside the room. Such a model acts here as a nonrealizable stochastic reference model. From the reference channel model, an efficient channel simulator with low realization expenditure can be derived by making use of the SOC principle [45].

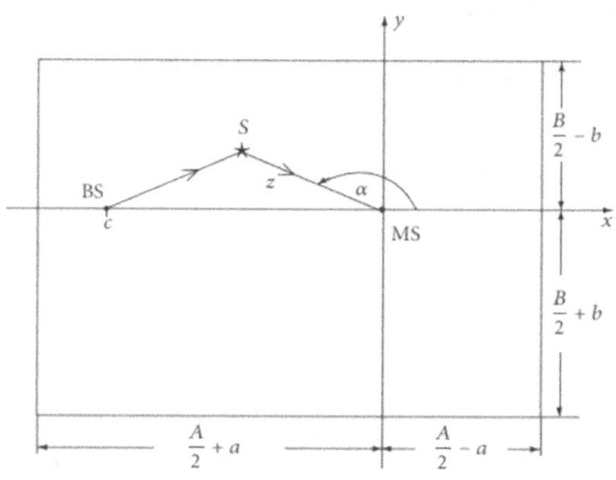

Figure 1: Geometrical indoor scattering model with local scatterers S, which are randomly distributed over the 2D horizontal plane of a room with length A and width B $(a \geq 0$ and $b \geq 0)$.

A straightforward assumption would be that the scatterers are uniformly distributed inside the room. For the modeling of indoor channels, however, it is more realistic to assume that the probability of the occurrence of a scatterer decays exponentially with the distance from the walls. This implies that it is more likely that the received scattered components are coming from the walls rather than from objects located in the room's centre. Instead of the uniform distribution, we use therefore a mixture of exponential functions to model the distribution of the scatterers inside the 2D horizontal plane of the room. The exponential functions are more general and include the uniform distribution as a special case, as explained in the next section.

STATISTICAL CHARACTERIZATION OF THE REFERENCE CHANNEL MODEL

This section studies the statistical properties of the proposed indoor wideband reference channel model. In this regard, the PDF of the AOA, the PDP, and the FCF are of special interest.

It is shown in Figure 1 that the BS is located at the position $(c, 0)$ $(c < 0)$, while the MS is located at the origin of the coordinate system. The position of the local scatterer S is described by (x,y), where x and y are independent random variables. As mentioned in Section 2, it is assumed that the scatterers are exponentially distributed within the 2D horizontal plane of the room. This implies that the random variables x and y must be exponentially distributed over the intervals $I_x = [-A/2-a, A/2-a]$ and $I_y = [-B/2-b, B/2-b]$, respectively. Thus, the PDF $p_x(x)$ of x and the PDF $p_y(y)$ of y are given by

$$p_x(x) = \begin{cases} P_1\left[e^{-w_{11}(x+A/2+a)} + e^{w_{12}(x-A/2+a)}\right], & x \in I_x, \\ 0, & \text{otherwise,} \end{cases} \tag{1a}$$

$$p_y(y) = \begin{cases} P_2\left[e^{-w_{21}(y+B/2+b)} + e^{w_{22}(y-B/2+b)}\right], & y \in I_y, \\ 0, & \text{otherwise,} \end{cases} \tag{1b}$$

where $w_{kl} \geq 0$ $(k, l = 1, 2)$ are real-valued parameters and

$$P_1 = \frac{w_{11}w_{12}}{w_{12}\left(1-e^{-w_{11}A}\right) + w_{11}\left(1-e^{-w_{12}A}\right)}, \tag{2a}$$

$$P_2 = \frac{w_{21} w_{22}}{w_{22}\left(1 - e^{-w_{21}B}\right) + w_{21}\left(1 - e^{-w_{22}B}\right)}. \tag{2b}$$

It should be mentioned that the distributions in (1a) and (1b) include the uniform distribution as a special case if $w_{kl} \to 0$ for $k, l = 1, 2$. With the rule of de l'Hospital, it can be shown that in this case P_1 and P_2 tend to $P_1 = 1/(2A)$ and $P_2 = 1/(2B)$, respectively. Owing to the assumption that the random variables x and y are statistically independent, we can express the joint PDF $p_{\mathbf{xy}}(x, y)$ of **x** and **y** as the product of the marginal PDFs $p_{\mathbf{x}}(x)$ and $p_{\mathbf{y}}(y)$; that is,

$$p_{\mathbf{xy}}(x, y) = p_{\mathbf{x}}(x) \cdot p_{\mathbf{y}}(y). \tag{3}$$

Next, we describe the position of the scatterers by using polar coordinates $(\mathbf{z}, \boldsymbol{\alpha})$, where z denotes the path length from the position of the scatterer S to the MS, and $\boldsymbol{\alpha}$ is the AOA (see Figure 1). Notice that z and $\boldsymbol{\alpha}$ are random variables. The joint PDF $p_{\mathbf{z\alpha}}(z, \alpha)$ of the position of the scatterers in polar coordinates $(\mathbf{z}, \boldsymbol{\alpha})$ can be written by using (3) as

$$\begin{aligned} p_{\mathbf{z\alpha}}(z, \alpha) &= z p_{\mathbf{xy}}(z\cos(\alpha), z\sin(\alpha)) \\ &= z p_{\mathbf{x}}(z\cos(\alpha)) \cdot p_{\mathbf{y}}(z\sin(\alpha)). \end{aligned} \tag{4}$$

Let D be the propagation path length, which is the plane wave's overall traveling distance from the BS via the scatterer S to the MS; then D is given by

$$\mathbf{D} = \mathbf{z} + \sqrt{\mathbf{z}^2 - 2c\mathbf{z}\cos(\boldsymbol{\alpha}) + c^2}. \tag{5}$$

Our next focus is on finding the joint PDF of the path length D and the AOA $\boldsymbol{\alpha}$ using (4). For this purpose, we introduce the auxiliary random variable $\boldsymbol{\theta} = \boldsymbol{\alpha}$. Solving the system of equations determined by $\mathbf{D} = \mathbf{z} + \sqrt{\mathbf{z}^2 - 2c\mathbf{z}\cos(\boldsymbol{\alpha}) + c^2}$ and $\boldsymbol{\theta} = \boldsymbol{\alpha}$ gives us the following two solutions:

$$\alpha = \theta, \tag{6a}$$

$$z = \frac{D^2 - c^2}{2(D - c\cos\alpha)}. \tag{6b}$$

By applying the fundamental theorem of transformation of random variables [46, p. 201], the joint PDF $p_{\mathbf{D}\theta}(D,\theta)$ of the path length D and the auxiliary random variable θ is given by

$$p_{\mathbf{D}\theta}(D,\theta) = |J(D,\theta)| \cdot p_{\mathbf{z}\alpha}\left(\frac{D^2 - c^2}{2(D - c\cos\theta)}, \theta\right), \tag{7}$$

Where

$$J(D,\theta) = \begin{vmatrix} \frac{\partial\alpha}{\partial D} & \frac{\partial\alpha}{\partial\theta} \\ \frac{\partial z}{\partial D} & \frac{\partial z}{\partial\theta} \end{vmatrix} = -\frac{D^2 - 2Dc\cos\theta + c^2}{2(D - c\cos\theta)^2} \tag{8}$$

denotes the Jacobian determinant of the inverse transformation. Substituting (8) into (7) and using (4), (6a), and (6b), we find the expression for the joint PDF $p_{\mathbf{D}\theta}(D,\theta)$. Owing to $\theta = \alpha$, the joint PDF $p_{\mathbf{D}\alpha}(D,\alpha)$ can be expressed directly using the joint PDF $p_{\mathbf{D}\theta}(D,\theta)$ by replacing θ in (7) by α. Thus, we obtain

$$p_{\mathbf{D}\alpha}(D,\alpha) = \frac{\left[D^2 - 2Dc\cos(\alpha) + c^2\right]\left(D^2 - c^2\right)}{4(D - c\cos(\alpha))^3} \cdot p_{\mathbf{x}}\left(\frac{\left(D^2 - c^2\right)\cos(\alpha)}{2(D - c\cos(\alpha))}\right) \cdot p_{\mathbf{y}}\left(\frac{\left(D^2 - c^2\right)\sin(\alpha)}{2(D - c\cos(\alpha))}\right), \tag{9}$$

where $p_{\mathbf{x}}(\cdot)$ and $p_{\mathbf{y}}(\cdot)$ are the densities given by (1a) and (1b), respectively.

The PDF of the AOA

In this subsection, we are concerned with the derivation of the PDF $p_{\alpha}(\alpha)$ of the AOA α. This PDF can be obtained from the joint PDF $p_{\mathbf{D}\alpha}(D,\alpha)$ in (9) by integrating over the range of D; that is,

$$p_{\alpha}(\alpha) = \int_{D\in\mathbf{I}_{\mathbf{D}}} p_{\mathbf{D}\alpha}(D,\alpha)\,dD, \tag{10}$$

$$D_{\max}(\alpha) = \begin{cases} \frac{A_-}{2\cos(\alpha)} + \sqrt{\left[\frac{A_-}{2\cos(\alpha)}\right]^2 - cA_- + c^2}, & \text{if } -\arctan\frac{B_+}{A_-} < \alpha \le \arctan\frac{B_-}{A_-}, \\ \frac{B_-}{2\sin(\alpha)} + \sqrt{\left[\frac{B_-}{2\sin(\alpha)}\right]^2 - cB_-\cot(\alpha) + c^2}, & \text{if } \arctan\frac{B_-}{A_-} < \alpha \le \pi - \arctan\frac{B_-}{A_+}, \\ -\frac{A_+}{2\cos(\alpha)} + \sqrt{\left[\frac{A_+}{2\cos(\alpha)}\right]^2 + cA_+ + c^2}, & \text{if } \pi - \arctan\frac{B_-}{A_+} < \alpha \le \pi,\ -\pi < \alpha \le -\pi + \arctan\frac{B_+}{A_+}, \\ -\frac{B_+}{2\sin(\alpha)} + \sqrt{\left[\frac{B_+}{2\sin(\alpha)}\right]^2 + cB_+\cot(\alpha) + c^2}, & \text{if } -\pi + \arctan\frac{B_+}{A_+} < \alpha \le -\arctan\frac{B_+}{A_-}, \end{cases} \tag{11}$$

where I_D represents the interval $(D_{min}, D_{max}(\alpha)]$ in which D_{min} and $D_{max}(\alpha)$ are the minimum and the maximum propagation path length, respectively. The minimum propagation path length D_{min} follows from (5) for $z = 0$; that is, $D_{min} = |c|$. Notice that the maximum propagation path length $D_{max}(\alpha)$ depends on $D_{max}(\alpha)$. It is shown that α is given as in (11), where

$$A_{\pm} = A \pm 2a, \tag{12a}$$

$$B_{\pm} = B \pm 2b. \tag{12b}$$

With $D_{min} = |c|$ and $D_{max}(\alpha)$, as given in (11), the PDF $p_\alpha(\alpha)$ can readily be obtained by solving the integral in (10) numerically.

The PDF of the Propagation Path Length

The PDF of the propagation path length D, denoted by $p_D(D)$, can be calculated by means of the relation

$$p_D(D) = \int_\alpha p_{D\alpha}(D, \alpha)\, d\alpha. \tag{13}$$

It is shown that the integral in (13) can be expressed by (14), which is presented as follows:

$$p_D(D) = \begin{cases} g(-\pi, \pi), & \text{if } |c| \le D \le D_4, \\ g(-\pi, \alpha_1) + g_2(\alpha_2, \pi), & \text{if } D_4 < D \le D_8, \\ g(-\pi, \alpha_3) + g_2(\alpha_4, \alpha_1) + g_2(\alpha_2, \pi), & \text{if } D_8 < D \le D_1, \\ g(-\pi, \alpha_3) + g_2(\alpha_4, \alpha_5) + g_2(\alpha_6, \alpha_1) + g_2(\alpha_2, \pi), & \text{if } D_1 < D \le D_2, \\ g(-\pi, \alpha_3) + g_2(\alpha_4, \alpha_5) + g_2(\alpha_2, \pi), & \text{if } D_2 < D \le D_3, \\ g(-\pi, \alpha_3) + g_2(\alpha_2, \pi), & \text{if } D_3 < D \le D_6, \\ g(\alpha_8, \alpha_3) + g_2(\alpha_2, \alpha_7), & \text{if } D_6 < D \le D_5, \\ g(\alpha_8, \alpha_3), & \text{if } D_5 < D \le D_7, \end{cases} \tag{14}$$

where the path lengths $D_1, D_2, \ldots, D_8$ are given, and $g(\alpha_m, \alpha_n)$ denotes the integral

$$g(\alpha_m, \alpha_n) = \int_{\alpha_m}^{\alpha_n} p_{D\alpha}(D, \alpha)\, d\alpha, \tag{15}$$

in which

$$\alpha_{1,2} = 2\arctan\left\{\frac{D-c}{B_-} \mp \sqrt{\left(\frac{D-c}{B_-}\right)^2 - \frac{D-c}{D+c}}\right\}, \tag{16a}$$

$$\alpha_{3,4} = -2\arctan\left\{\frac{D-c}{B_+} \pm \sqrt{\left(\frac{D-c}{B_+}\right)^2 - \frac{D-c}{D+c}}\right\}, \tag{16b}$$

$$\alpha_{5,6} = \mp\arccos\left\{\frac{DA_-}{D^2 - c^2 + cA_-}\right\}, \tag{16c}$$

$$\alpha_{7,8} = \pm\arccos\left\{\frac{-DA_+}{D^2 - c^2 - cA_+}\right\}. \tag{16d}$$

The PDF and PDP of the Propagation Delays

The objective of this subsection is to find the PDF $p_{\tau'}(\tau')$ of the propagation delays τ'. Here, the propagation delay τ' is defined as

$$\tau' = \frac{\mathrm{D} - |c|}{c_0}, \tag{17}$$

where c_0 denotes the speed of light. Taking this relation into account and applying the concept of transformation of random variables [46, p. 130], we can express the PDF $p_{\tau'}(\tau')$ of the propagation delays τ' as

$$p_{\tau'}\left(\tau'\right) = c_0 p_{\mathrm{D}}\left(c_0\tau' + |c|\right). \tag{18}$$

Here, $p_{\mathrm{D}}(c_0\tau' + |c|)$ can be obtained directly from the PDF in (14) by replacing the variable D by the term $c_0\tau' + |c|$. For brevity, the expression for $p_{\mathrm{D}}(c_0\tau' + |c|)$ is not presented here. Let the total power of the received multipath components be represented by $2\sigma_0^2$, and let us denote the PDP by $S_{\tau'}(\tau')$. Then, it follows that $\int_0^\infty S_{\tau'}(\tau')d\tau' = 2\sigma_0^2$ holds. By taking the relationship $S_{\tau'}(\tau') \sim p_{\tau'}(\tau')$ and the property $\int_0^\infty p_{\tau'}(\tau')d\tau' = 1$ into account, we can conclude that the following equation must hold for the PDP:

$$S_{\tau'}\left(\tau'\right) = 2\sigma_0^2 p_{\tau'}\left(\tau'\right) = 2c_0\sigma_0^2 p_{\mathrm{D}}\left(c_0\tau' + |c|\right). \tag{19}$$

Hence, the PDP $S_{\tau'}(\tau')$ can be obtained directly from the PDF $p_{\mathrm{D}}(D)$ in (14).

The FCF

According to the Wiener-Khinchin theorem, the PDP and the FCF form a Fourier transform pair. The FCF $r_{\tau'}(\nu')$ is defined as the Fourier transform of the PDP $S_{\tau'}(\tau')$; that is,

$$r_{\tau'}\left(\nu'\right) = \int_0^{\infty} S_{\tau'}\left(\tau'\right) e^{-j2\pi\nu'\tau'} d\tau'. \tag{20}$$

Substituting (19) in (20) results in the following expression for the FCF:

$$r_{\tau'}\left(\nu'\right) = 2c_0\sigma_0^2 \int_0^{\infty} p_D\left(c_0\tau' + |c|\right) e^{-j2\pi\nu'\tau'} d\tau'. \tag{21}$$

Since no closed-form solution exists, the integral above has to be solved numerically.

DESIGN OF AN SOC WIDEBAND INDOOR CHANNEL SIMULATOR

This section deals with the design of a stochastic SOC channel simulator for the proposed wideband indoor channel model. The time-variant impulse response of a wideband SOC channel simulator consisting of $\mathscr{L}$ discrete propagation paths can be expressed as follows [47, Eq. (7.55)]:

$$\hat{\mathbf{h}}\left(\tau', t\right) = \sum_{\ell=0}^{\mathscr{L}-1} a_\ell \hat{\boldsymbol{\mu}}_\ell(t)\, \delta\left(\tau' - \tau'_\ell\right). \tag{22}$$

Here, the path gains a_ℓ are determined by the square root of the PDP assigned to the ℓ th discrete propagation delay. By applying the procedure described in [47, p. 374], we have

$$a_\ell = \sqrt{\int_{\tau' \in I_\ell} S_{\tau'}\left(\tau'\right) d\tau'}, \quad \ell = 1, 2, \ldots, \mathscr{L}, \tag{23}$$

where I_ℓ are the intervals defined in [47, p. 374]. The stochastic complex process $\hat{\mu}_\ell(t)$ in (22) models the sum of all received scattered components having the same propagation delay τ'_ℓ. According to the SOC principle, such a stochastic process can be represented by a sum of N_ℓ cisoids of the form [45]

$$\hat{\mu}_\ell(t) = \sum_{n=1}^{N_\ell} c_{n,\ell} e^{j(2\pi f_{n,\ell} t + \theta_{n,\ell})}. \tag{24}$$

Here, the quantities $c_{n,\ell}$, $f_{n,\ell}$, and $\theta_{n,\ell}$ represent the gain, the Doppler frequency, and the Doppler phase of the n th component of the ℓ th discrete path, respectively. The gains $c_{n,\ell}$ and the Doppler frequencies $f_{n,\ell}$ are constant. They can be determined by a proper parameter computation method, for example, the modified method of equal areas (MMEA) [48], according to which the gains $c_{n,\ell}$ are given by

$$c_{n,\ell} = \sigma_0 \sqrt{\frac{2}{N_\ell}}. \tag{25}$$

The Doppler frequencies $f_{n,\ell}$ are determined by the AOAs $\alpha_{n,\ell}$ according to the well-known relation

$$f_{n,\ell} = f_{\max} \cos(\alpha_{n,\ell}), \tag{26}$$

where $f_{\max}$ stands for the maximum Doppler frequency. The AOAs $\alpha_{n,\ell}$ are constant parameters, which can be computed by means of the MMEA. This method requires solving the following equation:

$$\int_{-\pi}^{\alpha_{n,\ell}} p_{\alpha|\tau' \in I_\ell}(\alpha)\, d\alpha = \frac{1}{N_\ell}\left(n - \frac{1}{4}\right), \tag{27}$$

for all $n = 1, 2, \ldots, N_\ell$. The phases $\theta_{n,\ell}$ of the stochastic SOC channel simulator are independent and identically distributed (i.i.d.) random variables, each following a uniform distribution over $[0, 2\pi)$. Thus, the stochastic SOC channel simulator can be interpreted as a family of sample functions determined by the realizations of the phases $\theta_{n,\ell}$. A sample function can be obtained from the stochastic channel simulator by fixing all phases. Such a sample function, which can be interpreted as a waveform generated by a deterministic channel simulator, can be used to simulate the proposed wideband indoor channel.

NUMERICAL AND SIMULATION RESULTS

This section illustrates the theoretical results given by (10) and (21). The correctness of the theoretical results will be verified by simulations. We will also show that the SOC channel simulator matches the reference channel model with respect to the FCF. The validity of the proposed indoor channel model is also confirmed by indoor channel measurements by studying the mean delay and the RMS delay spread. In all simulations, we consider a room with length A=10 m and width B=5 m as our indoor environment. The parameters a and b have been chosen to be equal to 2 and 1, respectively. The SOC channel simulator is designed by applying the MMEA [48] using $N_\ell = 8$ $(\ell = 1, 2, \ldots, \mathscr{L})$ cisoids for each discrete propagation path.

The theoretical result of the PDF $p_{\boldsymbol{\alpha}}(\alpha)$ of the AOA $\boldsymbol{\alpha}$ [see (10)] of the wideband reference channel model is presented in Figure 2 for different BS locations (c=-2 and c=-4). It can be observed from this figure that the shape of the PDF of the AOA is independent of the position of the BS. The theoretical results illustrated in Figure 2 have been verified by experiments. In our experiments, we determined the horizontal (and vertical) locations of all scatterers in the Cartesian coordinate system as outcomes of a random generator, which generates uniformly distributed scatterers over $[-A/2 - a, A/2 - a]$ and $[-B/2 - b, B/2 - b]$ according to the distributions in (1a) and (1b), respectively. It should be mentioned that the results in Figure 2 correspond to the PDF of the AOA of the narrowband channel model (see [39, Eq. (9)]). In Section 6, we will graphically present the PDF of the AOA in case that the locations of the scatterers are exponentially distributed. The impact of the parameters a and b on the shape of the PDF $p_{\boldsymbol{\alpha}}(\alpha)$ of the AOA $\boldsymbol{\alpha}$ is illustrated in Figures 3 and 4, respectively. It can be observed from Figure 3 that an increase in the value of a from 0 to 4 results in a decrease in the PDF $p_{\boldsymbol{\alpha}}(\alpha)$ at small values of the AOAs, where they are confined to the interval $[-\arctan(B_+/A_-), \arctan(B_-/A_-)]$. In contrast, the PDF $p_{\boldsymbol{\alpha}}(\alpha)$ increases with increasing values of a at large values of the AOA; that is, $\boldsymbol{\alpha} \in [-\arctan(B_+/A_+), -\pi] \cup [\arctan(B_-/A_+), \pi]$.

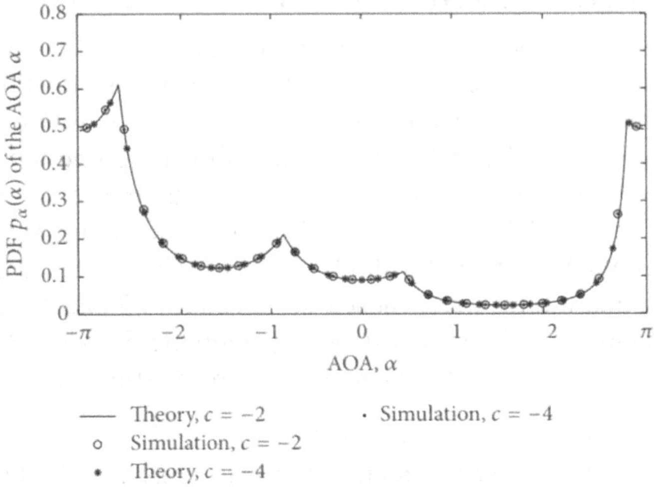

Figure 2: The PDF $p_\alpha(\alpha)$ of the AOA α for different locations of the BS.

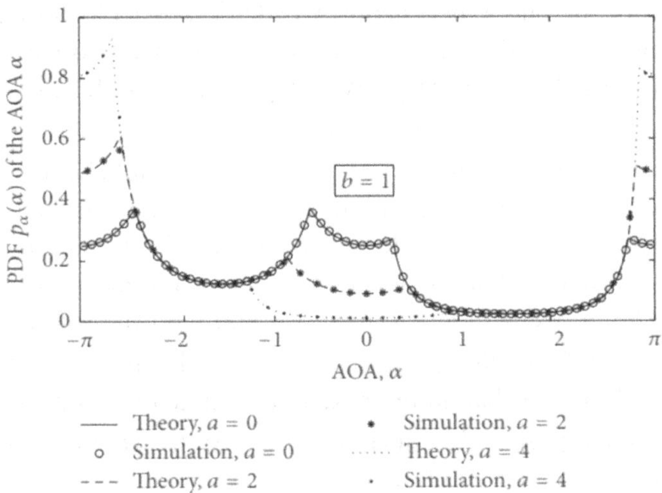

Figure 3: The PDF $p_\alpha(\alpha)$ of the AOA α for different locations of the MS.

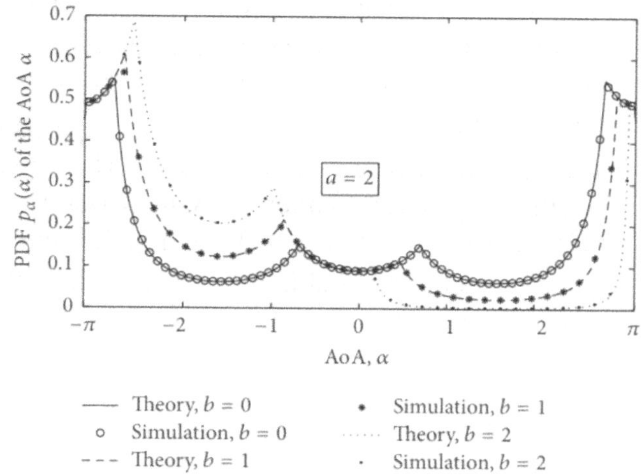

Figure 4: The PDF $p_\alpha(\alpha)$ of the AOA α for different locations of the MS.

Figure 4 reveals that an increase in b leads to an inverse effect on the shape of the PDF $p_\alpha(\alpha)$ compared to an increase in a. Figures 3 and 4 show that both the parameters a and b, which control the symmetry of the room, have impacts on the shape of the PDF of AOA. The main reason is that when a or b changes, it means the location of the mobile station relative to the room area moves (see the geometrical indoor scattering model in Figure 1). While a scatterer is fixed, the position of the mobile station determines the value of the AOA. If the mobile station moves closer to a wall, the AOA bounced by the scatterers in the triangle formed by the mobile station and two corners of the wall takes a larger range. Therefore, the PDF of one angle from that range decreases.

Figure 5 shows the absolute value of the FCF $r_{\tau'}(\nu')$ that has been calculated by using (21) for different values of the room length A. For comparison purposes, we also plot the theoretical curve for the absolute value of the FCF of the designed SOC simulation model by using Eq. (31) in [42]. We consider $\mathscr{L} = 20$ discrete propagation paths for the SOC channel simulator, where the corresponding delays τ'_ℓ are equally spaced between 0 and the maximum propagation delay $\tau'_{\max}$. The value of $\tau'_{\max}$ can be obtained from (18) as the maximum value of τ' such that $p_{\tau'}(\tau') \neq 0$. The path powers assigned to different propagation paths have been calculated according to the method described in [47, pp. 374–375]. The AOAs of the

SOC channel simulator are computed by employing the MMEA [48]. We observe from Figure 5 that the FCF of the SOC channel simulator can be brought into extremely good agreement with that of the reference model. The FCF decays faster with increasing the frequency separation ν' if the room length A increases (here from 10 m to 30 m). This means that the coherence bandwidth B_c becomes smaller as A increases (the coherence bandwidth B_c of a wideband mobile radio channel is the smallest positive value of the frequency separation variable $\upsilon' = B_c$ for which the condition $|r_{\tau'}(B_c)| = |r_{\tau'}(0)|/2$ is fulfilled).

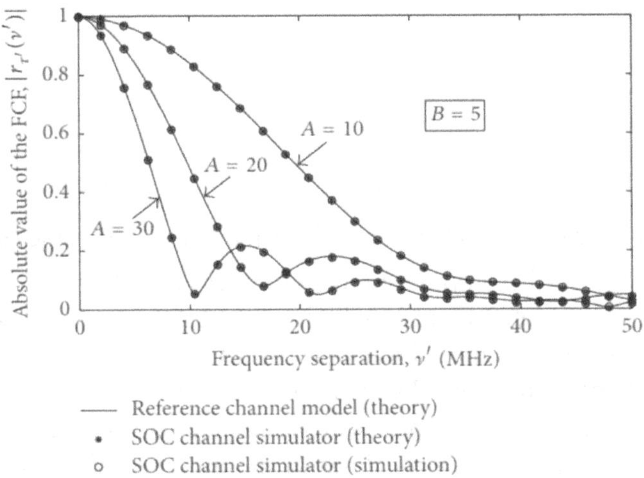

Figure 5: Absolute values of the FCFs $|r_{\tau'}(\nu')|$ (reference model) and $|\tilde{r}_{\tau'}(\nu')|$ (simulation model) for different values of the room length A.

The influence of the room width B on the FCF is presented in Figure 6 from which similar conclusions can be drawn as from Figure 5. When the dimensions of the room increase, the frequency correlation of time-variant transfer functions reduces at the same frequency separation. In other words, the time-variant transfer functions of the developed SOC channel simulator become independent at a smaller frequency separation, which leads to a smaller coherence bandwidth.

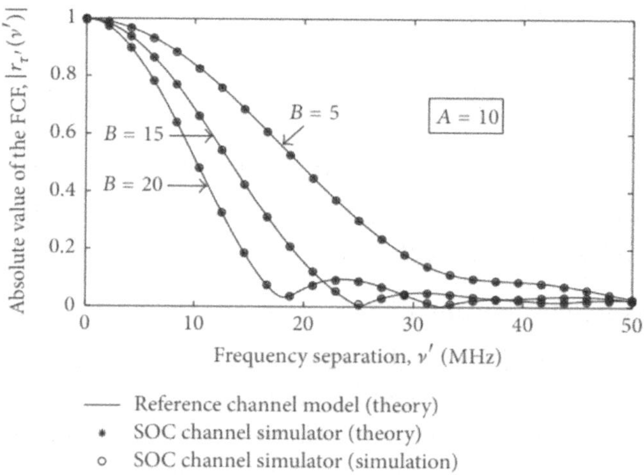

Figure 6: Absolute values of the FCFs $|r_{\tau'}(\nu')|$ (reference model) and $|\tilde{r}_{\tau'}(\nu')|$ (simulation model) for different values of the room width B.

EXPERIMENTAL VERIFICATION

To demonstrate the usefulness of the proposed geometrical-based indoor reference channel model, we will show how the statistics of the reference channel model can be fitted to the statistics of measured real-world channels by optimizing the key parameters of the reference model. Here, two of the most important characteristic quantities of wideband mobile radio channels are considered, namely, the mean excess delay and the RMS delay spread.

The mean excess delay $m_{\tau'}$ is defined as the first moment of the PDP [26, 49]; that is,

$$m_{\tau'} = \frac{\int_0^\infty \tau' \cdot S_{\tau'}(\tau')\, d\tau'}{\int_0^\infty S_{\tau'}(\tau')\, d\tau'}. \quad (28)$$

The RMS delay spread $\sigma_{\tau'}$ is defined as the square root of the second central moment of the PDP [26, 49]; that is,

$$\sigma_{\tau'} = \sqrt{\frac{\int_0^{\infty} \left(\tau' - m_{\tau'}\right)^2 \cdot S_{\tau'}\left(\tau'\right) d\tau'}{\int_0^{\infty} S_{\tau'}\left(\tau'\right) d\tau'}}. \tag{29}$$

The mean excess delay $m_{\tau'}$ and the RMS delay spread $m^{\star}_{\tau}$ of the reference model can easily be calculated by substituting the PDP $S_{\tau'}(\tau')$ [see (19)] into (28) and (29), respectively.

The measurement results of the mean excess delay $m^{\star}_{\tau}$ and the RMS delay spread $\sigma^{\star}_{\tau}$ considered in this section are taken from [16, 19, 23]. The measurement campaigns were carried out in laboratories, a conference room, and corridors. The measurements in [16] have been obtained at 2.4 GHz in a laboratory with dimensions 7.8 m × 9.95 m. Furthermore, the measurements reported in [19] have been obtained at 5 GHz in a conference room with dimensions 6.6 m × 5.9 m. Finally, the measurements in [23] have been obtained at 60 GHz in two 30 m × 1.75 m × 2.80 m corridors as well as inside a laboratory with dimensions 19.5 m × 7.5 m.

We combine all the relevant model parameters (except the parameters A and B) which control the mean excess delay $m_{\tau'}$ and the RMS delay spread $\sigma_{\tau'}$ of the reference channel model into a parameter vector denoted and defined by $\Omega := (a, b, c, w_{11}, w_{12}, w_{21}, w_{22})$. We also introduce the following error function:

$$E(\Omega) = W_1 \cdot \left|m^{\star}_{\tau'} - m_{\tau'}\right| + W_2 \cdot \left|\sigma^{\star}_{\tau'} - \sigma_{\tau'}\right|, \tag{30}$$

for measuring the deviations between the mean excess delays $m^{\star}_{\tau}$ and $m_{\tau'}$ as well as between the RMS delay spreads $\sigma^{\star}_{\tau}$ and $\sigma_{\tau'}$. In (30), W_1 and W_2 represent proper weighting factors.

The optimization of the parameter vector Ω has been carried out by minimizing numerically the above error function $E(\Omega)$ by means of the quasi-Newton optimization procedure [50]. The optimization results are presented in Tables 1, 2, and 3, for the laboratory, the conference room, and corridor scenarios. For comparison purpose, the measured mean assess delay $m^{\star}_{\tau'}$ and the measured RMS delay spread $\sigma^{\star}_{\tau}$, presented in [16, 19, 23], are also shown in Tables 1, 2, and 3. Note that the dimensions of the different rooms, which are controlled by the parameters A and B, have been chosen such that they are equal to the lengths and widths of the rooms in which the corresponding measurement campaigns have been conducted. Hence, A and B are fixed, and thus they have not been included in the optimization procedure. In Table 1, the measured mean assess delay $m^{\star}_{\tau'}$ is in the range from43.40 ns to 57.04 ns, while the RMS delay spread $\sigma^{\star}_{\tau}$ varies from 22.19 ns to 30.55 ns. Similarly, in Table 2, the measured mean assess delay $m^{\star}_{\tau'}$ is in the range from 16.40 ns to 24.0 ns, while the RMS delay spread $\sigma^{\star}_{\tau}$ varies from 15.5 ns to 20.7 ns. In Table 3, finally, the measured mean assess delay $m^{\star}_{\tau'}$ covers the range from 3.84 ns to 8.18 ns and holds for hallways, while the range from 3.52 ns to 14.69 ns refers to offices. In addition, the measured RMS delay spread $\sigma^{\star}_{\tau}$ varies from 12.34 ns to 15.04 ns for corridors and from 12.56 ns to 21.09 ns for laboratories. In our optimization procedure, the weighting factors W_1 and W_2 were chosen to be 0.35 and 0.65, respectively. For reasons of brevity, we have only presented the values of the optimized parameters in all tables with two decimals of precision. As shown in Tables 1, 2, and 3, the mean excess delay and the RMS delay spread of the proposed wideband indoor channel model are very close to the corresponding measured quantities for all considered indoor propagation scenarios, which demonstrates that the reference channel model is useful for characterizing real-world wideband indoor mobile radio channels.

Table 1: Optimized parameters of the reference models for laboratory scenarios at 2.4 GHz based on the measurements in [16]

Locations	$m^{\star}_{\tau'}$ (ns)	$\sigma^{\star}_{\tau'}$ (ns)	$m_{\tau'}$ (ns)	$\sigma_{\tau'}$ (ns)	A	B	a	b	c	w_{11}	w_{12}	w_{21}	w_{22}
Loc. 1	57.04	30.55	57.04	30.49	7.8	9.95	3.51	4.32	−0.01	701.53	198.76	3.71	34.05
Loc. 2	43.40	24.12	43.40	24.12	7.8	9.95	3.48	3.93	−1.02	10.17	10.10	10.16	9.94
Loc. 3	49.10	22.19	49.10	22.12	7.8	9.95	3.76	3.59	−1.01	10.04	12.04	7.34	12.59
Loc. 4	54.85	26.82	54.85	26.82	7.8	9.95	3.87	4.97	−0.86	12.01	13.18	4.60	8.79

Table 2: Optimized parameters of the reference models for the conference room at 5 GHz based on the measurements in [19]

Locations	$m_{\tau'}^{*}$ (ns)	$\sigma_{\tau'}^{*}$ (ns)	$m_{\tau'}$ (ns)	$\sigma_{\tau'}$ (ns)	A	B	a	b	c	w_{11}	w_{12}	w_{21}	w_{22}
Loc. 1	20.5	18.6	20.5	18.6	6.6	5.9	2.99	2.78	−0.88	12.78	7.83	13.12	3.70
Loc. 2	16.4	16.2	16.4	16.2	6.6	5.9	2.84	2.09	−0.66	5.65	2.79	14.15	1.61
Loc. 3	19.7	17.8	19.7	17.8	6.6	5.9	3.03	2.62	−0.68	8.53	1.33	7.25	5.38
Loc. 4	21.8	18.5	21.8	18.5	6.6	5.9	3.01	2.62	−0.80	5.83	2.07	6.93	4.82
Loc. 5	18.0	16.6	18.0	16.6	6.6	5.9	3.21	2.29	−0.84	6.85	1.23	7.21	4.21
Loc. 6	17.2	15.5	17.2	15.5	6.6	5.9	2.87	2.11	−0.96	7.44	1.62	6.84	3.67
Loc. 7	17.9	16.4	17.9	16.4	6.6	5.9	3.10	2.29	−0.87	6.84	1.20	7.19	4.37
Loc. 8	14.3	15.9	14.3	15.9	6.6	5.9	3.17	2.28	−0.95	7.93	1.33	7.42	2.58
Loc. 9	19.0	20.2	19.0	20.2	6.6	5.9	3.00	2.90	−0.17	18.97	4.84	17.05	7.14
Loc. 10	20.8	20.7	20.8	20.7	6.6	5.9	3.00	2.90	−0.14	18.02	6.70	16.74	6.51
Loc. 11	23.6	20.4	23.6	20.4	6.6	5.9	2.89	2.90	−0.15	19.31	7.13	12.38	7.29
Loc. 12	24.0	19.9	24.0	19.9	6.6	5.9	2.78	2.90	−0.15	19.90	7.33	12.18	7.45

Table 3: Optimized parameters of the reference models for the corridor and laboratory (Lab) scenarios at 60 GHz based on the measurements in [23]

Locations	$m_{\tau'}^{*}$ (ns)	$\sigma_{\tau'}^{*}$ (ns)	$m_{\tau'}$ (ns)	$\sigma_{\tau'}$ (ns)	A	B	a	b	c	w_{11}	w_{12}	w_{21}	w_{22}
Loc. 1	5.17	13.70	5.17	13.70	30	1.75	13.26	0.45	−15.64	9.61	0.23	14.50	1.67
Loc. 2	8.18	15.04	8.18	15.04	30	1.75	13.66	0.53	−16.08	31.23	1.17	11.89	1.02
Loc. 3	3.84	12.35	3.84	12.35	30	1.75	14.62	0.57	−19.17	30.88	1.02	11.67	1.05
Loc. 4	5.37	12.34	5.37	12.34	30	1.75	14.01	0.44	−17.95	31.85	0.98	14.31	1.13
Average corridor	*5.64*	*13.36*	*5.64*	*13.36*	30	1.75	12.98	0.56	−15.61	10.38	0.16	21.75	1.01
Loc. 5	8.42	14.72	8.42	14.72	19.5	7.5	9.00	1.34	−7.53	9.87	0.40	11.05	0.45
Loc. 6	3.52	12.56	3.77	12.56	19.5	7.5	9.02	3.42	−5.65	5.04	0.71	232.37	2.96
Average Lab (LOS)	*5.97*	*13.64*	*5.97*	*13.64*	19.5	7.5	9.02	2.29	−5.62	4.36	0.82	185.74	1.38
Loc. 7	12.81	19.94	12.51	19.28	19.5	7.5	7.85	2.98	−7.13	3.22	0.22	20.71	2.29
Loc. 8	14.69	21.09	14.67	21.08	19.5	7.5	9.21	0.72	−8.21	1.93	0.21	0.01	0.68
Average Lab (NLOS)	*13.75*	*20.52*	*13.72*	*20.51*	19.5	7.5	9.00	1.72	−9.05	1.48	0.21	19.28	0.33
Overall	*7.75*	*15.22*	*7.75*	*15.22*	19.5	7.5	9.00	2.26	−3.97	1.24	0.63	30.97	0.74

For completeness, we present in Figure 7 the nonuniform PDF $p_{\alpha}(\alpha)$ of the AOA α for three of the indoor propagation scenarios, namely, the Average Corridor, the Average Lab (NLOS), and the overall channel scenario. In this figure, we see that the PDF of the AOA has 4 peaks corresponding to the characteristics of the exponential distribution within the 2D horizontal plan of the room. The results in this figure have been obtained by evaluating (10) with the optimized channel parameters listed in Table 1.

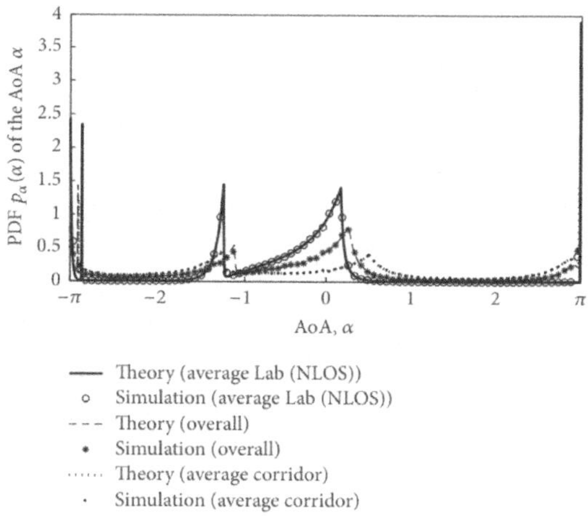

Figure 7: The PDF $p_\alpha(\alpha)$ of the AOA α for three different indoor propagation scenarios.

CONCLUSION

In this paper, we have developed a wideband mobile radio channel model for indoor propagation environments. The starting point for the derivation of the reference channel model was a geometrical scattering model, where we have assumed that an infinite number of scatterers are exponentially distributed within the 2D horizontal plane of a rectangular room. Analytical expressions have been derived for the PDF of the AOA, the PDF of the propagation delays, the PDP, and the FCF. We have shown that the shape of the PDF of the AOA is independent of the position of the transmitter (BS). Both the room length and width have a strong influence on the PDF of the AOA and the FCF. If the room length or width increases, the FCF decays faster if the frequency separation increases. The coherence bandwidth decreases with increasing the room size. The usefulness of the proposed reference channel model has been demonstrated by a close match between the channel statistics of the reference model and measured channels.

An efficient channel simulator has been derived from the reference channel model by applying the SOC principle. It has been shown that the SOC channel simulator matches closely the wideband reference model with respect to the FCF. The designed SOC channel simulator can be used for the performance evaluation of wideband indoor wireless communication systems under realistic propagation conditions.

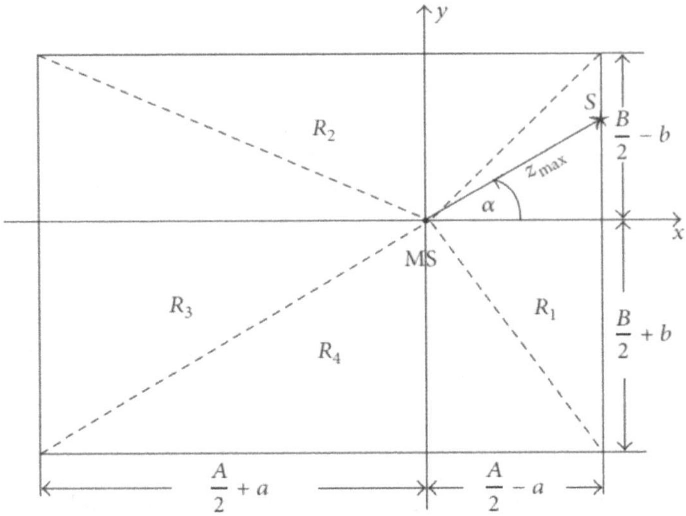

Figure 8: Partitioning of the scattering region inside a rectangular room into four subregions: R_1, R_2, R_3, and R_4 separated by dashed lines (---).

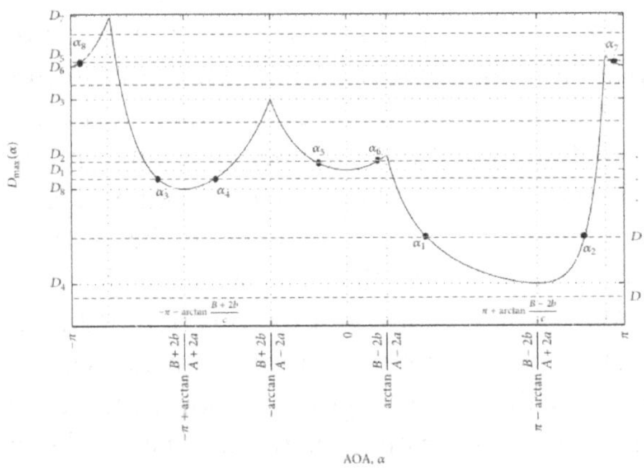

Figure 9: The maximum propagation path length $D_{\max}(\alpha)$ for different values of the AOA α ($A = 10\,\text{m}$, $B = 5\,\text{m}$, $a = 2\,\text{m}$, $b = 1\,\text{m}$, and $c = -2\,\text{m}$).

ACKNOWLEDGMENTS

This work was supported in part by the Key Scientific Research Team Cultivation Project founded by College of Information Science and

Engineering, Shandong Agricultural University, China. Also, this work was supported in part by the Spanish Ministry of Economy and Competitiveness through the RACHEL Project (TEC2013-47141-C4-2-R), the CARMEN Project (TEC2016-75067-C4-3-R), and the COMONSENS network (TEC2015-69648-REDC).

REFERENCES

1. J. E. Padgett, C. G. Gunther, and T. Hattori, "Overview of wireless personal communications," *IEEE Communications Magazine*, vol. 33, no. 1, pp. 28–41, 1995.
2. J. Karao�uz, "High-rate wireless personal area networks," *IEEE Communications Magazine*, vol. 39, no. 12, pp. 96–102, 2001.
3. A. F. Molisch, *Wireless Communications*, John Wiley & Sons, Chichester, England, 2005.
4. P. Ferré, A. Doufexi, J. Chung-How, A. R. Nix, and D. R. Bull, "Robust video transmission over wireless LANs," *IEEE Transactions on Vehicular Technology*, vol. 57, no. 4, pp. 2596–2602, 2008.
5. P. Lin and Y.-B. Lin, "Implementation and performance evaluation for mobility management of a wireless PBX network," *IEEE Journal on Selected Areas in Communications*, vol. 19, no. 6, pp. 1138–1146, 2001.
6. H. Elgala, R. Mesleh, and H. Haas, "Indoor optical wireless communication: Potential and state-of-the-art," *IEEE Communications Magazine*, vol. 49, no. 9, pp. 56–62, 2011.
7. K. Wang, A. Nirmalathas, C. Lim, and E. Skafidas, "High-speed optical wireless communication system for indoor applications," *IEEE Photonics Technology Letters*, vol. 23, no. 8, pp. 519–521, 2011.
8. T. Ohtsuki, "Multiple-subcarrier modulation in optical wireless communications," *IEEE Communications Magazine*, vol. 41, no. 3, pp. 74–79, 2003.
9. H. Elgala, R. Mesleh, and H. Haas, "Indoor broadcasting via white LEDs and OFDM," *IEEE Transactions on Consumer Electronics*, vol. 55, no. 3, pp. 1127–1134, 2009.
10. J. B. Carruthers, S. M. Carroll, and P. Kannan, "Propagation modelling for indoor optical wireless communications using fast multi-receiver channel estimation," *IEE Proceedings—Optoelectronics*, vol. 150, no. 5, pp. 473–481, 2003.
11. J. Kunthong and C. F. Bunting, "Statistical characterization of the 900 MHz and 1800 MHz indoor propagation using reverberation source stirring technique," in *Proceedings of the 2009 IEEE International Symposium on Antennas and Propagation and USNC/URSI National Radio Science Meeting, APSURSI 2009*, Charleston, SC, USA, June 2009.

12. E. Banner and A. B. Sesay, "Effects of antenna height, antenna gain, and pattern downtilting for cellular mobile radio," *IEEE Transactions on Vehicular Technology*, vol. 45, no. 2, pp. 217–224, 1996.
13. S. Y. Seidel and T. S. Rappaport, "914 MHz path loss prediction models for indoor wireless communications in multifloored buildings," *IEEE Transactions on Antennas and Propagation*, vol. 40, no. 2, pp. 207–217, 1992.
14. A. A. Saleh and R. A. Valenzuela, "A statistical model for indoor multipath propagation," *IEEE Journal on Selected Areas in Communications*, vol. 5, no. 2, pp. 128–137, 1987.
15. J. A. Wepman, J. R. Hoffman, and L. H. Loew, "Analysis of impulse response measurements for PCS channel modelling applications," *IEEE Transactions on Vehicular Technology*, vol. 44, no. 3, pp. 613–620, 1995.
16. T. A. Wysocki and H.-J. Zepernick, "Characterization of the indoor radio propagation channel at 2.4 GHz," *Journal of Telecommunications and Information Technology*, vol. 3-4, 2000.View at: Google Scholar
17. D. A. Hawbaker and T. S. Rappaport, "Indoor wideband radiowave propagation measurements at 1.3 GHz and 4.0 GHz," *IEEE Electronics Letters*, vol. 26, no. 21, pp. 1800–1802, 1990.
18. G. Santella and E. Restuccia, "Analysis of frequency domain wide-band measurements of the indoor radio channel at 1, 5.5, 10 and 18 GHz," in *Proceedings of the GLOBECOM '96*, pp. 1162–1166, London, UK.
19. M. Peter and W. Keusgen, "Analysis and comparison of indoor wideband radio channels at 5 and 60 GHz," in *Proceedings of the 3rd European Conference on Antennas and Propagation, (EuCAP '09)*, pp. 3830–3834, Berlin, Germany, 2009.
20. H. Droste and G. Kadel, "Measurement and analysis of wide band indoor propagation characteristics at 17 GHz and 60 GHz," in *Proceedings of the 9th International Conference on Antennas and Propagation*, vol. 2, pp. 288–291, Eindhoven, Netherlands, April 1995.
21. W. Fu, J. Hu, and S. Zhang, "Frequency-domain measurement of 60 GHz indoor channels: a measurement setup, literature data, and analysis," *IEEE Instrumentation & Measurement Magazine*, vol. 16, no. 2, pp. 34–40, 2013.
22. N. Moraitis and P. Constantinou, "Indoor channel measurements and characterization at 60 GHz for wireless local area network

applications," *IEEE Transactions on Antennas and Propagation*, vol. 52, no. 12, pp. 3180–3189, 2004.

23. N. Moraitis and P. Constantinou, "Measurements and characterization of wideband indoor radio channel at 60 GHz," *IEEE Transactions on Wireless Communications*, vol. 5, no. 4, pp. 880–889, 2006.
24. T. Zwick, T. J. Beukema, and H. Nam, "Wideband channel sounder with measurements and model for the 60 GHz indoor radio channel," *IEEE Transactions on Vehicular Technology*, vol. 54, no. 4, pp. 1266–1277, 2005.
25. H. Hashemi, "Impulse response modeling of indoor radio propagation channels," *IEEE Journal on Selected Areas in Communications*, vol. 11, no. 7, pp. 967–978, 1993.
26. D. Tholl and H. Hashemi, "Statistical modeling and simulation of the RMS delay spread of indoor radio propagation channels," *IEEE Transactions on Vehicular Technology*, vol. 43, no. 1, pp. 110–120, 1994.
27. K. Yu, M. Bengtsson, B. Ottersten, D. McNamara, P. Karlsson, and M. Beach, "Modeling of wide-band MIMO radio channels based on NLoS indoor measurements," *IEEE Transactions on Vehicular Technology*, vol. 53, no. 3, pp. 655–665, 2004.
28. S. Chamaani, S. A. Mirtaheri, Y. I. Nechayev, and P. S. Hall, "MICS band indoor channel modeling using ray tracing method," in *Proceedings of the 5th International Symposium on Telecommunications, (IST '10)*, pp. 126–131, Kish Island, Iran, December 2010.
29. S. Loredo, A. Rodríguez-Alonso, and R. P. Torres, "Indoor MIMO channel modeling by rigorous GO/UTD-based ray tracing," *IEEE Transactions on Vehicular Technology*, vol. 57, no. 2, pp. 680–692, 2008.
30. S.-H. Cheu and S.-K. Jeng, "All SBR/image approach to indoor radio propagation modeling," in *Proceedings of the IEEE Antennas and Propagation Society, AP-S international symposium (Digest)*, pp. 1952–1955, Newport Beach, Calif, USA, June 1995.
31. M. C. Lawton and J. P. McGeehan, "The application of a deterministic Ray launching algorithm for the prediction of radio channel characteristics in small-cell environments," *IEEE Transactions on Vehicular Technology*, vol. 43, no. 4, pp. 955–969, 1994.

32. Z. Ji, B.-H. Li, H.-X. Wang, H.-Y. Chen, and T. K. Sarkar, "Efficient ray-tracing methods for propagation prediction for indoor wireless communications," *IEEE Antennas and Propagation Magazine*, vol. 43, no. 2, pp. 41–49, 2001.
33. A. Ö. Kaya, L. J. Greenstein, and W. Trappe, "Characterizing indoor wireless channels via ray tracing combined with stochastic modeling," *IEEE Transactions on Wireless Communications*, vol. 8, no. 8, pp. 4165–4175, 2009.
34. G. Tiberi, S. Bertini, W. Q. Malik, A. Monorchio, D. J. Edwards, and G. Manara, "Analysis of realistic ultrawideband indoor communication channels by using an efficient ray-tracing based method," *IEEE Transactions on Antennas and Propagation*, vol. 57, no. 3, pp. 777–785, 2009.
35. G. E. Athanasiadou, A. R. Nix, and J. P. McGeehan, "Microcellular ray-tracing propagation model and evaluation of its narrow-band and wide-band predictions," *IEEE Journal on Selected Areas in Communications*, vol. 18, no. 3, pp. 322–335, 2000.
36. G. Liang and H. L. Bertoni, "A new approach to 3-D ray tracing for propagation prediction in cities," *IEEE Transactions on Antennas and Propagation*, vol. 46, no. 6, pp. 853–863, 1998.
37. T. Rautiainen, G. Wolfle, and R. Hoppe, "Verifying path loss and delay spread predictions of a 3D ray tracing propagation model in urban environment," in *Proceedings of the IEEE 56th Vehicular Technology Conference*, pp. 2470–2474, Vancouver, Canada, 2002.
38. L. Jiang and S. Y. Tan, "Geometrically based statistical channel models for outdoor and indoor propagation environments," *IEEE Transactions on Vehicular Technology*, vol. 56, no. 6, pp. 3587–3593, 2007.
39. Y. Ma and M. Patzold, "Design and simulation of narrowband indoor radio propagation channels under LOS and NLOS propagation conditions," in *Proceedings of the IEEE 71st Vehicular Technology Conference*, pp. 1–7, Taipei, Taiwan, May 2010.
40. Y. Ma and M. Pätzold, "Modeling and statistical characterization of wideband indoor radio propagation channels," in *Proceedings of the 2010 International Congress on Ultra Modern Telecommunications and Control Systems and Workshops, ICUMT 2010*, pp. 777–783, Moscow, Russia, October 2010.

41. Y. Ma, *Broadband wireless communication systems: Channel modeling and system performance analysis[Ph.D. thesis]*, Dept. of Inform. and Commun. Technol., University of Agder, Grimstad, Norway, 2011, https://brage.bibsys.no/xmlui/handle/11250/137621.
42. M. Pätzold and B. O. Hogstad, "A wideband space-time MIMO channel simulator based on the geometrical one-ring model," in *Proceedings of the 2006 IEEE 64th Vehicular Technology Conference, VTC-2006 Fall*, pp. 133–138, Montreal, Canada, September 2006.
43. M. Pätzold and B. O. Hogstad, "A wideband MIMO channel model derived from the geometric elliptical scattering model," in *Proceedings of the 3rd International Symposium on Wireless Communication Systems 2006, ISWCS '06*, pp. 138–143, Valencia, Spain, September 2006.
44. Y. Ma and M. Pätzold, "Wideband two-ring MIMO channel models for mobile-to-mobile communications," in *Proceedings of the 10th International Symposium on Wireless Personal Multimedia Communications, WPMC 2007*, pp. 380–384, Jaipur, India, 2007.
45. M. Pätzold and B. Talha, "On the statistical properties of sum-of-cisoids-based mobile radio channel simulators," in *Proceedings of the 10th International Symposium on Wireless Personal Multimedia Communications, WPMC 2007*, pp. 394–400, Jaipur, India, 2007.
46. A. Papoulis and S. U. Pillai, *Probability, Random Variables, and Stochastic Processes*, McGraw-Hill, New York, NY, USA, 4th edition, 2002.
47. M. Pätzold, *Mobile Radio Channels*, John Wiley & Sons, Chichester, UK, 2nd edition, 2011.View at: Publisher Site
48. C. A. Gutierrez-Diaz-De-Leon and M. Pätzold, "Sum-of-sinusoids-based simulation of flat fading wireless propagation channels under non-isotropic scattering conditions," in *Proceedings of the 50th Annual IEEE Global Telecommunications Conference, GLOBECOM 2007*, pp. 3842–3846, Washington, Wash, USA, November 2007.
49. H. Xu, V. Kukshya, and T. S. Rappaport, "Spatial and temporal characteristics of 60-GHZ indoor channels," *IEEE Journal on Selected Areas in Communications*, vol. 20, no. 3, pp. 620–630, 2002.
50. R. Fletcher and M. J. D. Powell, "A rapidly convergent descent method for minimization," *The Computer Journal*, vol. 6, no. 2, pp. 163–168, 1963.

15

A New Mathematical Analysis of the Probability of Detection in Cognitive Radio over Fading Channels

Omar Altrad[1] and Sami Muhaidat[2,3]

[1]Simon Fraser University, Burnaby, BC V5A 1S6, Canada
[2]Khalifa University, Abu Dhabi 127788, UAE
[3]Centre for Communication Systems Research (CCSR), University of Surrey, Surrey, GU2 7XH, Guildford, UK

ABSTRACT

Cognitive radio (CR) enriches wireless technology systems by harnessing spectrum white spaces. Such systems require continuous and reliable sensing to provide high-quality service to their users and to minimize the interference they may cause to legacy networks. As the simplicity of implementation of energy detectors and their incoherent requirements make them an ideal candidate for this type of application, this work provides a further mathematical analysis to the probability of detection over different

Citation: Altrad and Muhaidat: A new mathematical analysis of the probability of detection in cognitive radio over fading channels. EURASIP Journal on Wireless Communications and Networking 2013 2013:159. https://doi.org/10.1186/1687-1499-2013-159.

fading channels. We derive a tight closed-form expression for the probability of detection over Nakagami channels. In addition, we introduce an accurate recursive algorithm to compute the probability of detection for an odd degree of freedom over additive white Gaussian noise channels, which has been overlooked in the literature so far. Finally, we present the simulation results which concur with the numerical results and which are also a perfect match with the mathematical expressions derived.

INTRODUCTION

The rapid increase of wireless systems and applications raises spectrum demand. However, not all bands of the spectrum are fully utilized at specific times or at specific geographic locations. The Federal Communication Commission (FCC) reported that some of these bands (e.g., unlicensed bands at 2.4 and 5 Ghz) are overcrowded, while others (e.g., licensed bands such as the ultra-high frequency (UHF) band) are inefficiently used. Therefore, technology is emerging to reduce the spectrum scarcity issue by fully utilizing the unused portion of the spectrum. For example, IEEE 802.22 [1] proposes reusing the television (TV) UHF band without causing any interference to TV receivers. Another considers a cellular communication system that utilizes the wireless local area network system (*cf.* [2]). Researchers often refer to this technology as cognitive radio (CR) systems.

As the main objective of CR systems is the spectrum efficient utilization, an accurate design for a cognitive radio network (CRN) working under a licensed primary network (PN) needs to be considered. While the CRN may have its own frequency band of operation, it can also utilize the white spaces or spectrum holes in frequency bands of the PN to increase its performance and to provide a higher quality of service to its users. As a result, it is considered to be a secondary network relative to the primary network. One of the major challenge of CRN is spectrum sensing, i.e., a highly reliable sensing function must be implemented in the CRN's terminals. This arises from the fact that the CR receiver sensitivity must be as high as possible to detect the presence or absence of a primary user (PU) signal and to invoke other functions in the CR device which also depend mainly on sensing functionality. For example, in order to detect a primary signal, the CR system must have a sensitivity as much as 20 to 30 dB higher than that of the primary system [3]. Therefore, the core of CR systems is the spectrum sensing algorithm which determines the validity of a transmission opportunity.

In this paper, our study is limited to the energy sensing method [4–7]. In particular, for a local spectrum sensing scenario, i.e., the sensing is accomplished by a single cognitive radio. This detection method can be applied to any signal type with fewer computational requirements and a simpler implementation. Although several research papers have investigated the detection process using energy detector over a variety of fading channels (*cf.* [8–13]), the expressions derived for the probability of detection and the probability of false alarms were mainly evaluated for even degrees of freedom (e.g., [14, Eq.]10]). Therefore, we provide an algorithm to *compute* the detection probability in the case of odd degrees of freedom based on the suboptimal energy detector. Moreover, as spectrum sensing must detect a very low signal-to-noise ratio (SNR), which in turn requires a high degree of precision, the previously derived expressions mainly depend on the number of terms in the summation to get highly accurate results. In addition, they are numerically difficult and depend on other functions while their implementation is also susceptible to truncation errors. Therefore, closed-form expressions for the detection probability are derived. We summarize our contributions as follows:

- We provide a highly accurate recursive algorithm to compute the probability of detection for odd degrees of freedom. It should be noted that the mathematical derivation shows the steps of the algorithm when evaluating the detection probability in case of odd degrees of freedom, i.e., it is an algorithm rather than a mathematical derivation. An example of the algorithm importance is the Marcum function in Matlab which accepts only integer values in its third argument. Therefore, when the number of degrees of freedom is odd, the third argument is no longer accepted and the Marcum function cannot be used to evaluate the detection probability in this case. However, our algorithm solves this problem.
- We derive a closed-form expression over a Nakagami-*m* fading channel. Here, we use closed form in the sense that no summation and no integration are required. The accuracy of the closed form is very close to the previously reported expressions in which summation and integration are used to get highly accurate results. Our new expressions show how the ratio of the Nakagami parameter *m* and the average signal-to-noise ratio which affects the receiver operation characteristics (ROC) curves.

- We compare the derived expressions to the reported expressions in [14, 15] in which summation and integration are used. We also compare our derived expressions to other recently reported expressions, e.g., [16, 17], and we show that our new derived expressions can be used with no limitations. Moreover, the derived expressions are more accurate than the recently reported ones with less or almost the same computational complexity.

We also compare our simulation results with the analytical evaluation of the derived expressions.

The rest of the paper is organized as follows. In Section 2, the energy detector, system model, and the derivation of the recursive algorithm are introduced. We derive closed-form expressions for Nakagami channels in Section 3. Simulation and numerical results are introduced in Section 4, and we conclude our paper in Section 5.

ENERGY SENSING MODEL

The sensing process consists of two stages and is controlled by signals from the upper layers to sense a specific bandwidth B, as shown in Figure 1. In the first stage, the received signal $x(t)$ is filtered to the bandwidth of interest B to reject band noise and adjacent signals. It is then amplified using a low noise amplifier and is down-converted to an intermediate frequency. In the second stage, the received signal is sampled and quantized using an A/D converter. Next, a square-law device and an integrator with sensing interval T measures the received signal energy. Finally, the output of the integrator, represented by the test statistic Y, is compared to a predetermined threshold λ to determine the existence (H_1) or absence (H_0) of a PU.

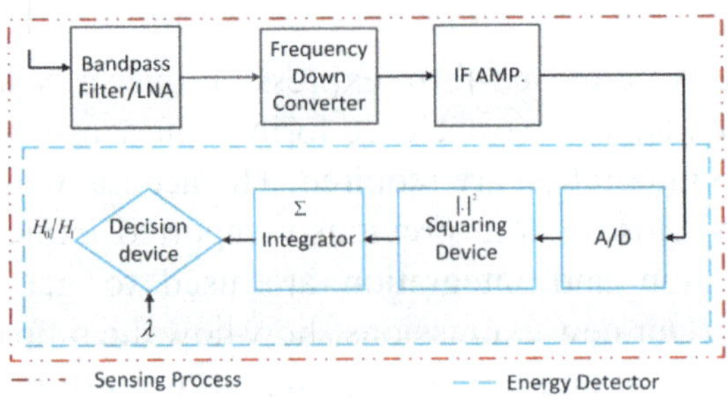

Figure 1: Schematic of sensing abstraction including an energy detector.

The existence or absence of a PU signal can be modeled as a binary hypotheses problem as originally proposed by [18, 19] and later followed by most researchers [20, 21]. This model differentiates between two hypotheses defined as:

$$x[n] = \begin{cases} w[n], & H_0 \\ hs[n] + w[n], & H_1 \end{cases}, n = 1, 2, \cdots N, \tag{1}$$

where $s[n]$ is the primary user signal component which is assumed to be an unknown deterministic signal, and $w[n]$ is the noise component which is assumed to be an additive white Gaussian noise (AWGN) with zero mean and variance σ^2. h is the channel coefficient which is assumed to be constant during the period of observation, i.e., for N samples, H_0 is the hypothesis test when noise only is present and H_1 is the hypothesis test when both noise and signal are present. We also assume that the noise samples are independent and identically distributed, and they are independent of the signal samples. The suboptimal energy detector is defined as:

$$Y = \sum_{N} |x[n]|^2 \tag{2}$$

Then, the distribution of the decision variable Y will be central chi-square χ_N^2 under H_0 and noncentral chi-square $\tilde{\chi}_N^2$ with N degrees of freedom under H_1. Notice that to reduce the overuse of notations, we distinguish between central and noncentral chi-square by the symbol (~). Thus, using this notation, the distribution can be expressed as [22]:

$$Y \sim \begin{cases} \chi_N^2, H_0 \\ \tilde{\chi}_N^2, H_1, \end{cases}$$

and its probability density function can be written as:

$$f_Y(y) = \begin{cases} \frac{1}{\sigma^N 2^{\frac{N}{2}} \Gamma\left(\frac{N}{2}\right)} y^{\left(\frac{N}{2}\right)-1} \exp\left(\frac{-y}{2\sigma^2}\right), H_0 \\ \frac{1}{2\sigma^2}\left(\frac{y}{\zeta}\right)^{(N-2)/4} \exp\left[\frac{-1}{2\sigma^2}(y+\zeta)\right] \times I_{\frac{N}{2}-1}\left(\frac{\sqrt{\zeta y}}{\sigma^2}\right), H_1 \end{cases} \tag{3}$$

where the noncentrality parameter $\zeta = \sum_{i=1}^{N} \mu_i^2,$ and μ_i is the mean of the i th Gaussian random variable of test Y. $I_r(.)$ is the r th-modified Bessel function of the first kind, which has a series representation [23]:

$$I_r(u) = \sum_{k=0}^{\infty} \frac{\left(\frac{1}{2}u\right)^{2k+r}}{k!\,\Gamma(r+k+1)} \tag{4}$$

Evaluating test Y by the decision device, which is shown in Figure 1, may result in two types of errors. We define the notation $P(H_i,H_j)$ to distinguish between these errors. When the decision device decides H_1 but H_0 is true, denoted as $P(H_1;H_0)$, this is called the probability of a false alarm (P_{fa}). When the device decides $P(H_0,H_1)$, this represents the probability of misdetection (P_{md}). The complementary to P_{md} is the probability of detection ($P_d = 1 - P_{md} = P(H_1;H_1)$). The performance of the energy detector can be characterized by the probability of detection in a low SNR regime. An alternative performance metric is the ROC curves which are generated by plotting P_{md} vs P_{fa}. Following the shorthand notation mentioned previously, the probability of detection and probability of a false alarm can be computed as:

$$\begin{aligned} P_d &= P(H_1;H_1) = P(y > \lambda; H_1) \\ &= \int_{\lambda}^{\infty} f_Y(y)dy, \qquad H_1 \end{aligned} \tag{5}$$

$$\begin{aligned} P_{fa} &= P(H_1;H_0) = P(y > \lambda; H_0) \\ &= \int_{\lambda}^{\infty} f_Y(y)dy, \qquad H_0 \end{aligned} \tag{6}$$

Probability of Detection and False Alarm under AWGN Channels

To derive the probability of false alarm using the right-tail probability of the central chi-squared density function $Q_{\chi_N^2}$, we define γ as $\gamma = \frac{h^2\zeta}{\sigma^2}$.[a] Substituting $t = y/\sigma^2$ and further integrating the probability density function in (3) under H_0 results in:

$$P_{fa} = \int_{\lambda/\sigma^2}^{\infty} \frac{1}{2^{\frac{N}{2}}\Gamma\left(\frac{N}{2}\right)} t^{\left(\frac{N}{2}\right)-1} \exp\left(\frac{-t}{2}\right) dt \quad \lambda/\sigma^2 \geq 0.$$

$$= Q_{\chi_N^2}\left(\lambda/\sigma^2\right) \tag{7}$$

where $Q_{\chi_N^2}$ can be written as ([24], Eq 26.4.4, Eq 26.4.5):

$$Q_{\chi_N^2}(\lambda/\sigma^2) = \begin{cases} 2Q\left(\sqrt{\lambda/\sigma^2}\right) & N=1 \\ 2Q\left(\sqrt{\lambda/\sigma^2}\right) + \frac{exp\left(-\frac{\lambda}{2\sigma^2}\right)}{\sqrt{\pi}} \sum\limits_{k=1}^{\frac{N-1}{2}} \frac{(k-1)!(2\lambda/\sigma^2)^{k-\frac{1}{2}}}{(2k-1)!} & N odd \\ exp\left(-\frac{\lambda}{2\sigma^2}\right) \sum\limits_{k=0}^{(\frac{N}{2})-1} \frac{\left(\frac{\lambda}{2\sigma^2}\right)^k}{k!} & N even \end{cases} \tag{8}$$

where $Q(.)$ is the complementary cumulative distribution function defined as $Q(x) = \int\limits_x^\infty \frac{1}{\sqrt{2\pi}} \exp\left(-\frac{1}{2}t^2\right)$. The same approach can be used to derive the probability of detection using the right tail of the noncentral chi-squared probability density function under H_1. By letting $t=y/\sigma^2$, the probability of detection is given by:

$$P_d = Q_{\tilde{\chi}_{N(\gamma)}^2}\left(\lambda/\sigma^2\right)$$

$$= \int\limits_{\lambda/\sigma^2}^{\infty} \left[\frac{1}{2}\left(\frac{t}{\gamma}\right)^{\frac{N-2}{4}} \exp\left[\frac{1}{2}(t+\gamma)\right] \times I_{\frac{N}{2}-1}\left(\sqrt{\gamma t}\right) dt\right] \tag{9}$$

We can rewrite (9) using ([25], Eq 2.45) for an even number of degrees of freedom as:

$$P_d = Q_{N/2}\left(\sqrt{\gamma}, \sqrt{\lambda'}\right) \tag{10}$$

where $\lambda'=\lambda/\sigma^2$, and $Q_m(.,.)$ is the m th generalized Marcum Q function which is given by:

$$Q_m(\alpha, \beta) = \frac{1}{\alpha^{m-1}} \int\limits_\beta^\infty x^m e^{-\left(\frac{x^2+\alpha^2}{2}\right)} I_{m-1}(\alpha x) dx \tag{11}$$

which is the same result as in [5].

For an odd number of degrees of freedom (10) cannot be directly evaluated. Therefore, we introduce an algorithm to solve this problem. To do so, we use the series expansion of the modified Bessel function defined in (4) and rewrite (9) as:

$$Q_{\tilde{\chi}^2_{N(\gamma)}}(\lambda/\sigma^2) = \sum_{k=0}^{\infty} \frac{\exp(-\gamma/2)(\gamma/2)^k}{k!}$$

$$\times \int_{\lambda'}^{\infty} \left(\frac{t^{\frac{N}{2}+k-1} \exp\left(-\frac{t}{2}\right)}{2^{\frac{N}{2}+k} \Gamma\left(\frac{N}{2}+k\right)} \right) dt$$

$$= \sum_{k=0}^{\infty} \frac{\exp(-\gamma/2)(\gamma/2)^k}{k!} \underbrace{Q_{\chi^2_{N+2k}}(\lambda')}_{\text{second term}} \qquad (12)$$

The second term of (12) represents the right-tail probability of a central chi-square with $l = N + 2k$ degrees of freedom. As a result, for N odd, $l = N + 2k$ is also odd. Thus, (12) can be rewritten using (8) for the odd case which results in (13), where $G_{\chi^2_{l-2}}(\lambda')$ is given by:

$$Q_{\tilde{\chi}^2_{N(\gamma)}}(\lambda') = \sum_{k=0}^{\infty} \frac{\exp(-\gamma/2)(\gamma/2)^k}{k!}$$

$$\times \left[2Q\left(\sqrt{\lambda'}\right) + \frac{\exp\left(-\frac{1}{2}\lambda'\right)}{\sqrt{\pi}} \right.$$

$$\left. \times \sum_{k=1}^{\frac{N-1}{2}} \frac{(k-1)!\,(2\lambda')^{k-\frac{1}{2}}}{(2k-1)!} \right]$$

$$= 2Q\left(\sqrt{\lambda'}\right) + \sum_{k=0}^{\infty} \frac{\exp(-\gamma/2)(\gamma/2)^k}{k!}$$

$$\times \left[\frac{\exp\left(-\frac{1}{2}\lambda'\right)}{\sqrt{\pi}} \sum_{j=1}^{\frac{l-2-1}{2}} \frac{(j-1)!\,(2\lambda')^{j-\frac{1}{2}}}{(2j-1)!} \right.$$

$$\left. + \frac{\exp\left(-\frac{1}{2}\lambda'\right)}{\sqrt{\pi}} \left(\frac{\left(\frac{l-1}{2}-1\right)!}{2\left(\frac{l-1}{2}-1\right)!} (2\lambda')^{\frac{l-1}{2}-\frac{1}{2}} \right) \right]$$

$$= 2Q\left(\sqrt{\lambda'}\right) + \sum_{k=0}^{\infty} \frac{\exp(-\gamma/2)(\gamma/2)^k}{k!}$$

$$\times \left[G_{\chi^2_{l-2}}(\lambda') + g(\lambda', l)\right] \tag{13}$$

$$G_{\chi^2_{l-2}}(\lambda') = \frac{\exp\left(-\frac{1}{2}\lambda'\right)}{\sqrt{\pi}} \sum_{j=1}^{\frac{l-3}{2}} \frac{(j-1)!\,(2\lambda')^{j-\frac{1}{2}}}{(2j-1)!} \tag{14}$$

and $g(\lambda',l)$ can be rewritten as:

$$g(\lambda', l) = \frac{\exp\left(-\frac{1}{2}\lambda'\right)}{\sqrt{\pi}} \left(\frac{\left(\frac{l-1}{2}-1\right)!}{2\left(\frac{l-1}{2}-1\right)!}(2\lambda')^{\frac{l-1}{2}-\frac{1}{2}}\right)$$

$$= \frac{\exp(\lambda'/2)}{\sqrt{\pi}} \left(\frac{((l-5)/2)!}{(l-4)!}(2\lambda')^{(l-4)/2}\right)$$

$$\times 2\lambda' \frac{((l-3)/2)}{(l-2)(l-3)} \tag{15}$$

(15) can also be reduced to

$$g(\lambda', l) = g(\lambda', l-2)\frac{\lambda'}{l-2} \tag{16}$$

where the initialization starts with $G_{\chi^2_{l-2}}(\lambda') = g(\lambda', 3) = \sqrt{\frac{2\lambda'}{\pi}}\exp(-\lambda'/2)$.

PROBABILITY OF DETECTION AND FALSE ALARMS UNDER NAKAGAMI FADING CHANNELS

To capture all different types of fading, the parameters of the Nakagami distribution can be adjusted to fit a variety of fading processes. If we define $\bar{\gamma} = E[h^2]\zeta/\sigma^2$ as the average signal-to-noise ratio, where $E(.)$ denotes the expectation operator, then the probability distribution of γ will be given as:

$$P_{\text{Nak}}(\gamma) = \left(\frac{m}{\bar{\gamma}}\right)^m \frac{\gamma^{m-1}}{\Gamma(m)} \exp\left[-\frac{m\gamma}{\bar{\gamma}}\right] \tag{17}$$

To compute the probability of detection, this must be averaged over the probability density function of the instantaneous value of γ, i.e., it can be written as:

$$P_{d_{\mathrm{Nak}}} = \int_0^\infty Q_{N/2}\left(\sqrt{\gamma}, \sqrt{\lambda'}\right) f(\gamma)\, d\gamma \tag{18}$$

Then, substituting (17) into (18) results in:

$$\begin{aligned} P_{d_{\mathrm{Nak}}} &= \int_0^\infty \left(\frac{m}{\bar{\gamma}}\right)^m \frac{\gamma^{m-1}}{\Gamma(m)} \exp\left(-\frac{m\gamma}{\bar{\gamma}}\right) Q_{N/2}\left(\sqrt{\gamma}, \sqrt{\lambda'}\right) d\gamma \\ &= \frac{2}{\Gamma(m)} \left(\frac{m}{\bar{\gamma}}\right)^m \int_0^\infty x^{2m-1} \exp\left(-\frac{\eta^2 x^2}{2}\right) Q_{N/2}\left(x, \sqrt{\lambda'}\right) dx \\ &= \alpha \int_0^\infty x^{2m-1} \exp\left(-\frac{\eta^2 x^2}{2}\right) Q_{N/2}\left(x, \sqrt{\lambda'}\right) dx \end{aligned} \tag{19}$$

where in the second step, we substitute $x = \sqrt{\gamma}$ and $\eta^2 = \frac{2m}{\bar{\gamma}}$, and in the last step, we substitute $\frac{2}{\Gamma(m)}\left(\frac{m}{\bar{\gamma}}\right)^m$ with α. Different combinations of m and $N/2$ lead to different results for the integration defined in the last step. In the following, the probability of detection is evaluated over both Rayleigh and Nakagami fading channels.

Special Case: Rayleigh fading

In the case of Raleigh fading, we set $m=1$ and use ([25], Eq B.53). Then the probability of detection can be written as:

$$\begin{aligned} P_{d_{\mathrm{Ray}}} = \exp\left(-\frac{\lambda'}{2}\right) \Bigg[& [1+\eta^2]^{u-1} \\ & \times \left\{ \exp\left(\frac{\lambda'}{2+2\eta^2}\right) - \sum_{k=0}^{u-2} \frac{1}{k!}\left(\frac{\lambda'}{2+2\eta^2}\right)^k \right\} \\ & + \sum_{k=0}^{u-2} \frac{1}{k!}\left(\frac{\lambda'}{2}\right)^k \Bigg] \end{aligned} \tag{20}$$

where $u=N/2$.

Nakagami Fading

In the case of Nakagami fading, we further simplify the expression in (19) using the series representation of the Marcum Q function [26], which is given by:

$$Q_u\left(\sqrt{\gamma}, \sqrt{\lambda'}\right) = 1 - \sum_{n \geq 0}\left[(-1)^n \exp\left(\frac{\gamma}{2}\right) \times \frac{L_n^{u-1}\left(\frac{\gamma}{2}\right)}{\Gamma(u+n+1)}\left(\frac{\lambda'}{2}\right)^{n+u}\right] \tag{21}$$

where L_j^k is the generalized Laguerre polynomial of degree j and order k. The absolute convergence of the series in (21) has been shown to be absolutely bounded by:

$$\sum_{n \geq 0}\left[(-1)^n \exp\left(\frac{\gamma}{2}\right) \frac{L_n^{u-1}\left(\frac{\gamma}{2}\right)}{\Gamma(u+n+1)}\left(\frac{\lambda'}{2}\right)^{n+u}\right] \leq \exp(-\gamma/4)\frac{1}{\Gamma(u)}\left(\frac{\lambda'}{2}\right)^{u-1}\left(\exp\left(\frac{\lambda'}{2}\right) - 1\right) \tag{22}$$

$$P_{d_{\text{Nak}}} = \alpha \int_0^\infty \left\{x^{2m-1} \exp(-\eta^2 x^2/2)\left[1 - \exp(-x^2/4)\frac{1}{\Gamma(u)}(\lambda'/2)^{u-1}(\exp(\lambda'/2) - 1)\right]\right\} dx$$

$$= \left[\begin{array}{l} \underbrace{\alpha \int_0^\infty x^{2m-1} \exp(-\eta^2 x^2/2)\, dx}_{\text{first term}} \underbrace{- \frac{\alpha}{\Gamma(u)}(\lambda'/2)^{u-1} \exp(\lambda'/2) \int_0^\infty x^{2m-1} \exp(-\eta^2 x^2/2) \exp(-x^2/4)\, dx}_{\text{second term}} \\ \underbrace{+ \frac{\alpha}{\Gamma(u)}(\lambda'/2)^{u-1} \int_0^\infty x^{2m-1} \exp(-\eta^2 x^2/2) \exp(-x^2/4) dx}_{\text{Third term}} \end{array}\right] \tag{23}$$

Then, substituting (22) into (19) results in (23). Next, by changing the variable M=2m-1, and further integrating, the first term of (23) can be reduced to:

$$\alpha \int_0^\infty x^{2m-1} \exp(-\eta^2 x^2/2)\, dx = \alpha \frac{\Gamma((M+1)/2)}{2\left(\frac{\eta^2}{2}\right)^{(M+1)/2}} = 1 \tag{24}$$

The second term of (23) can be further reduced to:

$$\frac{\alpha}{\Gamma(u)}(\lambda'/2)^{u-1}\exp(\lambda'/2)\int_0^\infty x^M \exp\left(-((\eta^2/2)+1/4)x^2\right)dx$$
$$=\frac{\exp(\lambda'/2)\,(\lambda'/2)^{u-1}}{\Gamma(u)}\frac{\left(\frac{m}{\bar{\gamma}}\right)^m}{\left(\frac{m}{\bar{\gamma}}+1/4\right)^m}, \quad (25)$$

while the third term can be reduced as:

$$\frac{\alpha}{\Gamma(u)}(\lambda'/2)^{u-1}\int_0^\infty x^M \exp\left(-((\eta^2/2)+1/4)x^2\right)dx$$
$$=\frac{1}{\Gamma(u)}(\lambda'/2)^{u-1}\frac{\left(\frac{m}{\bar{\gamma}}\right)^m}{\left(\frac{m}{\bar{\gamma}}+1/4\right)^m} \quad (26)$$

Then, the probability of detection under a Nakagami fading channel is the result of (24), (25), and (26) which can be written as:

$$P_{d_{\text{Nak}}}=1-\frac{1}{\Gamma(u)}(\lambda'/2)^{u-1}\frac{\left(\frac{m}{\bar{\gamma}}\right)^m}{\left(\frac{m}{\bar{\gamma}}+1/4\right)^m}\left[\exp(\lambda'/2)-1\right] \quad (27)$$

It is clear from (27) how changing various parameters affect the detection process. The new derived expression reveals the fact that the ratio of parameter m to parameter $\bar{\gamma}$ is an important consideration when evaluating the probability of detection over Nakagami fading channels. For example, at low $\bar{\gamma} < 2$ dB and when the degree of freedom u is fixed, the Nakagami parameter m has only a minor effect on the detection process. That means no matter how much m increases, the probability of detection stays almost the same. However, at high $\bar{\gamma} > 15$ dB, increasing m will greatly improve the probability of detection. This will be discussed further in the simulation section.

Another expression for the probability of detection over Nakagmai fading channels can easily be derived by rewriting the right-hand side of (21) as:

$$Q_u\left(\sqrt{\gamma},\sqrt{\lambda'}\right)=1-\sum_{n\geq 0}\exp(-\gamma/2)\left(\frac{\gamma}{2}\right)^n\left(\frac{\gamma^*(u+n,\lambda'/2)}{n!\,\Gamma(u+n)}\right) \quad (28)$$

where we use the notation $\gamma^*(.,.)$ to represent the lower incomplete gamma function. Equation (28) is the well-known canonical representation of the u th order generalized Marcum Q function. Then, with the help of (18) and using $\sum_{k=0}^{\infty} (a/2)^k/_{k!} = \exp(a/2)$ and after simple mathematical manipulation as shown in (24-26), the probability of detection over Nakagami fading channels can be approximated as:

$$P_{d_{\text{Nak}}} \cong 1 - \beta\left(\frac{\gamma^*(\lambda'/2, u)}{\Gamma(u)}\right) \tag{29}$$

where $\beta = [2m/(2m+\bar{\gamma})]$.

SIMULATION AND NUMERICAL RESULTS

A binary phase shift keying signal[b] with sampling frequency $f_s = 10 f_c$, where f_c is the carrier frequency, is used to investigate the detection probability for even/odd degrees of freedom. For the even degree of freedom with N arbitrarily chosen to be (10,20), the simulation results are compared to (10). As shown in Figure 2, it can be seen that increasing N or the SNR enhances the probability of detection. Therefore, this result is consistent with most of the reported results in the literature. We also notice that the simulation results coincide with the theoretical ones.

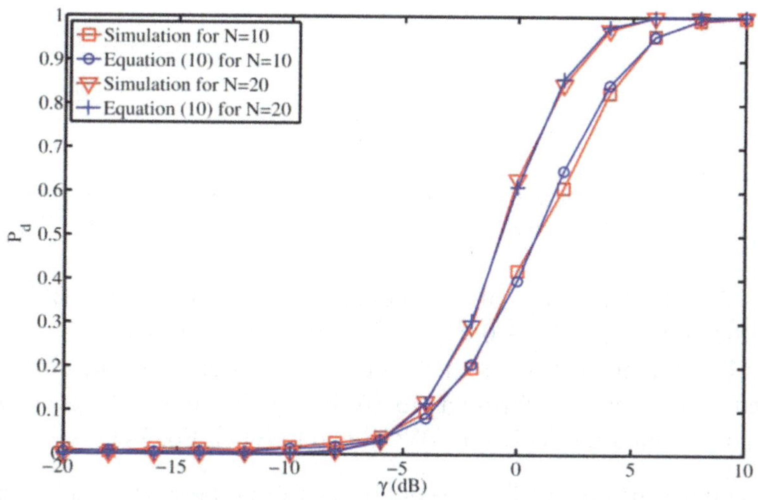

Figure 2: Probability of detection vs SNR for BPSK signal. With $f_s = 10f_c$, $P_{fa} = 0.01$, and different *even* numbers of degrees of freedom; the simulation is compared to (10) in an AWGN channel.

Figure 3 shows the detection probability using the recursive algorithm for odd/even degrees of freedom with N=31,32 for comparison. As shown, the recursive algorithm perfectly matches the simulation results. Moreover, the accuracy of the recursive algorithm goes up to 15 decimal places, which is the maximum number of digits that Matlab can support. Although there is a small effect on the detection probability when we compare the even/odd cases, in practice and since most current functions deals only with even degree of freedom, this algorithm becomes more beneficial. For example, when evaluating the detection probability using the Marcum function in Matlab with N=31 (odd), the third argument of the Marcum function will be 15.5; hence, the implemented Marcum function in Matlab cannot be used to evaluate the detection probability in this case as it accepts only integer numbers. However, our algorithm does.

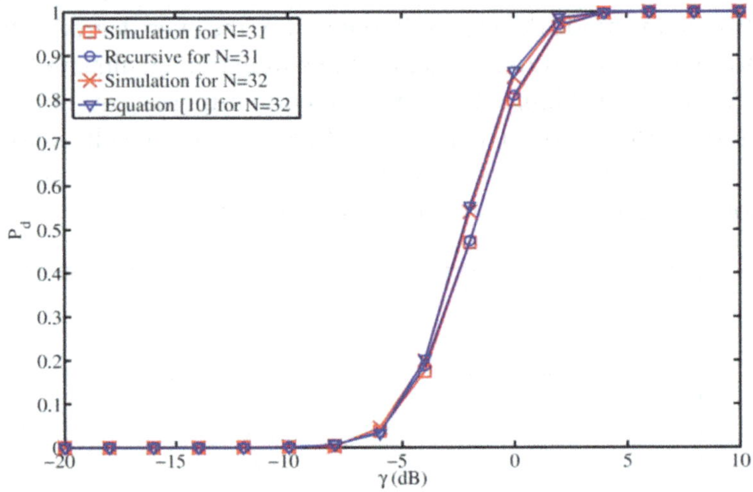

Figure 3: Probability of detection vs SNR for BPSK signal. With $f_s = 10f_c$, P_{fa} =0.01, and different odd with *N=31*, even with ***N=32*** number of degrees of freedom. In the odd case, the simulation is compared to the recursive formula (13,16) in AWGC, and in the even case, the simulation is compared to (10) in AWGC and to ([14], Eq. 14).

To evaluate the closed-form expressions derived for Nakagami channels, an extensive simulation has been performed using the ROC. The derived expressions are evaluated and compared with the numerical integration of (18) and with the expressions reported by [14–17].[c]

Comparison of the Derived Expressions with Equation 18

In Figure 4, we compare the derived expressions in (27) and (29) with the numerical integration of (18) for different values of $\bar{\gamma}$ and m. Figure 4 also shows the effect of varying the Nakagami parameter m on the misdetection probability at low and high values of $\bar{\gamma}$ which will be discussed in the following subsections.

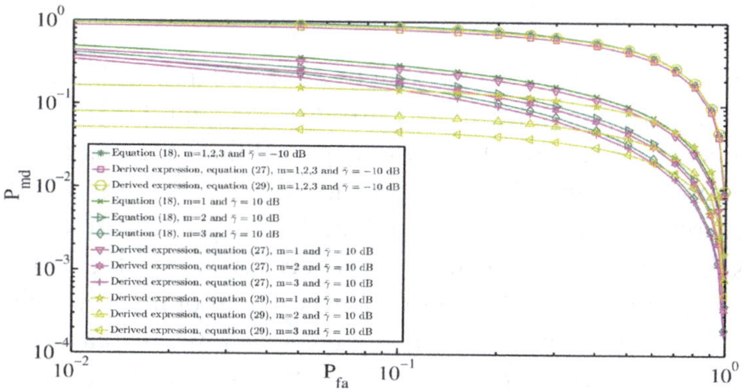

Figure 4: Comparison between the derived expressions, (27) and (29), and numerical integration of (18) for different values of $\bar{\gamma}$ and m .

Low value of $\bar{\gamma}$

At a low value of $\bar{\gamma}$, i.e., $\bar{\gamma}$ =−10 $\bar{\gamma}$ =-10 dB, it can be seen that increasing the value of m, (m=1,2,3), does not improve the misdetection probability for both derived expressions which concurs with the numerical integration of (18). We also note that (29) exactly matches (18); on the other hand, there is a minor discrepancy between (27) and (18).

High value of $\bar{\gamma}$

At a high value of $\bar{\gamma}$, i.e., $\bar{\gamma}$ =10 dB, increasing m will greatly improve the misdetection probability for both derived expressions (27 and 29), which also concurs with the numerical integration of (18) as can be seen in Figure 4. Further, we notice that at a very low false alarm probability, (29) is less accurate compared to (27). However, as the false alarm probability increases, the results for both expressions match that of the numerical integration of (18).

Comparison of the Derived Expressions with Related Works

In this subsection, we compare the new derived expressions with the previously reported expressions for high and low values of $\bar{\gamma}$ **and** $m = 1,2,3$.

Low value of $\bar{\gamma}$

In Figure 5, the new derived expressions are compared with the expressions of ([14]], Eq. 20), ([15]], Eq. 12), ([16]], Eq. 13), and ([17]], Eq. 13). It can be seen that at a very low false alarm probability, the results of the new derived expressions in (27) and (29) and all expressions in the previously mentioned references are a good match. As the probability of false alarm increases, however, a discrepancy arises between ([16]], Eq. 13) and ([17]], Eq. 13) when compared to the new expressions and/or the work of [14],[15]]. Moreover, as previously shown in Figure4, at low, increasing m did not enhance the probability of misdetection. This behavior can be seen from (27) and (29) which also concurs with the work of ([14]], Eq. 20) and ([15]], Eq. 12). However, the result of ([16]], Eq. 13) has some discrepancies when the false alarm probability or the value of m increases. Moreover, the result of ([17]], Eq. 13) is not consistent with increasing m. For example, the result at the value of $m=3$ is worse than the result at the value of $m=1,2$ as can be seen in Figure 5. We note also that the probability of misdetection approaches zero at the point of $(1 - P_{fa})$ for the expressions reported by [[16],[17]].

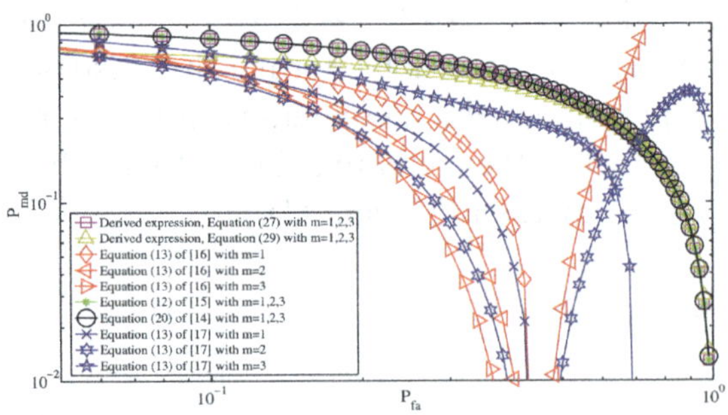

Figure 5: Comparison of the new derived expressions. Comparison of the new derived expressions (27) and (29), with the work of ([14]], Eq. 20), ([15]], Eq. 12), ([16]], Eq. 13), and ([17]], Eq. 13) with $\bar{\gamma}$ =–2 dB, $u=5$, and different values of m.

High value of $\bar{\gamma}$

Figure 6 shows the simulation results for all expressions when evaluated at a high value of $\bar{\gamma}$. The result of (27) is very close to that of ([14]], Eq. 20) and of ([15]] for the values of m=2,3. Moreover, although there are some discrepancies when evaluating (29) as compared to ([14]], Eq. 20) and ([15]], Eq. 12), at a high value of a false alarm probability, the expression (29) is a perfect match. To this end, we conclude that (29) works well at high values of $\bar{\gamma}$ and when $P_{fa}>0.2$. Moreover, (29) is accurate at low $\bar{\gamma}$ for all values of false alarm probability. On the other hand, the expressions ([[16]], Eq. 13) and ([17]], Eq. 13) are less accurate for all evaluated points of false alarm probability and the unpredictable behavior of these expressions still exists as discussed in the low value of $\bar{\gamma}$ case. Moreover, results from the expression ([[17]], Eq. 13) are also inconsistent as the value of m is increased.[d]

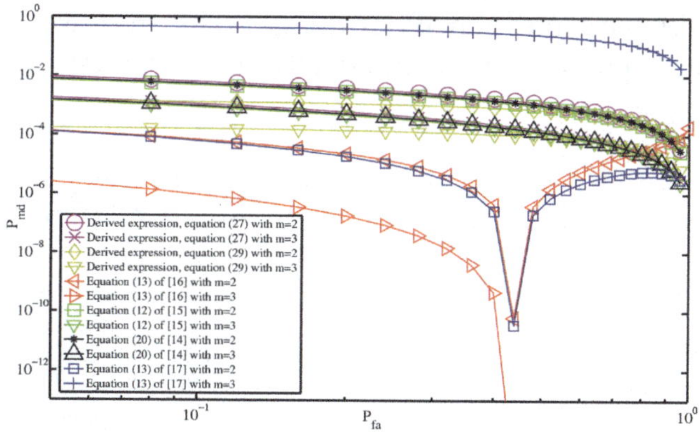

Figure 6: Simulation results for all expressions when evaluated at a high value of $\bar{\gamma}$. Comparison of the new derived expressions, (27) and (29), with the work of ([14]], Eq. 20), ([15]], Eq. 12), ([16]], Eq. 13), and ([17]], Eq. 13) with $\bar{\gamma}$ =20 dB, $u=5$, and $m=2,3$.

Computational Complexity

In the previous subsections, we discussed the accuracy of the derived expressions and compared them to the expressions in ([14]], Eq. 20) and ([15]], Eq. 12) that require summation and integration terms to get the needed accuracy. The derived expressions were also compared to the expressions of ([16]], Eq. 13) and ([17]], Eq. 13) that depend on evaluating the gamma function as does the derived expression in (29). To complete the picture, we need another performance metric that distinguishes between these expressions. To do so, a simple Matlab code was written to measure the computation time required of a central processing unit (CPU) to evaluate a point in the ROC domain. We ran the code on a computer equipped with a CPU with a speed of 3.07 GHz on which we cleared all background application processes that might run on the system.[e] We used 21 points of false alarm probability, i.e., (P_{fa} =0:0.05:1) with a step of 0.05. At each point, the code iterated 1,000 times and averaged afterward. The measured computation time of the CPU for all points used and for each expression is plotted in Figure7. It can be seen that more computation time is required to get highly accurate results. This is evident from the evaluated points of the expressions ([14]], Eq. 20) and ([15]], Eq. 12) as they have the highest computation time.[f] On the other hand, the derived expression in (27) and the reported expression in ([17]], Eq. 13) have almost the same computation time. Moreover, the expression of ([16]], Eq. 13) has the lowest computation time while the expression of ([14]], Eq. 20) has the highest.

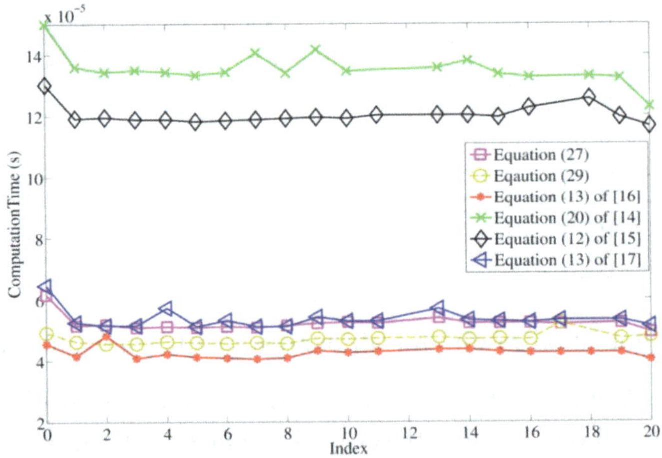

Figure 7: Comparison of the computation time, all expressions computed at $\bar{\gamma}$ =20 dB and $u=5$.

The derived expressions can also be compared in terms of the number of multiplications. For example, the derived expression in ([14], Eq. 20) is based on the summation of confluent hypergeometric functions, and such functions have a computational complexity of order $O\left[\log^2(n)\bar{M}(n)\right]$ for n-digit precision [27], where n means computing n digits, and $\bar{M}(n)$ is the bit complexity of multiplication. However, the reported expression of ([15],12) is based on an infinite series of gamma functions, and such functions have a computational complexity of $O\left[\sqrt{n}\bar{M}(n)\right]$. Using this notation, the derived expressions of (27) and (29) have a computational complexity of $O(\bar{M}(n))$. Since $O\left[\log^2(n)\bar{M}(n)\right] > O\left[\sqrt{n}\bar{M}(n)\right] > O(\bar{M}(n))$, which is also consistent with the simulation results of Figure 7, therefore, the derived expressions have a lower complexity than ([14], Eq. 20) and ([15], Eq. 12) and have the same computational complexity as ([16], Eq. 13) and ([17], Eq. 13).

CONCLUSIONS

Spectrum sensing using energy detectors under different fading channels was investigated. We derived tight closed-form expressions for the probability of detection in Nakagami channels. The closed-form expressions can easily be used for Rayleigh fading channels by setting $m=1$. The results of the closed-form formula as compared with other expressions based on summation and integration terms are very close. Furthermore, the derived expression of (27) can be used for all $\bar{\gamma}$; however, there is a minor limitation of using (29) specifically at high values of $\bar{\gamma}$. Moreover, the derived expressions have a lower computational complexity compared to other expressions with only a very small loss of accuracy. In addition, we introduced an accurate recursive algorithm to compute the probability of detection for an odd number of degrees of freedom under AWGN channels. Our simulation shows that the detection process for a binary phase shift keying signal using the recursive formula perfectly coincides with the recursive algorithm.

Endnotes

[a] In AWGN channels, there is no fading, i.e., $h^2=1$.

[b] Different modulation schemes could be used in the simulation since the derived expression is independent of the modulation used.

[c] For comparison purposes, we have used a value of $n=1$ in the expression of [17], Eq. 13), where n represents the number of nodes cooperating in the sensing process according to [17] notations.

[d] 'Equation (29) comes from another way of calculating the probability of detection over Nakagami fading channels in order to compare with results

from the state of the art, which seemed rather optimistic for low false alarm probability, and that for high SNR cases,' one of the anonymous reviewer's comment.

[e] The conducted simulations show that the computed time will only be scaled by a constant factor and that the calculated computation time will not be affected if we do or do not clear the processor cache of any background application processes.

[f] The number of terms used to calculate the summation of [15], Eq. 12) was 20.

REFERENCES

1. IEEE 802.22 *Working group on wireless regional area networks* . Accessed 1 May 2012 http://www.ieee802.org/22/
2. Allwright D, Heyes C, Hulbert P, Hurley S, Kennett M, Leese R, Mitchell S, Whittaker R, Wyman G: Improving the sharing of the radio spectrum: Final report. 2006.http://stakeholders.ofcom.org.uk/binaries/research/technology-research/final.pdf . Accessed 10 June 2012
3. 802.22-2011: IEEE Standard for information technology-telecommunications and information exchange between systems-wireless regional area networks WRAN,-specific requirements - part 22 cognitive wireless RAN medium access control MAC and physical layer PHY specifications: policies and procedures for operation in the TV bands. *IEEE P802.22/D2.0,* 2011, 1-698.
4. Cabric D, Mishra S, Brodersen R: Implementation issues in spectrum sensing for cognitive radios. In *Conference Record of the Thirty-Eighth Asilomar Conference on Signals, Systems and Computers*. Pacific Grove; 772-776. 7–10 Nov 2004
5. Digham F, Alouini MS, Simon M: On the energy detection of unknown signals over fading channels. *IEEE Trans. Commun* 2007, 55(1):21-24.
6. Pawelczak P, Janssen G, Prasad R: Performance measures of dynamic spectrum access networks. In *Global Telecommunications Conference,*. San Francisco; 2006:1-6. 27 Nov–1 Dec
7. Yuan Y, Bahl P, Chandra R, Chou P, Ferrell J, Moscibroda T, Narlanka S, Wu Y: Knows: Cognitive radio networks over white spaces. In *2nd IEEE International Symposium on New Frontiers in Dynamic Spectrum Access Networks*. Dublin; 416-427. 17–20 Apr 2007
8. Alam S, Odejide O, Olabiyi O, Annamalai A: Further results on area under the ROC curve of energy detectors over generalized fading channels. In *34th IEEE Sarnoff Symposium*. Princeton; 1-6. 3–4 May 2011
9. Herath S, Rajatheva N, Tellambura C: Energy detection of unknown signals in fading and diversity reception. *IEEE Trans. Commun* 2011, 59(9):2443-2453.
10. Kostylev VI: Characteristics of energy detection of quasideterministic radio signals. *Radiophys. Quantum Electron* 2000, 43(10):833-839.

11. Peh E, Liang YC, Guan YL, Pei Y: Energy-efficient cooperative spectrum sensing in cognitive radio networks. In *Global Telecommunications Conference (GLOBECOM 2011)*. Houston; 1-5. 5–9 Dec 2011
12. Peh E, Liang YC, Guan YL, Zeng Y: Cooperative spectrum sensing in cognitive radio networks with weighted decision fusion scheme. In *2010 IEEE 71st Vehicular Technology Conference (VTC 2010-Spring)*. Taipei; 1-5. 16–19 May 2010
13. Sofotasios P, Rebeiz E, Zhang L, Tsiftsis T, Cabric D, Freear S: Energy detection based spectrum sensing over κ-μ and κ-μ extreme fading channels. *IEEE Trans. Vehicular Tech* 2012, 62(3):1031-1040.
14. Digham F, Alouini MS, Simon MK: On the energy detection of unknown signals over fading channels. *IEEE Trans. Commun* 2003, 5: 3575-3579.
15. Annamalai A, Olabiyi O, Alam S, Odejide O, Vaman D: Unified analysis of energy detection of unknown signals over generalized fading channels. In *International Wireless Communications and Mobile Computing Conference (IWCMC)*. Istanbul; 636-641. 4–8 July 2011
16. Reisi N, Ahmadian M, Salari S: Performance analysis of energy detection-based spectrum sensing over fading channels. In *2010 6th International Conference on Wireless Communications Networking and Mobile Computing (WiCOM),*. Chengdu; 1-4. 23–25 Sept 2010
17. Horgan D, Murphy C: Fast and accurate approximations for the analysis of energy detection in nakagami-m channels. *IEEE Commun. Lett* 2012, 17(1):83-86.
18. Pridham R, Urkowitz H: Comment on energy detection of unknown deterministic signals. *Proc. IEEE* 1968, 56(8):1379-1380.
19. Urkowitz H: Energy detection of unknown deterministic signals. *Proc. IEEE* 1967, 55(4):523-531.
20. Axell E, Larsson E: Optimal and sub-optimal spectrum sensing of OFDM, signals in known and unknown noise variance. *IEEE J. Selec. Areas Commun* 2011, 29(2):290-304.
21. Gismalla E, Alsusa E: On the performance of energy detection using Bartlett's estimate for spectrum sensing in cognitive radio systems. *IEEE Trans. Signal Proces* 2012, 60(7):3394-3404.
22. Shnidman D: The calculation of the probability of detection and the generalized Marcum Q-function. *IEEE Trans. Inf. Theory* 1989, 35(2):389-400. 10.1109/18.32133

23. Proakis J, Salehi M: *Digital Communications*. McGraw Hill, New York; 2008.

24. Abramowitz M, Stegun IE, I.A: *Handbook of Mathematical Functions with Formulas, Graphs, and Mathematical Tables*. Dover, New York; 1972.

25. Simon K: *Probability Distribution Involving Gaussian Random Variables: A Handbook for Engineers and Scientists*. Springer, New York; 2002.

26. András S: The generalized Marcum Q-function: an orthogonal polynomial approach. *Acta Universitatis Sapientiae Mathematica* 2011, 3(1):60-76.

27. Borwein JM, Borwein PB: On the complexity of familiar functions and numbers. *SIAM Rev* 1988, 30(4):589-601. 10.1137/1030134

16

Signal-to-noise Ratio Estimation for M-QAM Signals in $\eta-\mu$ and $\kappa-\mu$ fading Channels

Wamberto José Lira de Queiroz[1], Danilo Brito Teixeira de Almeida[1], Francisco Madeiro[2] and Waslon Terllizzie Araújo Lopes[3]

[1]Federal University of Campina Grande and Institute for Advanced Studies in Communications, Campina Grande, Brazil
[2]Polytechnic School of Pernambuco, University of Pernambuco, Recife, Brazil
[3]Department of Electrical Engineering, Federal University of Paraíba, João Pessoa, Brazil

ABSTRACT

In this paper, signal-to-noise ratio (SNR) estimation is carried out by the method of moments (MOM) for fading channels modeled by probability distributions $\eta-\mu$ and $\kappa-\mu$, considering *M*-ary quadrature amplitude

Citation: Lira de Queiroz, W., Brito Teixeira de Almeida, D., Madeiro, F. et al. "Signal-to-noise ratio estimation for M-QAM signals in $\eta - \mu$ and $\kappa - \mu$ fading channels". EURASIP J. Adv. Signal Process. 2019, 20 (2019). https://doi.org/10.1186/s13634-019-0607-7

modulation (M-QAM) with constellation energy normalized to one. New expressions are presented for the SNR estimation and for the mean, variance, and normalized mean square error (NMSE) of the estimates, obtained by a statistical linearization argument. Additionally, it is shown how to obtain the SNR estimate for Nakagami-m channel from the estimation derived for the models η–μ and κ–μ. The results obtained from the analytical expressions are corroborated by simulation results and show that the MOM is a suitable alternative for scenarios in which the mathematical tractability does not suggest the application of other estimation techniques.

INTRODUCTION

Signal-to-noise ratio (SNR) estimation has been a recurrent research topic, due to the relevance of SNR for a variety of mobile communication systems. The a priori knowledge of the communication channel conditions is an important issue as long as those systems become more complex and widely required. In [1] for instance, the a priori knowledge of the channel, by means of the SNR, is proposed for evaluating the effective transmission rate (throughput) in a communication system with adaptive modulation and coding, while in [2] its use is considered in adaptive transmission systems.

In [3], the knowledge of the SNR is necessary for assessing the time-varying channel condition of an adaptive system with frequency hopping, and in [4] it is necessary for planning relay communication systems. In [5], SNR is a useful parameter in the scenario of turbo decoding systems, while in [6] it is useful in the context of low-density parity check (LDPC) codes.

Usually, the choice of the estimation technique depends on the complexity of the mathematical model of the signal in the receiver [4, 7]. Depending on that complexity, one can use a method belonging to the class of estimators that use a training sequence (that is, a data-aided method (DA)) [8, 9], or a method belonging to the class of estimators that do not have a priori knowledge of the transmitted sequence of symbols (that is, a non-data-aided method (NDA)) [10].

In [11], for example, a new SNR estimator is derived, DA and NDA, for a slow time-varying channel with impulse response characterized by a polynomial function of order L_c and phase-shift keying (PSK) signals. In the NDA scenario, the expectation maximization (EM) algorithm is proposed for the calculation of the estimates and improvement of performance is observed in relation to the estimator with channel considered constant throughout the time of observation, at the cost of a moderate increase in computational complexity. The difference of that approach in relation to the method of moments (MOM) NDA technique considered in our article is that the MOM is not efficient in reaching the Cramér-Rao bound.

In [12], the least-squares (LS) technique is proposed for estimation of SNR in a received signal model composed by M-QAM symbols (M=16) multiplied by samples of the impulse response of a time-varying channel added to samples of white Gaussian noise. The impulse response of the channel is approximated by a polynomial as a function of time with constant coefficients calculated by means of the LS technique, with the aid of a pilot sequence sent periodically and an array of antennas of N_a elements. The performance of the technique is evaluated by means of the normalized root mean square error (NRMSE), without comparison with the Cramér-Rao bound, and requires a much smaller number of samples than the required by the MOM without training sequence, in order to reach NRMSE values below 0.1 for an SNR above 10 dB.

In [13], one considers the maximum likelihood estimation (MLE) of the SNR in the demodulator of an orthogonal and non-coherent frequency-shift keying (FSK) signal, in which the transmitted signal is affected by Rayleigh fading and that this fading is constant in a block of k symbols, while in [14], the authors consider a binary frequency-shift keying (BFSK) affected by constant Rayleigh fading in a block of symbols whose BFSK carrier has frequency deviation. The MLE with and without pilot sequences are presented for the SNR, assisted by estimators by the MOM for the frequency deviation. Although the MLE is optimum in the sense of the minimum variance criterion, it depends on the expression of the probability density function (PDF) of the samples of the observed signal, and in the case of generalized fading models, such as those considered in this work, their PDFs make the PDFs of the received signal complicated for treatment by MLE.

In [15], the authors consider a model of received signal for a signal with frequency modulated differential chaos shift keying (FMDCSK) transmitted by a channel with multipath propagation and Rayleigh fading. The authors analyze the SNR estimators and their performance considering that the channel coefficients that characterize the fading are constant in a sequence of K symbols of the FMDCSK signal. The estimators are calculated for the data-aided (DA), non-data-aided (NDA), and joint DA-NDA cases, presenting good results in relation to the proximity of the Cramér-Rao Lower Bound (CRLB) for values of SNR above 20 dB.

In [16], an estimation method of SNR is presented for linearly modulated signals captured by an array of antennas in an environment with complex white additive Gaussian noise with spatial uncorrelation between the elements. In this NDA and single input multiple output (SIMO) estimator, based on the MOM, the performance is assessed in terms of the Normalized Mean Square Error (NMSE) for QAM signals and improves with the increase of the number of elements. The estimation of SNR in SIMO systems is also addressed in [17], in which the authors consider signal samples captured by an array of antennas in a channel model of constant gain by the time of observation of a sequence of symbols. A MLE based on the I/Q components of the received signal is evaluated, for the cases DA and NDA, that reach the CRLB for a wide range of SNR. In [18], the problem of estimation of the SNR is extended to a multiple input multiple output (MIMO) system in a channel model with block flat fading, in which the channel gain matrix is considered constant by a block of N symbols. From this consideration and reduction of the channel model to a Gaussian channel to each block of N symbols, the authors then present ML estimators for the DA and NDA cases, as well as the CRLB.

In [19], the main contribution of the authors is the analysis of the CRLB of SNR estimates of signals with minimum shift keying (MSK) modulation and QAM with turbo encoding. Even in a constant gain channel model over a K symbols window, the task of calculating the CRLB for the M-QAM symbols is the most laborious of the article and is solved from the analysis of the structure of Gray's mapping. The authors then show that the proposed DA ML estimator has a lower CRLB than the NDA CRLB case. The CRLB is also evaluated in [20] for a channel model of constant gain over the whole set of symbols observed under QAM modulation, for the cases DA and NDA. In the channel models of the aforementioned references, the channel gain is considered constant by a sequence of symbol intervals, which contributes to the mathematical treatment of estimation by maximizing the likelihood function. In our proposed model, in which fading can vary at each time interval in which a sample is obtained at the output of the receiver's matched filter, the PDF of the received signal modulus makes the likelihood function calculation more difficult.

In [21], the proposed solution consists of using the goodness-of-fit test of Kolmogorov-Smirnov (K-S) for the evaluation of the distance between empirical cumulative distribution function (CDF) generated from observed samples of the received signal and the theoretical curves of CDFs generated and stored in a file for different configurations of channel parameters. The proposed estimator was presented for a Gaussian channel, of constant gain by a sequence of symbol intervals, and was evaluated by means of the normalized root mean square error (NRMSE). For fading channels, such as the models considered in this manuscript, the method would depend on the calculation of the CDF of the envelope of the signal received, which would be a challenging task.

It is worth mentioning that in the paper by Bellili et al. [22], the authors do not make any other consideration about the channel coefficients than they are deterministic and unknown. In our article, the only consideration is that the fading is constant during the interval of the impulse response of the matched filter in the receiver. The main merit for our work is that the random behavior of the channel is taken into account by the fading probability distributions and the estimators obtained have simple expressions, encompass the estimators for other fading models, and are not aided by data sequence. In the flat fading channel model considered in [22], the NDA estimator needs the EM algorithm that converges for an optimum solution with a reduced number of iterations at cost of a greater computational complexity.

In propagation environments in which the fading is characterized by probability distributions such as $\eta-\mu$ and $\kappa-\mu$ [23], the SNR estimation is difficult by using estimators such as MLE, because the likelihood function for the problem becomes complicated. In these cases, the MOM is a good alternative.

Despite raising the mathematical complexity of the SNR estimation problem, general probability distributions, such as the $\eta-\mu$ and $\kappa-\mu$ distributions, model a wider variety of fading signals, such as signals received on Nakagami-*m*, Hoyt (Nakagami-*q*), Rice (Nakagami-*n*), unilateral Gauss, and Rayleigh channels. According to [23], the $\kappa-\mu$ distribution is better suited for line-of-sight applications and the $\eta-\mu$ distribution is more appropriate for non-line-of-sight applications. The SNR estimators obtained for the signal received under these two fading models encompass all the estimators obtained for the other fading models mentioned. The estimation of SNR by the MOM has been the considered technique, as shown, for example, in references [24, 25] and more recently in reference [26] in which the transmitted signal is modeled by a complex Gaussian random variable with zero mean by component. In [27], the MOM is proposed in a process of joint estimation of both the K parameter of the Rice fading distribution and SNR in a SIMO communication system.

In the present paper, the method of moments for SNR estimation is applied to channel models in which the fading is characterized by $\eta-\mu$ and $\kappa-\mu$ distributions and modulation scheme M-QAM [28] is considered.

Main Contributions

The main contributions of this paper are as follows:

- New expressions for the SNR estimates by NDAMOM for a received signal model for M-QAM signal under $\eta-\mu$ and $\kappa-\mu$ fading.
- New expressions for the evaluation of the mean, variance, and NMSE of the estimates, obtained from a statistical linearization procedure.
- Comparative analysis between the estimates obtained for the models $\eta-\mu$, $\kappa-\mu$, and Nakagami-m.
- Exact expressions for the moments of order k of the envelope of the observed signal.

The remaining of the paper is organized as follows. In Section 2, methods used in the work and the problem definition are presented. In Section 3, the derivation of the kth moment of the absolute value of the samples of the observed signal is presented. In Section 4, the derivation of the SNR estimates is presented. In Section 5, it is shown how to obtain the SNR estimates for the signal under Nakagami fading from the estimates for the models $\eta-\mu$ and $\kappa-\mu$. In Section 6, the derivation of the mean, variance, and NMSE of the estimates is presented. In Section 7, the moments of order 2, 4, 6, and 8 of M-QAM are presented. In Section 8, a proposal is presented for evaluating the CRLB. In Section 9, theoretical curves corroborated by curves obtained by simulations are presented, and in Section 10, the conclusions of the work are provided.

METHODS

The aim of this paper is to apply the MOM in the estimation of the SNR considering communication channels subject to $\eta-\mu$ and $\kappa-\mu$ generalized fading and M-QAM modulation schemes. Firstly, the channel model is described and the moments of the received signal are determined. Then, the estimation of the SNR is obtained from the ratio of the square of the second moment and the fourth moment considering the generalized fading models. The performance of the method is assessed by means of the normalized mean square error and the variance of the estimate.

Problem Definition

The mathematical model for a signal sample $r[n]$ observed at the output of the detector with a matched filter, in a discrete time n, from a transmission through a fading channel, modeled by a multiplicative gain $g[n]$ and by additive noise $w[n]$, can be written as

$$r[n] = g[n]\,s[n] + w[n]. \tag{1}$$

In this model of received signal, $s[n]$ represents the equiprobable symbols of a M-QAM constellation, normalized such that the average energy per symbol be unitary, and $w[n]$ is a Gaussian complex random variable of zero mean and variance $2\sigma_w^2$. In this paper, the modulus of the random variable $g[n]$ is characterized by the probability distributions $\eta-\mu$ and $\kappa-\mu$ [23], which have been used to characterize the envelope of the fading in mobile communication channels since both of them can represent a wide variety of fading models. The average power of the envelope of $g[n]$ is given by $\sigma_g^2 = E\left[|g[n]|^2\right]$, in which $E[\cdot]$ is the expected value operator.

The $\eta-\mu$ model is used to characterize small scale intensity variations in the faded signal in non-line-of-sight links and has two types, format I and format II. In format I, the parameter η varies in the interval $[1,\infty)$ and represents the ratio of the powers of the in-phase and quadrature components of the scattered electromagnetic wave, and in format II, the parameter η varies in the interval $(-1,1)$ and represents the correlation coefficient between the in-phase and quadrature components of the electromagnetic wave in each cluster of multipath. The parameter μ has the same physical meaning in both formats and represents an extension of the number of clusters of multipath to the real numbers [23]. The PDF of $|g[n]|$ can be written as

$$f_{|g[n]|}(|g[n]|) = \frac{4\sqrt{\pi}\mu^{\mu+\frac{1}{2}}h^{\mu}}{\Gamma(\mu)H^{\mu-\frac{1}{2}}}\frac{1}{\sqrt{\sigma_g^2}}\left(\frac{|g[n]|^2}{\sigma_g^2}\right)^{\mu}\times$$
$$\exp\left(-2\mu h\left(\frac{|g[n]|^2}{\sigma_g^2}\right)\right)\mathrm{I}_{\mu-\frac{1}{2}}\left(2\mu H\frac{|g[n]|^2}{\sigma_g^2}\right), \tag{2}$$

in which $h=(2+\eta^{-1}+\eta)/4$ and $H=(\eta^{-1}-\eta)/4$ for format I, while $h=(1-\eta^2)^{-1}$ and $H=\eta/(1-\eta^2)$ for format II. The term $\mathrm{I}_v(x)$ represents the modified

Bessel function of first kind and order v and $\Gamma(\mu)$ represents the Gamma function.

The probability distribution κ–μ, on the other hand, can be applied in the characterization of small scale fading (variations in the signal intensity in short time intervals from propagation by multiple paths) and with line-of-sight. The parameter $\mu \geq 0$ represents the number of clusters of multipath in the environment. The parameter κ is defined as the ratio of the total power of the dominant components to the total power of the scattered waves. For an envelope $|g[n]|$ of root mean square (rms) value $\hat{r} = \sqrt{E\left[|g[n]|^2\right]} = \sqrt{\sigma_g^2}$, the PDF of the envelope κ–μ can be written as

$$f_{|g[n]|}\left(|g[n]|\right) = \frac{2\mu\left(1+\kappa\right)^{\frac{\mu+1}{2}}}{\kappa^{\frac{\mu-1}{2}}\exp^{(\kappa\mu)}\sqrt{\sigma_g^2}}\left(\frac{|g[n]|}{\sigma_g^2}\right)^{\mu} \times \exp\left(-\mu\left(1+\kappa\right)\left(\frac{|g[n]|^2}{\sigma_g^2}\right)\right) \times \mathrm{I}_{\mu-1}\left(2\mu\sqrt{\kappa\left(1+\kappa\right)}\left(\frac{|g[n]|}{\sqrt{\sigma_g^2}}\right)\right). \tag{3}$$

Considering that the variables $g[n]$, $s[n]$, and $w[n]$ are independent, it follows that a sequence of samples of $r[n]$ can be seen as a sequence of the faded signal $g[n]s[n]$ affected by a white Gaussian noise sequence $w[n]$. In this model of $r[n]$, the average power can be written as

$$P_r = E[r[n]r^*[n]] = E\left[|g[n]|^2\right]E\left[|s[n]|^2\right] + E\left[|w[n]|^2\right] \tag{4}$$

and the SNR can be written as

$$\mathrm{SNR} = \frac{E\left[|g[n]|^2\right]E\left[|s[n]|^2\right]}{E\left[|w[n]|^2\right]} = \frac{\sigma_g^2}{2\sigma_w^2}\sum_{i=1}^{M} p_i|A_i|^2, \tag{5}$$

in which $\sigma_g^2 = E\left[|g[n]|^2\right]$ represents the average power of the fading, M represents the order of the QAM constellation, and $|A_i|$ and p_i represent the amplitude and probability, respectively, of the occurrence of the ith symbol.

Since the QAM constellation symbols considered in the study are equiprobable, it follows that $\sum_{i=1}^{M} p_i|A_i|^2 = 1$. Thus, if one assumes $\gamma = \sigma_g^2/\sigma_w^2$, it follows that the estimate of the SNR γ can be calculated from a function f(γ) defined as the ratio between the square of the second moment M_2 and the fourth moment M_4 of $|r[n]|$. If the rational function f(γ) is algebraically inverted in terms of the ratio between the square of the second sample moment and the fourth sample moment of $|r[n]|$, then this inverse obtained is the estimate of γ and will be denoted by $\hat{\gamma}$. The problem is then to find the expressions of the moments of the variable $|r[n]|$.

DERIVATION OF THE *K*TH MOMENT OF |*R*[*N*]|

For calculating the SNR estimate, $\hat{\gamma}$, one needs the moments of order 2 and 4 of $|r[n]|$. This is because the moment of order 2 of $|r[n]|$ is directly proportional to σ2wσw2, while the moment of order 4 is directly proportional to the square of σ2wσw2. Hence, the ratio of the square of the second moment and the fourth moment of $|r[n]|$ eliminates the dependence on the average power σ_w^2 and remains as a function of the SNR γ, which is the variable to be estimated in this study. In order to calculate the variance of the estimate $\hat{\gamma}$, the moments of order 2, 4, 6, and 8 of $|r[n]|$ are required. In this section, an exact expression is presented for the moment of order *k*, and then, the expressions of the individual moments are obtained in their simplified forms.

The *k*th moment of $|r[n]|$ can be calculated from the model $r[n]$ presented in (1) from the PDF of $|r[n]|$ conditioned on $|g[n]|$ and $|s[n]|$, written as [29]

$$f_{|r[n]|}(|r[n]|\,||g[n]|,|s[n]|) = \frac{|r[n]|}{\sigma_w^2} \times \exp\left(-\left(\frac{|r[n]|^2 + |g[n]|^2|s[n]|^2}{2\sigma_w^2}\right)\right) \times \mathrm{I}_0\left(|r[n]|\frac{|g[n]||s[n]|}{\sigma_w^2}\right), \tag{6}$$

in which $I_0(x)$ represents the zero order modified Bessel function. For the sake of simplicity, it is appropriate to simplify the notation and represent the variables $|r[n]|$ and $|g[n]|$, respectively, by *r* and *g*.

Taking the mean of $f_r(r|g,|s[n]|)$ with respect to probability distribution of the modulus of the QAM constellation symbols considered and substituting the resulting expression in the integral that defines the conditional expectation $E[r^k|g]$, it follows, after some simplifications, that

$$E\left[r^k|g\right] = \frac{1}{\sigma_w^2}\sum_{i=1}^{M} p_i \exp\left(-\frac{g^2|A_i|^2}{2\sigma_w^2}\right) \times \int_0^\infty r^{k+1}\exp\left(-\frac{r^2}{2\sigma_w^2}\right)\mathrm{I}_0\left(2\frac{g|A_i|}{2\sigma_w^2}r\right)dr. \tag{7}$$

The integral in (7) can be solved by Expression 6.643 of the table of integrals ([30] pp. 709). Representing the Whittaker's function resulting from the solution of this integral by the Kummer's confluent hypergeometric function ${}_1F_1(a;b;z)$ [30] and making the necessary algebraic simplifications, one can write the expression of kth moment of $|r[n]|$ conditioned on $|g[n]|$ as

$$E\left[r^k|g\right] = 2^{\frac{k}{2}}\left(\sigma_w^2\right)^{\frac{k}{2}+1}\Gamma\left(\frac{k}{2}+1\right) \times \sum_{i=1}^{M} p_i \exp\left(-\frac{g^2|A_i|^2}{2\sigma_w^2}\right){}_1\mathrm{F}_1\left(\frac{k}{2}+1;1;\frac{g^2|A_i|^2}{2\sigma_w^2}\right). \tag{8}$$

The kth moment of $|r[n]|$ can then be calculated taking the average of $E[r^k|g]$ with respect to PDF of the fading envelope, by the calculus of the integral

$$E\left[r^k\right] = \int_0^\infty E\left[r^k|g\right] f_g(g)dg. \tag{9}$$

Moments of $|r[n]|$ for the Distribution η−μ

The kth moment of $|r[n]|$ for η−μ fading can be obtained by substituting the expressions (2) and (8) in (9), so that one can write

$$E\left[|r[n]|^k\right] = \frac{\sqrt{2\pi}2^{\frac{k}{2}}\Gamma\left(\frac{k}{2}+1\right)}{\Gamma(\mu)(2H)^{\mu}}\left(\frac{H}{h}\right)^{\frac{1}{2}}\left(\sigma_w^2\right)^{\frac{k}{2}} \times \sum_{i=1}^{M}\frac{p_i}{(1+\gamma_i)^{\mu+\frac{1}{2}}}\int_0^\infty v^{\mu-\frac{1}{2}}\exp^{(-v)}\mathrm{I}_{\mu-\frac{1}{2}}\left[\frac{H}{h}\frac{v}{1+\gamma_i}\right] \times {}_1F_1\left(\frac{k}{2}+1;1;\frac{v}{1+\frac{1}{\gamma_i}}\right)dv, \tag{10}$$

in which

$$\gamma_i = \frac{|A_i|^2}{4h\mu}\frac{\sigma_g^2}{\sigma_w^2}. \tag{11}$$

The expressions of the moments of order 2, 4, 6, and 8, denoted by M_2, M_4, M_6 and M_8, respectively, can then be obtained from (10) by writing the Kummer's confluent hypergeometric function in terms of Laguerre polynomials [31],

$${}_1F_1(a;1;z) = e^z L_{a-1}(-z), \tag{12}$$

in which the polynomials $L_n(x)$ can be generated from the Olinde Rodrigues formula [31], given by

$$L_n(x) = \frac{e^x}{n!}\frac{d^n}{dx^n}\left(e^{-x}x^n\right). \tag{13}$$

Using the representation (12) and simplifying the resulting integrals, one obtains, after the procedure of simplification, that the moments M_2, M_4, M_6, and M_8 can be written as polynomials as a function of the SNR γ as

$$\begin{aligned}
M_2 &= \left(\sigma_w^2\right)^1\left(\alpha_{21}\gamma^1 + 2\right),\\
M_4 &= \left(\sigma_w^2\right)^2\left(\alpha_{42}\gamma^2 + \alpha_{41}\gamma + 8\right),\\
M_6 &= \left(\sigma_w^2\right)^3\left(\alpha_{63}\gamma^3 + \alpha_{62}\gamma^2 + \alpha_{61}\gamma + 48\right),\\
M_8 &= \left(\sigma_w^2\right)^4\left(\alpha_{84}\gamma^4 + \alpha_{83}\gamma^3 + \alpha_{82}\gamma^2 + \alpha_{81}\gamma + 384\right),
\end{aligned} \tag{14}$$

in which the parameters α_{ij} can be described by the group of expressions (15) and by the relations (16),

$$\begin{cases}
\alpha_{21} = C_2,\\
\alpha_{42} = \delta\left[2\mu + \xi^2 + 1\right]C_4,\\
\alpha_{63} = \delta^2\left[2\mu^2 + 3\left(\xi^2+1\right)\mu + 3\xi^2 + 1\right]2C_6,\\
\alpha_{84} = \delta^3\left[4\mu^3 + 12(\xi^2+1)\mu^2 + \left(3\xi^4 + 30\xi^2 + 11\right)\mu\right.\\
\qquad \left. + 3\xi^4 + 18\xi^2 + 3\right]2C_8
\end{cases} \tag{15}$$

and

$$\begin{cases}
\alpha_{41} = 8\alpha_{21},\\
\alpha_{61} = 72\alpha_{21}, \quad \alpha_{62} = 18\alpha_{42},\\
\alpha_{81} = 768\alpha_{21}, \ \alpha_{82} = 288\alpha_{42}, \ \alpha_{83} = 32\alpha_{63},
\end{cases} \tag{16}$$

in which δ=1/(2μ), ξ=H/h, and C_k represent the kth moment of the constellation M-QAM.

Moments of $|r[n]|$ for the Distribution κ–μ

Similarly, substituting expressions (3) and (8) in (9), from the kth conditioned moment, one can write the kth moment of $|r[n]|$ under κ–μ fading as

$$E\left[|r[n]|^k\right] = \frac{2^{\frac{k}{2}}(\sigma_w^2)^{\frac{k}{2}}\Gamma\left(\frac{k}{2}+1\right)}{\kappa\mu^{\frac{\mu-1}{2}}\exp(\kappa\mu)}\sum_{i=1}^{M} p_i\left(\frac{1}{1+\gamma_i}\right)^{\frac{\mu}{2}+\frac{1}{2}}$$
$$\times\int_0^\infty w^{\frac{\mu}{2}-\frac{1}{2}}\exp(-w)\,{}_1F_1\left(\frac{k}{2}+1,1;\frac{w}{1+\frac{1}{\gamma_i}}\right)$$
$$\times\, \mathrm{I}_{\mu-1}\left(2\sqrt{\frac{\kappa\mu w}{1+\gamma_i}}\right)dw, \tag{17}$$

in which

$$\gamma_i = \frac{\sigma_g^2}{\sigma_w^2}\frac{|A_i|^2}{2\mu(1+\kappa)}. \tag{18}$$

Expressions of the moments of order 2, 4, 6, and 8, denoted by M_2, M_4, M_6 and M_8, respectively, can be obtained from (17) writing the Kummer confluent hypergeometric function in terms of polynomials of Laguerre [31], so that the moments can be written as a function of the SNR γ,

$$M_2 = \left(\sigma_w^2\right)^1\left(\beta_{21}\gamma^1+2\right),$$
$$M_4 = \left(\sigma_w^2\right)^2\left(\beta_{42}\gamma^2+\beta_{41}\gamma+8\right),$$
$$M_6 = \left(\sigma_w^2\right)^3\left(\beta_{63}\gamma^3+\beta_{62}\gamma^2+\beta_{61}\gamma+48\right),$$
$$M_8 = \left(\sigma_w^2\right)^4\left(\beta_{84}\gamma^4+\beta_{83}\gamma^3+\beta_{82}\gamma^2+\beta_{81}\gamma+384\right) \tag{19}$$

in which the relations between the coefficients β_{ij} can be written as in the groups of expressions (20) and (21).

$$\begin{cases} \beta_{21} = C_2, \\ \beta_{42} = \zeta^2\left[(\kappa+1)^2\mu^2+(2\kappa+1)\mu\right]C_4, \\ \beta_{63} = \zeta^3\left[(\kappa+1)^3\mu^3+3(2\kappa^2+3\kappa+1)\mu^2\right. \\ \qquad \left.+2(3\kappa+1)\mu\right]C_6, \\ \beta_{84} = \zeta^4\left[\left(\kappa^4+64\kappa^2+4\kappa+1\right)\mu^4+6\left(2\kappa^3+5\kappa^2\right.\right. \\ \qquad \left.+4\kappa+1\right)\mu^3 \\ \qquad \left.+(36\kappa^2+44\kappa+11)\mu^2+6(4\kappa+1)\mu\right]C_8, \end{cases} \tag{20}$$

and

$$\begin{cases} \beta_{41} = 8\beta_{21}, \\ \beta_{61} = 72\beta_{21}, \quad \beta_{62} = 18\beta_{42}, \\ \beta_{81} = 768\beta_{21}, \ \beta_{82} = 288\beta_{42}, \ \beta_{83} = 32\beta_{63}, \end{cases} \tag{21}$$

in which $\zeta=1/(\mu(1+\kappa))$.

DERIVATION OF SNR ESTIMATES

In Section 3, exact expressions were obtained for the moments of order 2, 4, 6, and 8 of the received signal envelope in discrete time, $|r[n]|$, for the fading models $\eta-\mu$ and $\kappa-\mu$. The estimation of the SNR can then be obtained from the ratio between the square of the second moment and the fourth moment for these models. This ratio can be expressed in terms of the SNR $\gamma = \sigma_g^2/\sigma_w^2$ and is denoted f(γ) for formats I and II of fading $\eta-\mu$ and for $\kappa-\mu$. Thus, if the ratio

$$\mathrm{f}(\gamma) = \frac{E\left[|r[n]|^2\right]^2}{E\left[|r[n]|^4\right]} = \frac{\mathrm{p}(\gamma)}{\mathrm{q}(\gamma)} \tag{22}$$

can be written as a function of γ and this function can be inverted in terms of the ratio of the square of the second sample moment and the fourth sample moment of $|r[n]|$, denoted by sample ratio and represented by $\hat{\mathrm{f}}_1$, then the inverse obtained is an estimate $\hat{\gamma}$ of the SNR γ.

In the first case, in which the fading is characterized by format I of the distribution $\eta-\mu$, for which $h=(2+\eta^{-1}+\eta)/4$ and $H=(\eta^{-1}-\eta)/4$, the function f(γ) can be written as the ratio of two polynomials of order 2 in γ,

$$\mathrm{f}(\gamma) = \frac{4 + 4C_2\gamma + C_2^2\gamma^2}{8 + 8C_2\gamma + 4\varrho C_4\gamma^2}, \tag{23}$$

in which ϱ is given in (24).

When inverted in terms of the sample ratio $\hat{\mathrm{f}}_1$, the estimate $\hat{\gamma}$ under the format I of fading $\eta-\mu$ may be written as

$$\text{Format I } \eta-\mu \begin{cases} \hat{\gamma} = \dfrac{2(1-2\hat{f})+2\sqrt{2\hat{f}(1-2\hat{f})(\epsilon-1)}}{C_2(2\epsilon\hat{f}-1)} \\ \epsilon = 2\varrho\dfrac{C_4}{C_2^2} \\ \varrho = \dfrac{\mu(1+\eta)^2+(1+\eta^2)}{4\mu(1+\eta)^2}. \end{cases} \quad (24)$$

In the second case, in which the fading is characterized by format II of distribution η−μ, for which h=1/(1−η²) and H=η/(1−η²), the parameter ϱ of the function f(γ) presented in (23) is given in (25). When inverted in terms of the sample ratio $\hat{f}$, the estimate $\hat{\gamma}$ under format II of fading η−μ can be written as

$$\text{Format II } \eta-\mu \begin{cases} \hat{\gamma} = \dfrac{2(1-2\hat{f})+2\sqrt{2\hat{f}(1-2\hat{f})(\epsilon-1)}}{C_2(2\epsilon\hat{f}-1)} \\ \epsilon = 2\varrho\dfrac{C_4}{C_2^2} \\ \varrho = \dfrac{1+2\mu+\eta^2}{8\mu} \end{cases} \quad (25)$$

In the third case, in which the fading is characterized by the distribution κ−μ, the parameter ϱ of the function f(γ) is given in (26). When inverted in terms of the sample ratio $\hat{f}$, the estimate $\hat{\gamma}$ under κ−μ fading can be written as

$$\text{Model } \kappa-\mu \begin{cases} \hat{\gamma} = \dfrac{2(1-2\hat{f})+2\sqrt{2\hat{f}(1-2\hat{f})(\epsilon-1)}}{C_2(2\epsilon\hat{f}-1)} \\ \epsilon = 2\varrho\dfrac{C_4}{C_2^2} \\ \varrho = \dfrac{(\kappa\mu(\kappa\mu+2\mu+2)+\mu^2+\mu)}{4\mu^2(1+\kappa)^2}. \end{cases} \quad (26)$$

As the generalized distributions encompass the distribution of Nakagami-*m* as a particular case and this distribution is one of the most used to characterize the fading in mobile communication channels, the following section shows the relationship between the estimators obtained in this article and the SNR estimator for the received signal model under Nakagami-*m* fading.

RELATION WITH THE ESTIMATOR FOR NAKAGAMI FADING CHANNEL

According to [23], the Nakagami-m distribution can be obtained from the distribution $\eta-\mu$ making $\mu=m$ and $\eta\to0$ or $\eta\to\infty$ in format I or $\mu=m$ and $\eta\to\pm1$ in format II. It can still be obtained making $\mu = \frac{m}{2}$ and $\eta\to1$ in format I or $\mu = \frac{m}{2}$ and $\eta\to0$ in format II. The Nakagami-m distribution can also be obtained from the $\kappa-\mu$ distribution making $\kappa=0$ and $\mu=m$.

Using the estimates summarized in Table 1, one can obtain the parameters

$$\varrho = \frac{(m+1)}{4m}$$

and

$$\epsilon = \frac{(m+1)}{2m}\frac{C_4}{C_2^2}$$

for the Nakagami fading from the following possible substitutions

- For $\eta-\mu$ fading format I: making $\mu=m$ and $\eta\to0$ in the expression of ϱ.

Table 1: SNR estimates under generalized fading

fading	Estimate of SNR γ
Nakagami-m	$\hat{\gamma} = \frac{2(1-2\hat{f})+2\sqrt{2\hat{f}(1-2\hat{f})(\epsilon-1)}}{C_2(2\epsilon\hat{f}-1)}$ $\epsilon = 2\varrho\frac{C_4}{C_2^2}$ $\varrho = \frac{(m+1)}{4m}$
Format I $\eta-\mu$	$\hat{\gamma} = \frac{2(1-2\hat{f})+2\sqrt{2\hat{f}(1-2\hat{f})(\epsilon-1)}}{C_2(2\epsilon\hat{f}-1)}$ $\epsilon = 2\varrho\frac{C_4}{C_2^2}$ $\varrho = \frac{\mu(1+\eta)^2+(1+\eta^2)}{4\mu(1+\eta)^2}$
Format II $\eta-\mu$	$\hat{\gamma} = \frac{2(1-2\hat{f})+2\sqrt{2\hat{f}(1-2\hat{f})(\epsilon-1)}}{C_2(2\epsilon\hat{f}-1)}$ $\epsilon = 2\varrho\frac{C_4}{C_2^2}$ $\varrho = \frac{1+2\mu+\eta^2}{8\mu}$
$\kappa-\mu$	$\hat{\gamma} = \frac{2(1-2\hat{f})+2\sqrt{2\hat{f}(1-2\hat{f})(\epsilon-1)}}{C_2(2\epsilon\hat{f}-1)}$ $\epsilon = 2\varrho\frac{C_4}{C_2^2}$ $\varrho = \frac{(\kappa\mu(\kappa\mu+2\mu+2)+\mu^2+\mu)}{4\mu^2(1+\kappa)^2}$

1. For η−μ fading format I: making $\mu = \frac{m}{2}$ and η=1 in the expression of ϱ.
2. For η−μ fading format II: making μ=m and η→±1 in the expression of ϱ.
3. For η−μ fading format II: making $\mu = \frac{m}{2}$ and η→0 in the expression of ϱ.
4. For κ−μ: making κ=0 and μ=m in the expression of ϱ.

EVALUATION OF THE ESTIMATE BY MEANS OF THE NMSE

The normalized mean square error (NMSE) of a parameter θ, considering an estimate $\hat{\theta}$, is defined as [26]

$$\text{NMSE} = \frac{E\left[(\hat{\theta} - \theta)^2\right]}{\theta^2}, \tag{27}$$

and the variance of the estimate $\hat{\theta}$ is given by

$$\text{Var}\left[\hat{\theta}\right] = E\left[\hat{\theta}^2\right] - E^2\left[\hat{\theta}\right]. \tag{28}$$

Using $E\left[\hat{\theta}^2\right]$ from (25) and applying in (24), one can write

$$E\left[(\hat{\theta} - \theta)^2\right] = \text{Var}\left[\hat{\theta}\right] + \left(E\left[\hat{\theta}\right] - \theta\right)^2. \tag{29}$$

Hence, the NMSE of the estimate $\hat{\gamma}$ can be written as

$$\text{NMSE} = \frac{\text{Var}[\hat{\gamma}] + (E[\hat{\gamma}] - \gamma)^2}{\gamma^2}, \tag{30}$$

in which $\text{Var}[\hat{\gamma}]$ and $E[\hat{\gamma}]$ as a function of γ can be computed by a linearization process, as presented in [32]. According to that method, if an estimate $\hat{\theta}$ can be written as a function of the observed data vector **x**,

$$\hat{\theta} = h^{-1}(\hat{\boldsymbol{\mu}}) = g(\mathbf{x}), \tag{31}$$

in which $\hat{\boldsymbol{\mu}}$ is the vector of sample moments of **x**, it is possible to obtain approximate expressions for the mean and variance of the estimate to evaluate the performance of the estimator by means of a Taylor's expansion of $g(\mathbf{x})$ about a vector of statistics $\boldsymbol{\mu}=E[\boldsymbol{T}]$ [32]. The mean and variance values given by these expressions approximate the exact values as the number of samples observed in the vector **x** increases. The exact mean and variance, however, can only be determined by means of computational simulation.

In the context of the study presented in this paper, the estimated parameter is scalar, $\hat{\theta} = \hat{\gamma}$, and can be written in terms of the relation (28). Calculating approximate expressions for the mean and the variance of $\hat{\theta}$ can then be performed assuming that this estimate depends on the vector $\boldsymbol{T}$ of two statistics of the observed samples, $T=[T_1(x)T_2(x)]$, respectively, the second sample moment and fourth sample moment of $|r[n]|$.

The Taylor's expansion of an estimate $\hat{\theta}$, about a vector $\boldsymbol{\mu}$ of two statistics, can be approximated by

$$\hat{\theta} \approx g(\boldsymbol{\mu}) + \left[\left. \frac{\partial g}{\partial \boldsymbol{T}} \right|_{\boldsymbol{T}=\boldsymbol{\mu}} \right]^T (\boldsymbol{T} - \boldsymbol{\mu}), \tag{32}$$

and

$$\hat{\theta}^2 \approx g^2(\boldsymbol{\mu}) + 2g(\boldsymbol{\mu}) \left[\left. \frac{\partial g}{\partial \boldsymbol{T}} \right|_{\boldsymbol{T}=\boldsymbol{\mu}} \right]^T (\boldsymbol{T} - \boldsymbol{\mu}) + \left[\left. \frac{\partial g}{\partial \boldsymbol{T}} \right|_{\boldsymbol{T}=\boldsymbol{\mu}} \right]^T (\boldsymbol{T} - \boldsymbol{\mu})(\boldsymbol{T} - \boldsymbol{\mu})^T \left[\left. \frac{\partial g}{\partial \boldsymbol{T}} \right|_{\boldsymbol{T}=\boldsymbol{\mu}} \right]. \tag{33}$$

It follows that $\mathrm{Var}(\hat{\theta})$ can be written as

$$\mathrm{Var}(\hat{\theta}) = \left. \frac{\partial g}{\partial \boldsymbol{T}} \right|_{\boldsymbol{T}=\boldsymbol{\mu}}^T C_{\mathrm{T}} \left. \frac{\partial g}{\partial \boldsymbol{T}} \right|_{\boldsymbol{T}=\boldsymbol{\mu}}, \tag{34}$$

in which $C_T=E[(\boldsymbol{T}-\boldsymbol{\mu})(\boldsymbol{T}-\boldsymbol{\mu})^T]$ represents the covariance matrix of the statistics vector $\boldsymbol{T}$.

In the context of SNR estimation, it is observed that all estimates $\hat{\gamma}$ summarized in Table 1 can be written from a single function, in terms of the parameters that characterize the type of fading and sample moments of $|r[n]|$, so that

$$\hat{\gamma} = g(\mathbf{T}) = 2(1 - 2q(T_1, T_2))(2\epsilon q(T_1, T_2) - 1)^{-1}$$
$$+ 2\sqrt{2(\epsilon - 1)q(T_1, T_2)(1 - 2q(T_1, T_2))}(2\epsilon q(T_1, T_2) - 1)^{-1}, \tag{35}$$

in which $q(T_1, T_2) = \hat{\mathrm{f}} = \frac{T_1^2}{T_2}$,

$$T_1 = \frac{1}{N}\sum_{n=1}^{N} |r[n]|^2$$
$$T_2 = \frac{1}{N}\sum_{n=1}^{N} |r[n]|^4. \tag{36}$$

and

$$E[T_1] = \frac{1}{N}\sum_{n=1}^{N} E\left[|r[n]|^2\right] = M_2$$
$$E[T_2] = \frac{1}{N}\sum_{n=1}^{N} E\left[|r[n]|^4\right] = M_4. \tag{37}$$

If the constellation of the modulation scheme used has equiprobable symbols and with energy per symbol such that the second moment of the entire constellation is unitary, then $C_2=1$. Hence, the mean value of the estimate, $E[\hat{\gamma}] = g(\boldsymbol{\mu})$, can be written as

$$E[\hat{\gamma}] = 2\left(1 - 2\frac{M_2^2}{M_4}\right)\left(2\epsilon\frac{M_2^2}{M_4} - 1\right)^{-1}$$
$$+ 2\left(2(\epsilon - 1)\frac{M_2^2}{M_4}\left(1 - 2\frac{M_2^2}{M_4}\right)\right)^{1/2}$$
$$\times \left(2\epsilon\frac{M_2^2}{M_4} - 1\right)^{-1}. \tag{38}$$

For the evaluation of the variance of the estimates, it is necessary, according to the expression (31), the calculation of the partial derivatives of $\hat{\gamma}$ in relation to the statistics T_1 and T_2. These derivatives can be written as

$$\frac{\partial g}{\partial T_1} = \frac{\partial g}{\partial q}\frac{\partial q}{\partial T_1}$$
$$\frac{\partial g}{\partial T_2} = \frac{\partial g}{\partial q}\frac{\partial q}{\partial T_2}, \tag{39}$$

and the partial derivative $\frac{\partial g}{\partial q}$ can be written as

$$\frac{\partial g}{\partial q} = h(q(T_1, T_2)), \tag{40}$$

in which the function $h(x)$ is given by

$$\begin{aligned} h(x) &= -4\epsilon(1-2x)(2\epsilon x-1)^{-2} - 4(2\epsilon x-1)^{-1} \\ &- 4\epsilon(2(\epsilon-1))^{1/2}(2\epsilon x-1)^{-2}x^{1/2}(1-2x)^{1/2} \\ &+ 2(\epsilon-1)(2(\epsilon-1))^{-1/2}(1-4x)x^{-1/2}(1-2x)^{-1/2} \\ &\times (2\epsilon x-1)^{-1}. \end{aligned} \tag{41}$$

Using the fact that

$$\begin{aligned} \frac{\partial}{\partial T_1} q(T_1, T_2) &= 2\frac{T_1}{T_2} \\ \frac{\partial}{\partial T_2} q(T_1, T_2) &= -\frac{T_1^2}{T_2^2}, \end{aligned} \tag{42}$$

one can write the vector $\left.\frac{\partial g}{\partial T}\right|_{T=M}$ as

$$\left.\frac{\partial g}{\partial T}\right|_{T=M} = \begin{bmatrix} 2h(g(M_2, M_4))\frac{M_2}{M_4} \\ -h(g(M_2, M_4))\frac{M_2^2}{M_4^2} \end{bmatrix}. \tag{43}$$

The elements of the covariance matrix C_T, presented in (31), can be obtained from the expressions of the statistics T_1 and T_2, given in (33), and written as

$$\begin{aligned} \text{Cov}(T_1, T_1) &= \frac{1}{N}\left(M_4 - M_2^2\right) \\ \text{Cov}(T_1, T_2) &= \text{Cov}(T_2, T_1) = \frac{1}{N}(M_6 - M_2 M_4) \\ \text{Cov}(T_2, T_2) &= \frac{1}{N}\left(M_8 - M_4^2\right). \end{aligned} \tag{44}$$

Substituting (40) and (41) in the expression of the variance of $\hat{\gamma}$, one can write Var($\hat{\gamma}$) as

$$\begin{aligned} \text{Var}[\hat{\gamma}] &= \frac{h^2(g(M_2, M_4))}{N}\left(\frac{M_2}{M_4}\right)^2 \begin{bmatrix} 2 \\ -\frac{M_2}{M_4} \end{bmatrix}^T \\ &\times \begin{bmatrix} M_4 - M_2^2 & M_6 - M_2 M_4 \\ M_6 - M_2 M_4 & M_4 - M_8 - M_4^2 \end{bmatrix} \begin{bmatrix} 2 \\ -\frac{M_2}{M_4} \end{bmatrix}, \end{aligned} \tag{45}$$

which can be written as

$$\mathrm{Var}[\hat{\gamma}] = \frac{1}{N}h^2(f_M)\left[4f_M - f_M^2 - \frac{4M_6}{M_2M_4}f_M^2 + \frac{M_8}{M_4^2}f_M^2\right], \tag{46}$$

in which

$$f_M = \frac{M_2^2}{M_4}.$$

MOMENTS OF THE *M*-QAM CONSTELLATIONS

In this section, the moments of order 2, 4, 6, and 8 are determined for the M-QAM constellation. The moments of order 2 and 4 are necessary to calculate the SNR estimate, while the higher order moments are necessary to evaluate the variance of the estimate.

In this calculation, it is assumed that the symbols are equiprobable and normalized from the total energy of the constellation, obtained from the sum of the square modulus of M symbols. Writing these symbols in terms of the corresponding coordinates in the plane $\mathbb{R}^2$, $(2i - 1 - \sqrt{M}, 2j-1-\sqrt{M})$ for $i, j = 0, 1, \cdots, \sqrt{M}-1$, one has that the total energy is given by $E_t = \sum_{n=0}^{M-1} |s[n]|^2 = 2M(M-1)/3$, while the average energy per symbol, obtained by dividing the total energy by the number of symbols, can be written as $E_{av}=2(M-1)/3$. The constellation can then be normalized by the constant $C = 1/\sqrt{E_{av}}$.

The moment of order $2k$ of the normalized constellation can then be obtained by multiplying the symbols by the constant of normalization C and calculating the sum

$$C_{2k} = \sum_{n=0}^{M-1} p_n(|s[n]|)^{2k}. \tag{47}$$

Writing $s[n]$ in terms of its coordinates in the plane R2R2 and developing the sum of (44) for k equal to 1, one has that the moment of order 2, C_2, can be written as

$$C_2 = \sum_{n=0}^{M-1} p_n|s[n]|^2 = \frac{1}{M}\sum_{n=0}^{M-1} |s[n]|^2 = \frac{E_t}{M} = 1. \tag{48}$$

Making k equal to 2, 3, and 4, one can obtain, respectively, the moments of order 4, 6, and 8, denoted by C_4, C_6, and C_8. Developing the summation (44) and making the necessary simplifications, one has

$$C_4 = \frac{(7M^2 - 20M + 13)}{5(M-1)^2},$$
$$C_6 = \frac{9(9M^2 - 40M + 51)}{35(M-1)^2},$$
$$C_8 = \frac{9(83M^3 - 637M^2 + 1897M - 2183)}{175(M-1)^3}. \quad (49)$$

DISCUSSION ON THE CRAMÉR-RAO BOUND

In this section, a discussion is presented about the calculation of the Cramér-Rao bound for the η−μ fading model. This proposal can be extended to κ−μ fading following the same procedure. The PDF of the samples modulus of the received signal $|r[n]|$ under η−μ fading can be obtained by taking the expected value of the conditioned PDF $f_{|r[n]|}(|r[n]|||g[n]|,|s[n]|)$ by the distributions of $|g[n]|$ and $|s[n]|$. Thus, the PDF can be written as

$$f_{|r[n]|}(x) = \frac{\beta_{\eta-\mu}}{M}\frac{x}{\sigma_w^2}\exp\left(-\frac{x^2}{2\sigma_w^2}\right)\sum_{i=1}^{M}\int_0^{\infty} v^{\mu-\frac{1}{2}}\exp\left(-\left(2\mu h + \frac{|A_i|^2}{2}\frac{\sigma_g^2}{\sigma_w^2}\right)v\right) \times \mathrm{I}_{\mu-\frac{1}{2}}(2\mu vH)\mathrm{I}_0\left(x|A_i|\frac{\sigma_g}{\sigma_w^2}\sqrt{v}\right)dv, \quad (50)$$

in terms of the average powers of the noise, σ_w^2, and fading, σ_g^2, for the normalized M-QAM constellation of equiprobable symbols. After searching in classical Laplace integrals and transform tables involving Bessel functions, it was not possible to find an exact solution to the integral in (47) because one of the Bessel functions has as argument a square root of the integration variable and this became the main constraint to calculating the Cramér-Rao bound.

A possible solution to this problem is to consider the average power, σ_g^2, of the fading of unit value, so that the SNR is $\gamma = 1/\sigma_w^2$. This consideration is appropriate from the point of view of computational simulation because it makes the generation of variables η−μ faster by the method of acceptance-rejection employed in this study, since the variation of σ_g^2 influences the PDF format of the η−μ distribution. In addition, this consideration allows to calculate only the Fisher information, instead of the Fisher information matrix, which is four times more laborious. Thus, by considering σ_g^2

$=1, f_{|r[n]|}(x)$ can be written as

$$f_{|r[n]|}(x;\gamma) = \frac{\beta_{\eta-\mu}}{M} x\gamma \exp\left(-\frac{\gamma}{2}x^2\right) Q_{\eta-\mu}(x,\gamma), \tag{51}$$

in which

$$Q_{\eta-\mu}(x,\gamma) = \sum_{i=1}^{M} \int_0^\infty v^{\mu-\frac{1}{2}} \exp\left(-\left(2\mu h + \frac{|A_i|^2}{2}\gamma\right) v\right) \times \mathrm{I}_{\mu-\frac{1}{2}}(2\mu v H) \mathrm{I}_0\left(x|A_i|\gamma\sqrt{v}\right) dv \tag{52}$$

and

$$\beta_{\eta-\mu} = \frac{2\sqrt{\pi}\,\mu^{\mu+\frac{1}{2}} h^{\mu}}{\Gamma(\mu) H^{\mu-\frac{1}{2}}}. \tag{53}$$

Considering the N observed samples of $|r[n]|$ independent and identically distributed with PDF $f_{|r[n]|}(|r[n]|;\gamma)$, then the Fisher information $I(\gamma)$ can be written as

$$\begin{aligned} I(\gamma) &= -NE\left[\frac{\partial^2}{\partial\gamma^2} \ln f_{|r[n]|}(|r[n]|;\gamma)\right] \\ &= N\int_0^\infty \frac{\partial}{\partial\gamma} f_{|r[n]|}(x;\gamma) \frac{\partial}{\partial\gamma} \ln f_{|r[n]|}(x;\gamma) dx. \end{aligned} \tag{54}$$

in which the expected value was taken by the PDF $f_{|r[n]|}(|r[n]|;\gamma)$.

The integral in (51) can be written as

$$\begin{aligned} &\int_0^\infty \frac{\partial}{\partial\gamma} f_{|r[n]|}(x,\gamma) \frac{\partial}{\partial\gamma} \ln f_{|r[n]|}(x,\gamma) dx \\ &= \frac{1}{\gamma}\frac{\beta_{\eta-\mu}}{M} \underbrace{\int_0^\infty x Q_{\eta-\mu}(x,\gamma) \exp\left(-\frac{\gamma}{2}x^2\right) dx}_{I_1} \\ &\quad - \frac{\beta_{\eta-\mu}}{M} \underbrace{\int_0^\infty x^3 Q_{\eta-\mu}(x,\gamma) \exp\left(-\frac{\gamma}{2}x^2\right) dx}_{I_2} \\ &\quad + \frac{\gamma}{4}\frac{\beta_{\eta-\mu}}{M} \underbrace{\int_0^\infty x^5 Q_{\eta-\mu}(x,\gamma) \exp\left(-\frac{\gamma}{2}x^2\right) dx}_{I_3} \\ &\quad + 2\frac{\beta_{\eta-\mu}}{M} \underbrace{\int_0^\infty x Q'_{\eta-\mu}(x,\gamma) \exp\left(-\frac{\gamma}{2}x^2\right) dx}_{I_4} \\ &\quad - \gamma\frac{\beta_{\eta-\mu}}{M} \underbrace{\int_0^\infty x^3 Q'_{\eta-\mu}(x,\gamma) \exp\left(-\frac{\gamma}{2}x^2\right) dx}_{I_5} \\ &\quad \underbrace{+ \gamma\frac{\beta_{\eta-\mu}}{M} \int_0^\infty x \frac{(Q'_{\eta-\mu}(x,\gamma))^2}{Q_{\eta-\mu}(x,\gamma)} \exp\left(-\frac{\gamma}{2}x^2\right) dx,}_{I_6} \end{aligned} \tag{55}$$

in which

$$\frac{\partial}{\partial \gamma} Q_{\eta-\mu}(x,\gamma) = \sum_{i=1}^{M} \int_0^{\infty} v^{\mu-\frac{1}{2}} \exp(-2\mu h v) \mathrm{I}_{\mu-\frac{1}{2}}(2\mu H v) \times \frac{\partial}{\partial \gamma}\left[\exp\left(-\frac{|A_i|^2}{2} v\gamma\right) \mathrm{I}_0(x|A_i|\gamma\sqrt{v})\right] dv. \tag{56}$$

Calculation of the expressions I_1, I_2, I_3, I_4, and I_5 could be performed and simplified without great difficulties because when $Q_{\eta-\mu}(x,\gamma)$ was considered, the change of order of integration in the resulting double integrals led to integrals that could be calculated as Laplace transforms involving the modified Bessel functions. Therefore, one has the following solutions.

$$I_1 = \frac{M}{\gamma} \frac{\Gamma(\mu) H^{\mu-\frac{1}{2}}}{2\sqrt{\pi}\mu^{\mu+\frac{1}{2}}} \frac{1}{[h^2 - H^2]^{\mu}}. \tag{57}$$

$$I_2 = \frac{1}{\gamma} \sum_{i=1}^{M} |A_i|^2 \left(\frac{\Gamma(\mu) H^{\mu-\frac{1}{2}}}{2\sqrt{\pi}\mu^{\mu+\frac{1}{2}}} \frac{h}{[h^2 - H^2]^{\mu+1}} \right) + \frac{2}{\gamma^2} \sum_{i=1}^{M} \left(\frac{\Gamma(\mu) H^{\mu-\frac{1}{2}}}{2\sqrt{\pi}\mu^{\mu+\frac{1}{2}}} \frac{1}{[h^2 - H^2]^{\mu}} \right). \tag{58}$$

$$I_3 = \frac{1}{2\gamma} \sum_{i=1}^{M} |A_i|^4 \left(\frac{\Gamma(\mu) H^{\mu-\frac{1}{2}}}{2\sqrt{\pi}\mu^{\mu+\frac{1}{2}}} \frac{(H^2 + (2\mu+1)h^2)}{\mu [h^2 - H^2]^{\mu+2}} \right) + \frac{8}{\gamma^2} \sum_{i=1}^{M} |A_i|^2 \left(\frac{\Gamma(\mu) H^{\mu-\frac{1}{2}}}{2\sqrt{\pi}\mu^{\mu+\frac{1}{2}}} \frac{h}{[h^2 - H^2]^{\mu+1}} \right) + \frac{8}{\gamma^3} \sum_{i=1}^{M} \left(\frac{\Gamma(\mu) H^{\mu-\frac{1}{2}}}{2\sqrt{\pi}\mu^{\mu+\frac{1}{2}}} \frac{1}{[h^2 - H^2]^{\mu}} \right). \tag{59}$$

$$I_4 = \frac{1}{2\gamma} \sum_{i=1}^{M} |A_i|^2 \left(\frac{\Gamma(\mu) H^{\mu-\frac{1}{2}}}{2\sqrt{\pi}\mu^{\mu+\frac{1}{2}}} \frac{h}{[h^2 - H^2]^{\mu+1}} \right). \tag{60}$$

$$I_5 = \frac{1}{4\gamma} \sum_{i=1}^{M} |A_i|^4 \left(\frac{\Gamma(\mu) H^{\mu-\frac{1}{2}}}{2\sqrt{\pi}\mu^{\mu+\frac{1}{2}}} \frac{(H^2 + (2\mu+1)h^2)}{\mu [h^2 - H^2]^{\mu+2}} \right) + \frac{3}{\gamma^2} \sum_{i=1}^{M} |A_i|^2 \left(\frac{\Gamma(\mu) H^{\mu-\frac{1}{2}}}{2\sqrt{\pi}\mu^{\mu+\frac{1}{2}}} \frac{h}{[h^2 - H^2]^{\mu+1}} \right). \tag{61}$$

The integral I_6 cannot be written in closed-form. It is rewritten as

$$I_6(\gamma) = \gamma \frac{\beta_{\eta-\mu}}{M} \int_0^{\infty} x \frac{(Q'_{\eta-\mu}(x,\gamma))^2}{Q_{\eta-\mu}(x,\gamma)} \exp\left(-\frac{\gamma}{2} x^2\right) dx. \tag{62}$$

In both Format I and Format II of the model η−μ, one has $h/(h^2-H^2)=1$. Thus, after substituting the results I_k, $1\leq k\leq 5$ and $I_6(\gamma)$ in (52), Fisher information $I(\gamma)$ can be written, from (51), as

$$I(\gamma) = N\left[I_6(\gamma) + \frac{1}{\gamma^2} - \frac{C_2}{\gamma} - C_4\lambda\right]. \quad (63)$$

Therefore, one has

$$\text{CRLB}[\hat{\gamma}] = \frac{1}{N\left[I_6(\gamma) + \frac{1}{\gamma^2} - \frac{C_2}{\gamma} - C_4\lambda\right]}, \quad (64)$$

in which

$$\lambda = \frac{1}{8\lambda}\left[\left(\frac{H}{h}\right)^2 + (2\mu + 1)\right], \quad (65)$$

$C_2=1$ for the normalized M-QAM constellation and C_4 is given by (44).

RESULTS AND DISCUSSION

In this section, results are presented for numerical evaluation of the mathematical expressions obtained for the estimates of the SNR and its NMSE, corroborated by simulations performed by the Monte Carlo method. For each SNR value, the variance and the mean of each estimate were obtained from a mean of 10 realizations of N samples of $r[n]$. In Figs. 1 and 2, curves are presented for the rational function f(γ) obtained from the ratio between the square of the second moment and the fourth moment of the magnitude of the received signal $|r[n]|$, for the fading models η−μ and κ−μ, considering the 64-QAM modulation scheme.

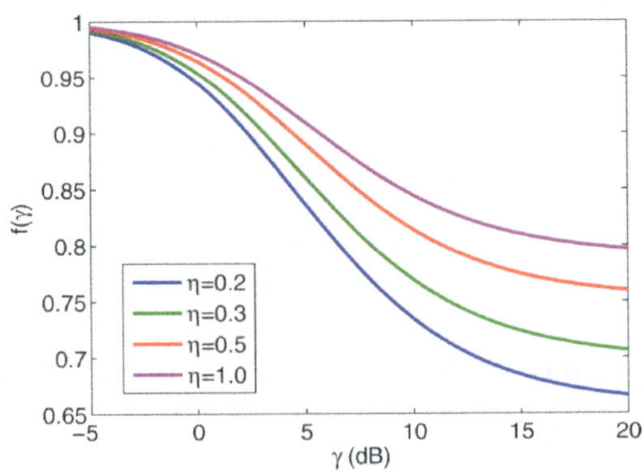

Figure 1: Curves of the rational function f(γ), as a function of SNR γ (dB), for fading model η−μ with μ=0.6 for 64-QAM and different values of η.

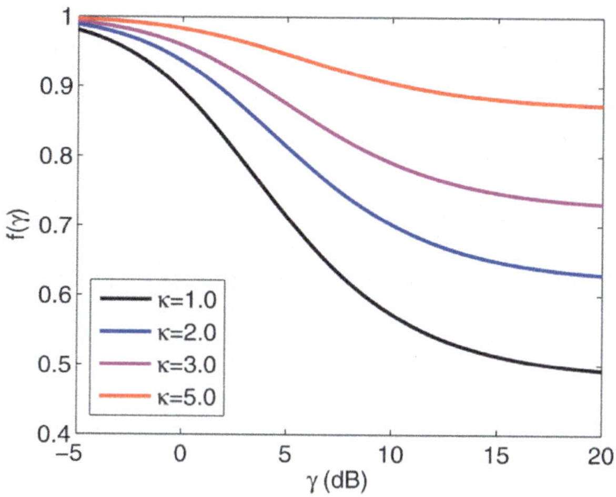

Figure 2: Curves of the rational function f(γ), as a function of SNR γ (dB), for fading model κ−μ with μ=0.5 with 64-QAM modulation and different values of κ.

The importance of these curves is that they allow to determine the interval within which the ratio between the square of the second sample moment and the fourth sample moment of $|r[n]|$ must lie within so that f(γ) can be inverted and provide the estimate $\hat{\gamma}$ of the SNR. In both figures, one observes that the length of this interval increases as κ and η decrease. When these parameters decrease, for μ fixed, the probability of the normalized envelope of the fading take values close to zero, for example, is larger, and this behavior can be confirmed by means of the CDF of the distributions η−μ and κ−μ presented in [23].

The next group of curves, in Figs. 3 and 4, show the behavior of the NMSE and the mean of the estimate $\hat{\gamma}$ for the received signal model $r[n]$ under fading η−μ as a function of the SNR γ in dB. The theoretical curves of the NMSE were plotted for different values of η from the expressions (27), (35), (43), and the moments obtained for $|r[n]|$ under fading η−μ. The modulation used was 64-QAM, and the number of samples N considered was 5×10^4, for the plot of the theoretical curves of the NMSE and for the the mean of the estimate obtained by simulation.

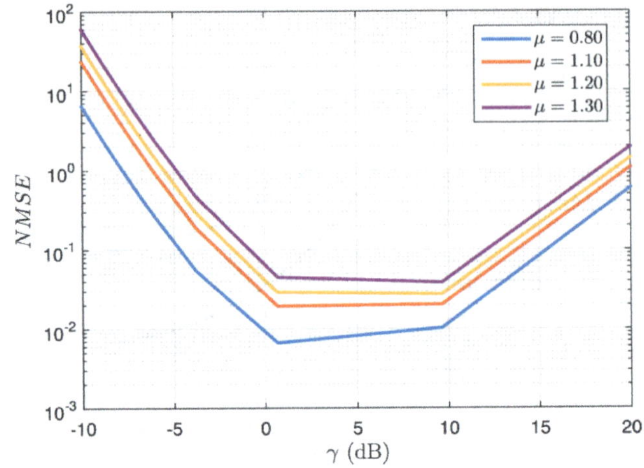

Figure 3: NMSE of the estimate $\hat{\gamma}$ as a function of SNR in dB for fading η−μ with η=0.1, different values of μ, N=5×10^4 and 64-QAM modulation.

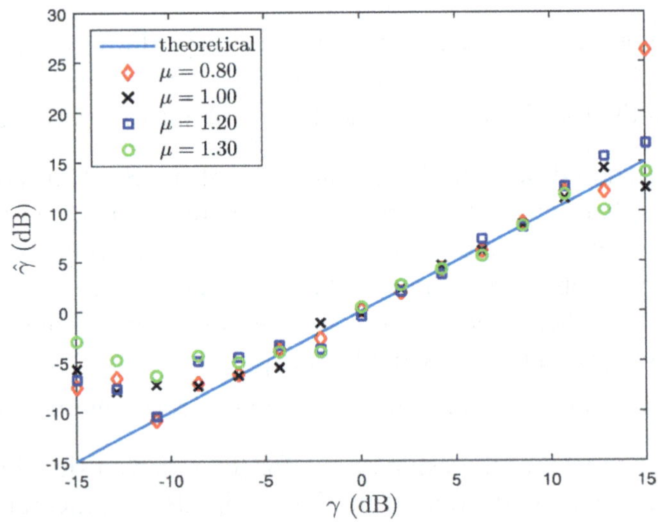

Figure 4: Mean of the estimate $\hat{\gamma}$ as a function of SNR in dB for fading η−μ with η=0.1, different values of μ, N=5×10^4 and 64-QAM modulation.

Figures 3 and 4, together, show the interval of SNR values for which the estimator presents its better performance. While the mean of the estimates approximates the desired values in the interval from 0 to 10 dB for all values of the parameter μ considered, the NMSE remains bellow 10^{-1} from –5 to 15 dB only for μ=0.8. It can be seen from Fig. 4 that the estimates adhere well to the expected average behavior in the range of SNR values between 0 dB and approximately 15 dB for all values of the parameter μ, while the NMSE of the estimate is more sensitive to the variation of intensity of fading characterized by the different values μ. For η fixed, the probability of the intensity of the fading envelope decreases below a fixed value decreases as μ increases. This behavior of the fading envelope η−μ can be verified by means of its CDF, presented in [23], and means that the increase of parameter μ models fading situations of lesser intensity.

Figures 5 and 6 present curves of the variance of the estimate $\hat{\gamma}$ as a function of the number of samples N of the signal envelope $|r[n]|$ on a channel under fadings κ−μ and η−μ. It was considered, both in the simulation and in the plot of theoretical curves, fading κ−μ with κ=1.1 and μ=0.3 and fading η−μ with η=0.1 and μ=0.8 for different values of γ (dB) and 64-QAM modulation scheme.

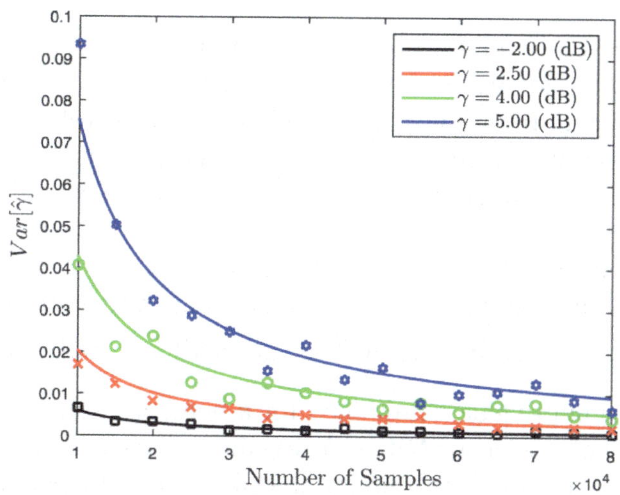

Figure 5: Variance of the estimate $\hat{\gamma}$ as a function of the number of samples N for different values of SNR γ and 64-QAM modulation considering κ−μ fading with κ=1.1 and μ=0.3.

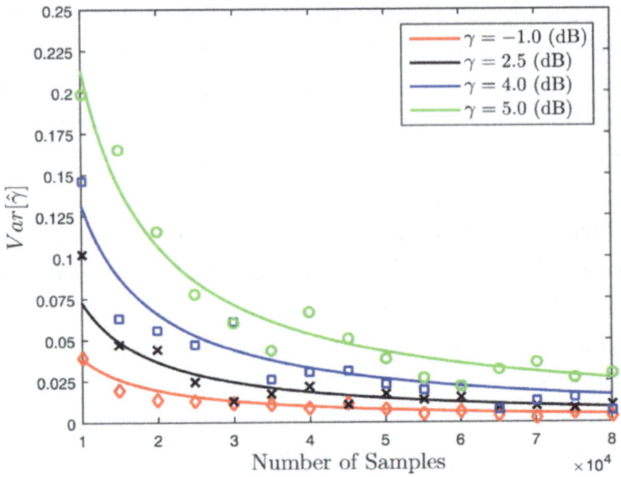

Figure 6: Variance of the estimate $\hat{\gamma}$ as a function of the number of samples N for different values of SNR γ and 64-QAM modulation considering η−μ fading with η=0.1 and μ=0.8.

It is observed from Fig. 5 that the variance of $\hat{\gamma}$ is below 0.015 for $N=8\times10^4$, for all values of SNR considered. The theoretical curves were all obtained from the expression (43) and corroborated with the points of the simulations carried out. By the curves, it is observed that the estimator performs better, in terms of the variance, for smaller values of SNR. For γ=−2 dB, the variance remains below 0.01 for the entire range of N values considered, indicating that with 10^4 samples, it is possible to get the estimate with a variance of less than 0.01 around the average.

From Fig. 6, it can be seen that for the SNR of −1 dB the variance of $\hat{\gamma}$ takes maximum value below 0.05 for a number of samples N of $|r[n]|$ equals 10^4, whereas for a SNR of 5 dB, the variance is only below 0.05 for $N=4\times10^4$. It is also worth noting that the setting of parameters η=0.1 and μ=0.8 models a case of more intense fading than configurations with values greater than μ. Therefore, the γ estimator of the η−μ model, presented in the second and third lines of Table 1, presents better performance for this more severe channel condition than the case where the SNR is larger and the fading is less intense. Although these MOM NDA estimators require much more samples of the received signal, compared to MLE estimators, it is still appropriate, due to its simplicity, to be implemented in hardware systems in which the processor capacity is reduced.

The group of curves of Figs. 7 and 8 shows the behavior of the NMSE and the mean of the estimate $\hat{\gamma}$ for different configurations of parameters of the κ−μ fading model. Each simulated value of the SNR estimate was obtained from N=5×10^4 samples of the envelope of the received signal $|r[n]|$, and the theoretical curves were obtained from the expressions (27), (35), and (43) and from the expressions of moments of $|r[n]|$ for the κ−μ model.

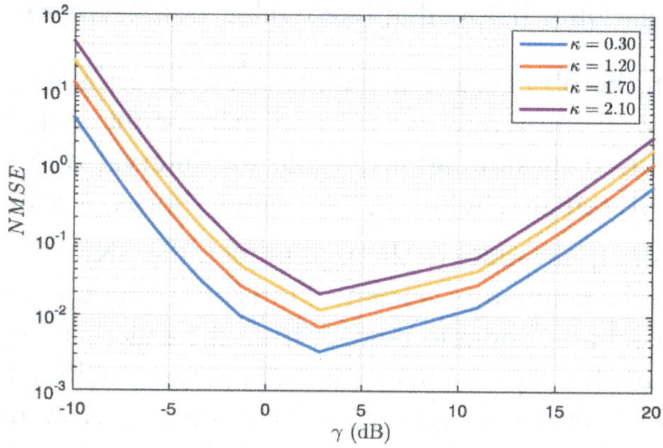

Figure 7: NMSE of the estimate $\hat{\gamma}$ as a function of SNR in dB for fading κ−μ for N=5×10^4 and 64-QAM modulation considering different values of κ and μ=0.8.

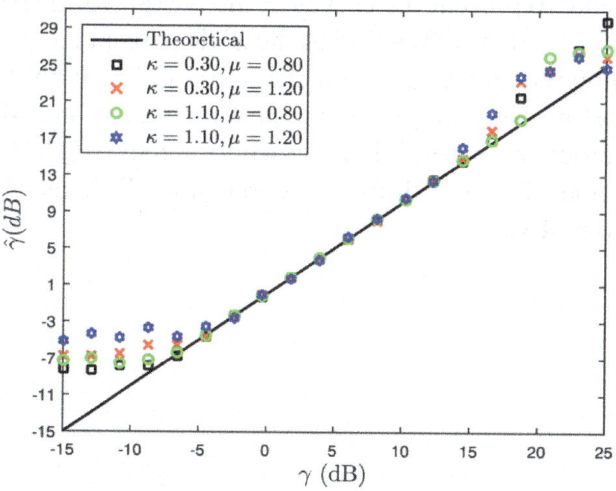

Figure 8: Mean of the estimate $\hat{\gamma}$ as a function of SNR in dB for fading κ−μ for N=5×10^4 and 64-QAM modulation considering different values of κ and μ.

Regarding Fig. 7, it is observed that for μ=0.8 and fixed SNR γ, the NMSE of the estimate $\hat{\gamma}$ decreases with the decreasing of κ parameter. It can be observed, by means of the CDF of the envelope of κ–μ fading presented in [23], that with smaller values of κ, the distribution κ–μ models cases of greater fading depth. This conclusion can be obtained by fixing a value of envelope of the κ–μ fading and verifying the probability of the envelope taking values below that fixed value as κ decreases. With respect to Fig. 8, it can be observed that the mean of the estimate $\hat{\gamma}$ is close to the target value in the range of –5 to 15 dB, for the different parameter configurations used in simulations, while the interval of SNR values for which the NMSE remains below 10^{-1} becomes shorter as κ increases.

Either for the η–μ model or the κ–μ model, the SNR range for the best performance of the estimators is from –2 to 12 dB. Above 12 dB, the average values of the estimates begin to deviate from the exact value. Nevertheless, the simplicity of the expressions of the estimators justifies the number of samples of the observed signal greater than the necessary for the optimal MLE estimator. As can be seen in Section 8, the greater complexity of the fading model leads to a greater complexity of the likelihood function and the difficulty for obtaining both a MLE estimator and an exact expression for the CRLB.

Figures 9 and 10 present the curves of the NMSE for different values of the constellation order, as a function of the SNR γ, respectively, on channels under η–μ fading with η=0.1 and μ=0.6 and κ–μ fading with κ=3 and μ=0.5. In both channel models, the estimator performance is not affected as the constellation order increases. This happens for the M-QAM constellation because the value of its fourth moment changes a little as M increases, as can be seen from (46).

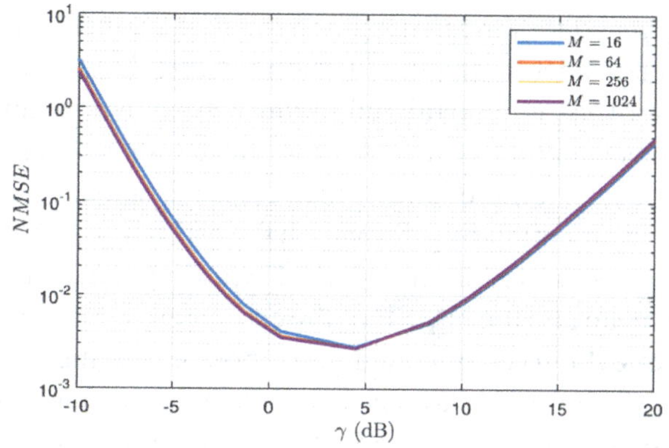

Figure 9: NMSE of the estimate $\hat{\gamma}$ as a function of the SNR in dB, N=5×10^4, η−μ fading for η=0.1 and μ=0.6, and different orders of the M-QAM constellation.

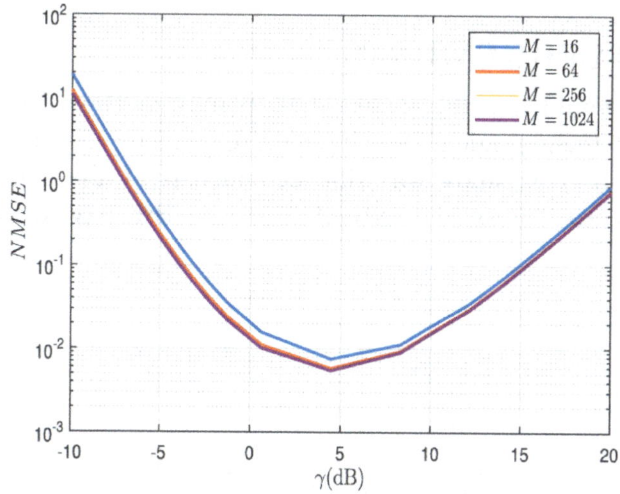

Figure 10: NMSE of the estimate $\hat{\gamma}$ as a function of the SNR in dB, N=5×10^4, κ−μ fading for κ=3 and μ=0.5, and different orders of the M-QAM constellation.

CONCLUSION

In this article, new and exact expressions are presented for the estimation of SNR in generalized fading channels characterized by the probability distributions $\eta-\mu$ and $\kappa-\mu$, by using the method of moments. In the model of received signal considered, the modulation scheme M-QAM was used. The main advantage of using generalized models, such as $\eta-\mu$ and $\kappa-\mu$, is that their distributions encompass several other distributions commonly used to model fading. As an example, one of the scenarios considered in this article was that of the Nakagami fading, for which the estimate of SNR was obtained from expressions presented in this work. Regarding the performance of the estimators, the mean and the NMSE of the estimates were used. It is observed from the theoretical curves obtained, corroborated by the simulations, that the performance of the estimators is satisfactory in the range of SNR values in which the mean of the estimate adheres to the target SNR value. It is also noticed, from the curves presented, that the performance of the estimators is better for the parameter settings that characterize more intense fading and low SNR. As future works on SNR estimation, the authors will investigate the use of goodness-of-fit tests and kernel density estimation.

ACKNOWLEDGEMENTS

The authors would like to express their thanks to Brazilian Council for Scientific and Technological Development (CNPq) for the financial support to the research. This study was financed in part by the Coordenação de Aperfeiçoamento de Pessoal de Níível Superior – Brazil (CAPES) – Finance Code 001.

REFERENCES

1. C. Seo, S. Portugal, S. Malik, C. You, T. Jung, H. Liu, I. Hwang, Throughput performance analysis of AMC based on a new SNR estimation algorithm using preamble. Wirel. Pers. Commun. **68**(4), 1225–1240 (2013). https://doi.org/10.1007/s11277-012-0505-x.
2. S.S Chauhan, S. Kumar, Channel capacity of TAS/MRC system with adaptive transmission and channel estimation errors. Telecommun. Syst. **63**(3), 347–355 (2016). https://doi.org/10.1007/s11235-015-0124-4.
3. W. Aihua, Y.Y Shujing, in *Proceedings of the 2nd International Conference on Future Control and Automation*, ed. by W. Deng. The SNR estimation and prediction for, AFH/MSK system, vol 1 (Springer Berlin HeidelbergBerlin, Heidelberg, 2012), pp. 317–324. https://doi.org/10.1007/978-3-642-31006-5-38.
4. Y. Zhou, Y. Chen, Performance analysis of end-to-end SNR estimators for AF relaying. Telecommun. Syst. **67**(2), 269–280 (2017). https://doi.org/10.1007/s11235-017-0327-y.
5. Elezabi, A. Gomaa, Diversity combining and SNR estimation for turbo-coded frequency-hopped spread-spectrum in partial-band interference and fading channels. Wirel. Pers. Commun. **50**(3), 351–365 (2009). https://doi.org/10.1007/s11277-008-9593-z.
6. J.Y Park, K.S Chung, An adaptive low-power LDPC decoder using SNR estimation. EURASIP J. Wirel. Commun. Netw. **2011**(1), 48 (2011). https://doi.org/10.1186/1687-1499-2011-48.
7. D.R Pauluzzi, N.C Beaulieu, A comparison of SNR estimation techniques for the AWGN channel. IEEE Trans. Commun. **48**(10), 1681–1691 (2000). https://doi.org/10.1109/PACRIM.1995.519404.
8. H. Abeida, T.Y Al-Nafouri, Al-Ghadhban S., in *IEEE International Workshop on Signal Processing Advances in Wireless Communications - SPAWC*. Data-aided SNR estimation in time-variant Rayleigh fading channels (IEEEMarrakech, 2010), pp. 1–5. https://doi.org/10.1109/TSP.2010.2063429.
9. F. Qin, X. Dai, J.E Mitchell, Effective-SNR estimation for wireless sensor network using Kalman filter. Ad Hoc Netw. **11**(3), 944–958 (2013). https://doi.org/10.1016/j.adhoc.2012.11.002.
10. T. Salman, A. Badawy, T. M Elfouly, T. Khattab, Mohamed A., in *IEEE International Conference on Wireless and Mobile Computing,*

Networking and Communications - WiMob. Non-data-aided SNR estimation for QPSK modulation in AWGN channel (IEEELarnaca, 2014), pp. 611–616. https://doi.org/10.1109/WiMOB.2014.6962233.

11. Wiesel, J. Goldberg, H.M. Yaron, SNR estimation in time-varying fading channels. IEEE Trans. Commun. **54**(5), 841–848 (2006). https://doi.org/10.1109/TCOMM.2006.873995.
12. F. Bellili, A. Stéphenne, Affes S., in *IEEE International Symposium on Wireless Communication Systems*. SNR estimation of QAM-modulated transmissions over time-varying SIMO channels (IEEEReykjavik, 2008), pp. 199–203. https://doi.org/10.1109/ISWCS.2008.4726046.
13. S.A Hassan, M.A Ingram, SNR estimation for M-ary non-coherent frequency shift keying systems. IEEE Trans. Commun. **59**(10), 2786–2795 (2011). https://doi.org/10.1109/TCOMM.2011.080111.090627.
14. S.A Hassan, M.A Ingram, SNR estimation in a non-coherent BFSK receiver with a carrier frequency offset. IEEE Trans. Signal Process. **59**(7), 3481–3486 (2011). https://doi.org/10.1109/TSP.2011.2131137.
15. G. Cai, L. Wang, L. Kong, G. Kaddoum, in *Vehicular Technology Conference (VTC Spring)*. SNR estimation for FM-DCSK system over multipath Rayleigh fading channels (IEEENanjing, 2016), pp. 1–5. https://doi.org/10.1109/VTCSpring.2016.7504478.
16. Stéphenne, F. Bellili, S. Affes, Moment-based SNR estimation over linearly-modulated wireless SIMO channels. IEEE Trans. Wirel. Commun. **9**(2), 714–722 (2010). https://doi.org/10.1109/TWC.2010.02.081719.
17. M.A Boujelben, F. Bellili, S. Affes, A. Stéphenne, SNR estimation over SIMO channels from linearly modulated signals. IEEE Trans. Sig. Process. **58**(12), 6017–6028 (2010). https://doi.org/10.1109/TSP.2010.2074197.
18. Das, B. D. Rao, SNR and noise variance estimation for MIMO systems. IEEE Trans. Signal Process. **60**(8), 3929–3941 (2012). https://doi.org/10.1109/TSP.2012.2194707.
19. F. Bellili, A. Methenni, S. Affes, Closed-form CRLBs for SNR estimation from turbo-coded BPSK, MSK and square-QAM-modulated signals. IEEE Trans. Signal Process. **62**(15), 4018–4033 (2014). https://doi.org/10.1109/TSP.2014.2328328.

20. F. Bellili, A. Stéphenne, S. Affes, Cramer-Raó lower bounds for NDA SNR estimates of square QAM modulated transmissions. IEEE Trans. Commun. **58**(11), 3211–3218 (2010). https://doi.org/10.1109/TCOMM.2010.093010.080568.

21. Y. Fu, J. Zhu, S. Wang, H. Zhai, Robust non-data-aided SNR estimation for multilevel constellations via Kolmogorov-Smirnov test. IEEE Commun. Lett. **18**(10), 1707–1710 (2014). https://doi.org/10.1109/LCOMM.2014.2356473.

22. F. Bellili, R. Meftehi, S. Affes, A. Stéphenne, Maximum likelihood SNR estimation of linearly-modulated signals over time-varying flat-fading SIMO channels. IEEE Trans. Signal Process. **63**(2), 441–456 (2015). https://doi.org/10.1109/TSP.2014.2364017.

23. M.D Yacoub, The κ−μ distribution and the η−μ distribution. IEEE Antennas Propag. Mag. **49**(1), 68–81 (2007). https://doi.org/10.1109/MAP.2007.370983.

24. M.A Diaz, R. L. Valcarce, C. Mosquera, SNR estimation for multilevel constellations using higher-order moments. IEEE Trans. Signal Process. **58**(4), 1515–1526 (2010). https://doi.org/10.1109/TSP.2009.2036069.

25. S.A Dianat, *SNR estimation in Nakagami fading channels with arbitrary constellation, vol. 2* (IEEE, Honolulu, 2007). https://doi.org/10.1109/ICASSP.2007.366238.

26. M. Hafez, T. Khattab, H.M Shalaby, Blind SNR estimation of Gaussian-distributed signals in Nakagami fading channels. IEEE Trans. Wirel. Commun. **14**(7), 3509–3518 (2015). https://doi.org/10.1109/TWC.2015.2403325.

27. M.B.B Salah, A. Samet, Moment-based joint estimation of Ricean K-factor and SNR over linearly-modulated wireless SIMO channels. Wirel. Pers. Commun. **91**(2), 903–918 (2016). https://doi.org/10.1007/s11277-016-3503-6.

28. K. Cho, D. Yoon, On the general BER expression of one and two-dimensional amplitude modulations. IEEE Trans. Commun. **50**(7), 1074–1080 (2002). https://doi.org/10.1109/TCOMM.2002.800818.

29. J.G Proakis, M. Salehi, *Digital Communications*, 5th edn (McGraw-Hill Higher Education, New York, 2001).

30. I.S Gradshteyn, I.M Ryzhik, *Table of Integrals, Series and Products*, 7th edn (Elsevier Science, Burlington, Massachusetts, 2007).

31. M. Abramowitz, I. Stegun, *Handbook of Mathematical Functions, 1st edn. (National Bureau of Standards Applied Mathematics Series)* (Dover Publications, New York, 1965).
32. S. M. Kay, *Fundamentals of Statistical Signal Processing: Estimation Theory* (Prentice Hall, Upper Saddle River, New Jersey, 2013).

17

An Efficient MIMO Scheme with Signal Space Diversity for Future Mobile Communications

Zhanji Wu and Xiang Gao

School of Telecommunication Engineering, Beijing University of Posts and Telecommunications, Xitucheng Road 10, Beijing, China

ABSTRACT

An efficient wireless transmission scheme with the signal space diversity (SSD) is proposed to improve the performance of multiple-input multiple-output (MIMO) systems in fading channels. By introducing the rotated modulation and space-time component interleaver, the proposed scheme jointly optimizes channel coding, modulation, and MIMO and can improve the link reliability and energy efficiency. An optimum spatial component interleaver is proposed to maximize the MIMO achievable rate. Based on the average mutual information (AMI)-maximization criterion, the optimal

Citation: Wu, Z., Gao, X. An efficient MIMO scheme with signal space diversity for future mobile communications. J Wireless Com Network 2015, 87 (2015). https://doi.org/10.1186/s13638-015-0301-x

rotation angles of real-valued signal and complex-valued QAM signal are investigated for the MIMO scheme. For the iterative demapping and decoding (ID) scheme, a simple genetic algorithm (GA) to search binary convolution code (BCC) is also put forward to match the rotated modulation. Simulation results show that the optimized BCC-coded MIMO scheme with SSD-ID outperforms the turbo-coded MIMO scheme with bit-interleaved coded modulation (BICM)-ID by 1.4 dB signal-to-noise ratio (SNR) gain, while the new scheme has much lower complexity. So, the proposed scheme is simple, efficient, and promising for future wireless communication systems.

INTRODUCTION

Wireless communications have made a great progress in the recent few years. By introducing more advanced technology, 5G will provide higher spectral efficiency, more spectrum resources, and more reliability to meet the growing demand for mobile traffic [1].

Bit-interleaved coded modulation (BICM) is a bandwidth-efficient coded modulation scheme which increases the time diversity in fading channels [2,3]. For its iterative version, BICM with iterative demapping and decoding (BICM-ID), the extrinsic information is transferred between the channel decoder and the soft-in-soft-out demapper, which is like the serial turbo decoder. Multiple-input multiple-output (MIMO) scheme is the extension of the coding theory on the space domain, so it is also named space-time coding (STC) [4]. Foschini proposed a layered space-time (LST) architecture to process multidimensional signals in the space domain [5]. The BICM-LST is a conventional spectral-efficient spatial multiplexing technology to deal with MIMO fading channels, and the BICM-threaded layered space-time (TLST) with a cyclic-shift spiral spatial interleaver is regarded as the most efficient method, because the cyclic-shift spatial interleaver introduces effective space diversity for the codeword on each layer [6]. In general, the BICM-LST can be viewed as the serial concatenation of the channel coding, modulation, and spatial layered multiplexing. Because BICM-LST exhibits a robust diversity performance on fading channels, it is widely deployed by wireless communication standards.

As for the bandwidth-efficient quadrature amplitude modulation (QAM), uncoded rotated multidimensional modulation schemes over independent Rayleigh fading channels were studied in [7] for the single-input single-output (SISO) scheme. Different from the other well-known diversity (time, frequency, code, space), it has an intrinsic modulation diversity, which is named signal space diversity (SSD). Through the combination of constellation rotation and component interleaver, the schemes can achieve very high modulation diversity, and the error performance over fading channels can approach that on the additive white Gaussian noise (AWGN) channels. SSD schemes for SISO system have been extensively researched. In [8], SSD is introduced to the BICM by the means of modifications to the QAM constellation mapper and demapper so as to improve the BICM performance of QAM constellations for broadcasting applications. In [9], a LDPC-coded SSD scheme for multi-level modulation was presented. N.F. Kiyani and J.H. Weber studied the rotated-MPSK SISO BICM-ID system [10,12], which focused on two-dimensional multiphase shift keying (MPSK) scheme. In [11], the performance analysis of BICM-ID with SSD in fading channels is presented. In [13], the extension of BICM-SSD schemes with a non-binary code was proposed. We also proposed coded orthogonal frequency division multiplexing (OFDM) systems with SSD in [14,15]. In [16], the schemes combining SSD with SISO-coded BICM and BICM-ID systems were investigated. It provided a new criterion for determining the optimal rotation angle by maximizing the average mutual information (AMI). For the optimization of BICM-ID system, it proved that SSD can mitigate the different-slope problem of the demapper's extrinsic information transfer (EXIT) curve under different channels. However, finding well-matched channel codes for given labeling in BICM-ID system with SSD is still a big challenge.

The combination of signal rotation and space-time coding in MIMO system can effectively improve the diversity gain. In order to achieve full diversity, the quasi-orthogonal space time block codes (QOSTBC) with rotating the constellations of half of the complex symbols has been widely discussed in [17-20]. Some specific optimal rotation angles and corresponding optimization criterions for QAM and phase shift keying (PSK) constellations are provided. A rotation-based method that aims at maximizing the minimum distance in the space-time constellation is proposed in [17]. The proposed scheme shows good improvement of the codes compared to their non-rotated counterparts. In [18], the authors considered the design of rotated QOSTBC for the MISO system. The code designs are based mainly on the rank and the determinant criteria, and the optimal rotation angle $\pi/6$ can provide full diversity and the optimal coding gain. In [19], the authors proposed to design the signal constellations properly to ensure that the resulting quasi-orthogonal STBCs can guarantee to achieve the full diversity. The optimal rotation angles are determined by maximizing the diversity product. A novel method to exactly derive the coding gain of QSTBC as a function of the rotation angle and the minimum Euclidean distance of two-dimensional constellations is proposed in [20]. A coded MIMO scheme for block-fading channels was proposed in [21], which consists of a channel code and a space-time code. The space-time code is designed based on SSD technique, which allows full spatial multiplexing MIMO transmission and achieves full space diversity. In [22], the uncoded SSD scheme was extended to V-BLAST MIMO systems in order to achieve the maximum diversity gain without additional power or bandwidth consumption. An improved turbo-coded SSD scheme was proposed for MIMO-OFDM BICM system in [23], and the linear minimum mean square error (LMMSE) equalization is utilized for the non-ID MIMO detection. In general, the research of SSD technique in coded MIMO systems is still on the original stage. There are still many open problems. For instance, the optimal rotation angles in current research mainly depend on the maximum product distance introduced in [7]. Unfortunately, this criterion is only valid for the SISO system in high signal-to-noise ratio (SNR) region. As for the coded MIMO scheme, when powerful forward error-correction codes (FECs) are considered, actual SNR can be quite low. Hence, the angle values applied to uncoded SISO system do not lead to the best error performance for the coded modulation MIMO scheme. What is more, current research works mainly focus on local optimizations. For example, most proposed MIMO systems with SSD are only an extension of SISO-SSD system, and all are based on the conventional non-precoding transmitter. The channel coding, QAM modulation, and STC are independent with each other, which is just a straightforward serial concatenation. Hence, the performance of the BICM-LST is still rather far away from the MIMO fading channel capacity. For example, a near-capacity BICM-LST scheme was proposed in [24], which allows the iterative processing of the list sphere detection (LSD) and turbo decoding, but simulation results indicate that the gaps to the MIMO capacity are still more than 2 dB. As each individual optimization becomes mature so far, from the philosophy, it is high time to optimize these key technical elements jointly so as to improve the overall performance.

An improved coded MIMO system based on SSD is proposed for the jointly optimization of constellation rotation angle, spatial component interleaver, and the matching of channel coding and labeling, which is named joint coding and modulation diversity (JCMD), where the terminology 'coding' refers to both the channel coding and space-time coding. Firstly, in order to maximize the MIMO achievable rate, an optimum spatial component interleaver is proposed. Secondly, based on the AMI-maximization criterion, the optimal rotation angles of real-valued signal and complex-valued QAM signal are investigated for MIMO schemes, which are different from the SISO scheme. Thirdly, for the JCMD-ID scheme, a simple genetic algorithm (GA) to search binary convolution code (BCC) is put forward to match the rotated QAM modulation. Simulation results show that the optimized BCC-coded JCMD-ID MIMO scheme outperforms the turbo-coded BICM-ID MIMO scheme in [24] by 1.4 dB SNR gain, while the new scheme has much lower complexity.

Throughout this paper, we use bold letters to represent vectors or matrices. $(\cdot)^T$ and $(\cdot)^H$ represent transposition and conjugate transposition, respectively. SNR $=E_s/N_0$, where E_s denotes the average symbol energy per receive antenna and $N_0=2\sigma^2$ denotes the variance of the complex Gaussian noise.

The paper is organized as follows. An improved JCMD MIMO scheme is proposed in Section 2. Theoretical analysis about the achievable rate of JCMD-MIMO for rotated real-valued signals is given in Section 3. Based on the AMI analysis, the optimal rotation angles for JCMD and JCMD-ID MIMO systems are presented in Section 4. Section 5 introduces an outer convolutional code search method for the optimization of JCMD-ID system with optimal rotation angle. Simulation results are presented in Section 6 on fast fading channels. Concluding remarks are offered in Section 7.

SYSTEM MODEL

Based on the BICM-TLST scheme, a system model of N_L-layer MIMO-JCMD-ID is shown in Figure 1. Perfect channel state information (CSI) is assumed to be known at both the transmitter and the receiver. In Figure 1, the iterative feedback processing is depicted in dashed lines. Without loss of generality, a rank-L $N_R \times N_T$ MIMO system with L nonzero eigenvalues is assumed, where N_R and N_T are the number of receive and transmit antennas, respectively, and $N_L \leq L \leq \min\{N_R, N_T\}$. In the transmitter, K information bits $\mathbf{B}=(b_1,b_2,\ldots,b_K)^T$ are encoded and interleaved to yield the coded bit sequence $\mathbf{C}=(c_1,c_2,\ldots,c_N)^T$. Afterwards, m-tuple coded bits are mapped to a complex symbol $x_k=x_k(I)+\mathrm{j}\cdot x_k(Q)$, which is chosen from a 2^m-ary rotated QAM constellation set $\chi = \{\hat{x}_1, \hat{x}_2, \ldots, \hat{x}_{2^m}\}$ according to some optimal angle. Each symbol x_k has one Q-component $x_k(Q)$ and one I-component $x_k(I)$. The rotated mapped symbol sequence is first mapped onto N_L layers in a round robin manner. Afterwards, the conventional cyclic-shift spatial interleaver of TLST is used for different symbols on N_L layers to exploit both the space and time diversity as the following, because the cyclic-shift spatial interleaver allows the codewords to be distributed on all layers:

$$w_k^l = x_k^i, \quad l = (i + k - 2) \bmod N_L + 1, \tag{(1)}$$

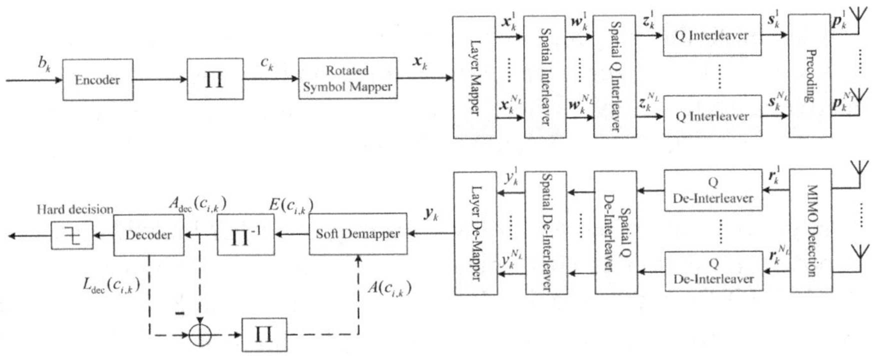

Figure 1: System model for MIMO-JCMD-ID with perfect CSI. The iterative feedback processing is depicted in dashed lines.

where w_k^l denotes the kth ($k \in \mathbf{N}^+$) symbol at the lth ($l \in [1, N_L]$) layer after the spatial interleaver, and x_k^i denotes the kth symbol at the ith ($i \in [1, N_L]$) layer before the interleaver. Then, a spatial Q-component interleaver is applied for the Q components of N_L symbols at the same instant as the following:

$$z_k^n(Q) = w_k^l(Q), \quad n = N_L - l + 1, \tag{(2)}$$

where z_k^n denotes the kth symbol at the nth ($n \in [1, N_L]$) layer after the spatial Q interleaver. Thus, I components keep the same layer order as before, and just Q components change the layer order. The spatial Q-component interleaver is used to make the fading of I component and that of the Q component as uncorrelated as possible in the space domain. Thus, the modulation diversity of the proposed scheme is further extended to the spatial dimension. Actually, the Q-component spatial interleaver can be different from the reverse interleaver in Equation 2. For example, it can be a cyclic-shift interleaver as the following,

$$z_k^n(Q) = w_k^l(Q), \quad n = (l \bmod N_L) + 1. \qquad ((3))$$

If perfect CSI is known, we prove that the reverse interleaver is better than other interleavers through the later theoretical analysis and computer simulations. If CSI is unknown at the transmitter, the cyclic-shift interleaver in Equation 3 can be used.

In order to make the fading of I component and that of the Q component as uncorrelated as possible in the time domain, after the spatial Q interleaving, Q components of the mapped symbols in each layer are interleaved through a time-domain pseudo S-random interleaver to reconstruct a new symbol vector $\mathbf{s}_k = \left[s_k^1 \cdots s_k^{N_L}\right]^T$, where s_k^l denotes the *k*th symbol at the *l*th layer after the component interleaving. Afterwards, the symbols are mapped onto N_T transmit antennas via the spatial precoding and then transmitted.

The ideal precoding matrix can be obtained by singular value decomposition (SVD), which can divide MIMO channel into parallel independent SISO channels. According to the SVD criterion, the $N_R \times N_T$ MIMO channel matrix $\mathbf{H}_k$ can be decomposed as

$$\mathbf{H}_k = \mathbf{U}_k \mathbf{D}_k \mathbf{V}_k{}^H, \qquad ((4))$$

where the $N_R \times N_R$ matrix $\mathbf{U}_k$ and the $N_T \times N_T$ matrix $\mathbf{V}_k$ are unitary matrices. $\mathbf{D}_k$ is a $N_R \times N_T$ non-negative diagonal matrix with N_l nonzero descending-order singular values, $\sqrt{\rho_1} \geq \sqrt{\rho_2} \geq \ldots \geq \sqrt{\rho_{N_L}} > 0$, where ρ_i is the *i*th largest eigenvalue of $\mathbf{H}_k \cdot \mathbf{H}_k{}^H$. Thus, the SVD-based linear precoding is performed as the following:

$$\mathbf{p}_k = \mathbf{V}_k \cdot \mathbf{s}_k. \qquad ((5))$$

In the receiver, the corresponding detection matrix is $\mathbf{U}_k^H$. The precoding and detection process can be expressed as linear transformations as shown in Equation 6.

$$\begin{aligned}\mathbf{r}_k &= \mathbf{U}_k^H \cdot (\mathbf{H}_k \cdot \mathbf{p}_k + \mathbf{n}_k) \\ &= \mathbf{U}_k^H \cdot \mathbf{H}_k \cdot \mathbf{V}_k \cdot \mathbf{s}_k + \mathbf{U}_k^H \cdot \mathbf{n}_k \\ &= \mathbf{D}_k \cdot \mathbf{s}_k + \mathbf{n}'_k,\end{aligned} \qquad ((6))$$

where $\mathbf{r}_k = \left[r_k^1 \cdots r_k^{N_R}\right]^T$ denotes the received symbol vector, $\mathbf{n}_k$ and $\mathbf{n}'_k$ are column vectors of N_p complex Gaussian random variables with mean zero and variance $\sigma^2 = \frac{N_0}{2}$. Thus, due to SVD, the MIMO channel can be viewed as N_L parallel fading channels, and for lth layer, the kth received symbol that corresponds to s_k^l in the transmitter can be expressed as

$$r_k^l = \sqrt{\rho_l} \cdot s_k^l + n'^l_k. \qquad ((7))$$

After the corresponding Q-component de-interleaving in time domain for each layer and spatial Q-component de-interleaving, the kth received symbol on lth layer is reconstructed as y_k^l that corresponds to x_k^l in the transmitter. For y_k^l, the fading coefficients of I-component $\lambda_k(I)$ and Q-component $\lambda_k(Q)$ are different, which can be expressed as

$$\begin{aligned} y_k^l(I) &= \lambda_k^l(I)x_k^l(I) + n_k^l(I) \\ y_k^l(Q) &= \lambda_k^l(Q)x_k^l(Q) + n_k^l(Q). \end{aligned} \qquad ((8))$$

Assuming that the Q interleaver is long enough, each coordinate of the symbol after the Q de-interleaving can be regarded as suffering from independent fading coefficients. The equivalent coded modulation (CM) channel for JCMD system before sending to the soft demapper can be modeled as

$$Y_k^l(\eta) = \Lambda_k^l(\eta)X_k^l(\eta) + N_k^l(\eta), \eta \in \{I, Q\}, l \in [1, N_L], \qquad ((9))$$

where $N_k^l(I)$ and $N_k^l(Q)$ are identically independently distributed (i.i.d.) Gaussian noise random variables with zero mean and variance of $\sigma^2 = \frac{N_0}{2}$. For MIMO fading channels, $\Lambda_k^l(I)$ and $\Lambda_k^l(Q)$ are singular values of corresponding sub-channels. This is in net contrast with respect to SISO scheme where the fading coefficients are Rayleigh distributed. That means the modulation diversity of proposed scheme is further extended to the spatial dimension. By denoting $\mathbf{X} = \left[\mathbf{X}_1, \ldots, \mathbf{X}_{N_L}\right]^T$ $\mathbf{Y} = \left[\mathbf{Y}_1, \ldots, \mathbf{Y}_{N_L}\right]^T, \mathbf{N} = \left[\mathbf{N}_1, \ldots, \mathbf{N}_{N_L}\right]^T,$ and $\Lambda=$

$\operatorname{diag}(\Lambda_1, \ldots, \Lambda_{N_L})$ representing a ($2N_L \times 2N_L$) diagonal matrix, the channel model in Equation 9 can be written in the matrix form as **Y**=**Λ** **X**+**N**, where $\mathbf{X}_l = \left[X_k^l(I), X_k^l(Q)\right]$, $\mathbf{Y}_l = \left[Y_k^l(I), Y_k^l(Q)\right]$, $\mathbf{N}_l = \left[N_k^l(I), N_k^l(Q)\right]$, and $\Lambda_l = \operatorname{diag}\left(\Lambda_k^l(I), \Lambda_k^l(Q)\right)$.

After that, symbols on multiple layers are reassembled as symbol streams **y**=$[y_1, y_2 \cdots]^T$ through the layer demapping. A serial concatenation of a soft-in-soft-out rotated symbol demapper and a channel decoder are employed to approach the maximum likelihood (ML) receiver performance. For JCMD-ID, the iterative demapping and decoding scheme is an application of the turbo decoder principle. The soft demapper calculates the extrinsic value $E(c_{i,k})$ of bit $c_{i,k}$ which corresponds to the *i*th (*i*=1,2,…,*m*) bit of the received symbol y_k as follows:

$$E(c_{i,k}) = L(c_{i,k}) - A(c_{i,k}), \qquad ((10))$$

where $L(c_{i,k}) \triangleq \ln \frac{P(c_{i,k}=0|y_k)}{P(c_{i,k}=1|y_k)}$ is the log-likelihood ratio (LLR) and $A(c_{i,k}) \triangleq \ln \frac{P(c_{i,k}=0)}{P(c_{i,k}=1)}$ is *a priori* L-value. Based on Bayes' theorem, we can write

$$E(c_{i,k}) = \ln \frac{\sum\limits_{\substack{\hat{x}\in\chi,\\ \hat{c}_i=0}} P(y_k|x_k=\hat{x}) \exp\left\{\sum\limits_{\substack{j=1,\\ j\neq i}}^{m}\left[(-1)^{\hat{c}_j}\frac{A(c_{j,k})}{2}\right]\right\}}{\sum\limits_{\substack{\hat{x}\in\chi,\\ \hat{c}=1}} P(y_k|x_k=\hat{x}) \exp\left\{\sum\limits_{\substack{j=1,\\ j\neq i}}^{m}\left[(-1)^{\hat{c}_j}\frac{A(c_{j,k})}{2}\right]\right\}}, \qquad ((11))$$

where $\hat{c}_i$ is the *i*th bit corresponding to symbol $\hat{x}$. For the fading channel, the conditional probability is given by

$$P(y_k|x_k=\hat{x}) = \frac{1}{2\pi\sigma^2}\exp\left(-\frac{\left(y_k^I - \lambda_k^I\hat{x}^I\right)^2 + \left(y_k^Q - \lambda_k^Q\hat{x}^Q\right)^2}{2\sigma^2}\right). \qquad ((12))$$

To the complexity, by applying Max-Log-MAP algorithm, Equation 11 can be simplified as

$$E(c_{i,k}) = \max_{\substack{\hat{x}\in\chi,\\ \hat{c}_{i,k}=0}} \left\{ \Omega_k\left(\hat{x}\right) - \sum_{\substack{j=1,\\ j\neq i}}^{m} \left[(-1)^{\hat{c}_{j,k}} \frac{A\left(c_{j,k}\right)}{2} \right] \right\} - \max_{\substack{\hat{x}\in\chi,\\ \hat{c}_{i,k}=1}} \left\{ \Omega_k\left(\hat{x}\right) - \sum_{\substack{j=1,\\ j\neq i}}^{m} \left[(-1)^{\hat{c}_{j,k}} \frac{A\left(c_{j,k}\right)}{2} \right] \right\}, \quad ((13))$$

where $\Omega_k\left(\hat{x}\right) = -\frac{\left(y_k^I-\lambda_k^I\hat{x}^I\right)^2+\left(y_k^Q-\lambda_k^Q\hat{x}^Q\right)^2}{2\sigma^2}$.

For the JCMD system without the iterative demapping and decoding, $A(c_{i,k})$=0. Finally, the decoder can utilize the extrinsic values to decode information bits.

In the transmitter, compared with the conventional BICM, the JCMD scheme introduces extra constellation rotation and Q-component interleavers. Constellation rotation does not increase the complexity, because the rotated symbol mapping can be implemented through look-up table operations as the same as the conventional modulation without rotation. Q-component interleavers also can be implemented by the low-complexity index-based look-up table operations.

In the receiver, the soft rotated demapping operation of JCMD system is the same as that of BICM system, which is shown in Equation 11. Q-component de-interleavers also can be implemented by the simple reverse index-based look-up table operations.

THEORETICAL ANALYSIS OF THE ACHIEVABLE RATE FOR ROTATED REAL-VALUED SIGNALS

Firstly, since real-valued signals are elementary, we analyze the real-valued transmit signals in MIMO systems.

Lemma 1.

For any constant rank-2 MIMO fading channel with real-valued transmit signals, the constant achievable rate of JCMD-MIMO is not less than that of BICM-MIMO with the conventional uniform power allocation. If and only if the constant eigenvalues are identical, both of them are equal, otherwise the former is greater than the latter.

Proof.

A simple rank-2 MIMO case with two eigenvalues ρ_1 and ρ_2 is illustrated in Figure 2. Due to the well-known SVD, BICM-MIMO can be viewed as two parallel fading channels with two eigenvalues ρ_1 and ρ_2 for spatial layer 1 and layer 2, respectively. Two fading amplitude coefficients of layer 1 and layer 2 are the corresponding singular values $\sqrt{\rho_1}$ and $\sqrt{\rho_2}$, respectively, as shown in the left half of Figure 2. Thus, given the same transmit power $\frac{P}{2}$ on each layer for one real-valued symbol, the received symbol power of layer 1 and that of layer 2 are $\frac{\rho_1 P}{2}$ and $\frac{\rho_2 P}{2}$, respectively, where P is the total transmit power for two layers. So, according to Shannon's theory, the achievable rate of BICM-MIMO is shown as the following:

$$
\begin{aligned}
C_1 &= \frac{W}{2} \cdot \log_2 \left[\left(1 + \frac{\rho_1 P}{2\sigma^2}\right) \left(1 + \frac{\rho_2 P}{2\sigma^2}\right) \right] \\
&= \frac{W}{2} \cdot \log_2 \left(1 + \frac{\rho_1 + \rho_2}{2\sigma^2} P + \frac{\rho_1 \rho_2}{4\sigma^4} P^2 \right),
\end{aligned}
\tag{(14)}
$$

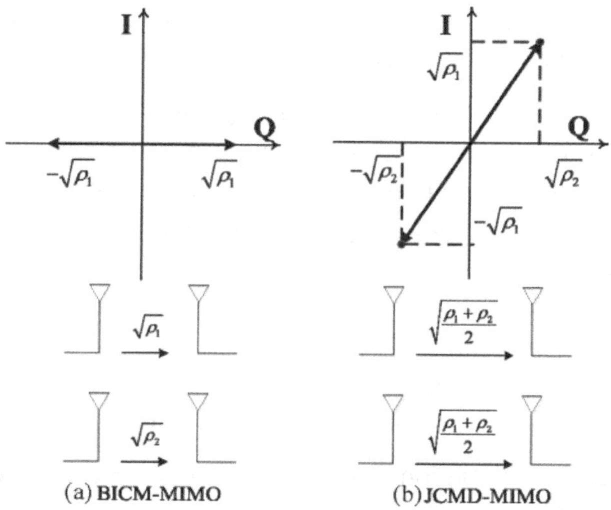

Figure 2: Achievable rates of conventional BICM-MIMO and JCMD-MIMO system. (a) BICM-MIMO can be viewed as two parallel fading channels with two eigenvalues ρ_1 and ρ_2. (b) JCMD-MIMO can be viewed as two parallel fading channels with identical fading amplitude coefficient ρ1+ρ22——√ρ1+ρ22 for both layers.

where W is the channel bandwidth, σ^2 is the variance of AWGN. As we know, in order to achieve the capacity in the Equation 14, the transmit signals should be Gaussian distributed. Note that the rotation does not change the achievable rate for the BICM-MIMO scheme without the I/Q-component interleaver.

For the JCMD-MIMO scheme, we consider the $\frac{\pi}{4}$-rotated real-valued transmit signal, which is rotated by $\frac{\pi}{4}$ compared with the conventional real-valued signal. Due to the orthogonal modulation, the transmit powers of I component and that of Q component on each layer are both $\frac{P}{4}$. For JCMD-MIMO, after the spatial Q-component de-interleaver at the receiver, the fading amplitude coefficient of I component is different from that of Q component in one symbol, that is to say, one is $\sqrt{\rho_1}$, and another is $\sqrt{\rho_2}$. . Thus, the received power of I component is also different from that of Q component, that is to say, one is $\frac{\rho_1 P}{4}$, and another is $\frac{\rho_2 P}{4}$. So, the total received symbol power is $\frac{(\rho_1+\rho_2)P}{4}$ both for layer 1 and layer 2. Thus, JCMD-MIMO can be viewed as two parallel fading channels with identical fading amplitude coefficient $\sqrt{\frac{\rho_1+\rho_2}{2}}$ for both layer 1 and layer 2, as shown in the right half of Figure 2. In the receiver, after the phase compensation, the received signal also can be viewed as the real-valued signal with the power $\frac{(\rho_1+\rho_2)P}{4}$. Therefore, the achievable rate of JCMD-MIMO is shown as the following:

$$\begin{aligned} C_2 &= \frac{W}{2} \cdot \log_2 \left[1 + \frac{(\rho_1+\rho_2)}{4\sigma^2} P \right]^2 \\ &= \frac{W}{2} \cdot \log_2 \left[1 + \frac{\rho_1+\rho_2}{2\sigma^2} P + \frac{(\rho_1+\rho_2)^2}{16\sigma^4} P^2 \right]. \end{aligned} \qquad ((15))$$

So, comparing Equation 14 with 15, it is easy to come to the conclusion: $C_2 \geq C_1$

If and only if $\rho_2 = \rho_1$, $C_2 = C_1$.

Furthermore, we will prove that $\frac{\pi}{4}$ is the optimum rotation angle for real-valued transmit signals in the JCMD-MIMO scheme. Let us consider a general θ-rotated real-valued transmit signal, which is rotated by θ compared with the conventional real-valued signal. Thus, the transmit power of I component and that of Q component on layer 1 are $\frac{P}{2}\cos^2\theta$ and $\frac{P}{2}\sin^2\theta$, respectively, and the total transmit power on layer 1 is also $\frac{P}{2}$. Therefore, the received power of I component and that of Q component are $\frac{\rho_2 P}{2}\cos^2\theta$ and $\frac{\rho_1 P}{2}\sin^2\theta$, respectively. So, the total received symbol power on layer 1 is $\frac{(\rho_1\sin^2\theta+\rho_2\cos^2\theta)}{2}P$. Likewise, the total received symbol power on layer 2 is $\frac{(\rho_2\sin^2\theta+\rho_1\cos^2\theta)}{2}P$. Thus, we can get the following achievable rate:

$$
\begin{aligned}
C(\theta) &= \frac{W}{2}\cdot\log_2\left\{\left[1+\frac{(\rho_1\sin^2\theta+\rho_2\cos^2\theta)}{2\sigma^2}P\right]\cdot\left[1+\frac{(\rho_1\cos^2\theta+\rho_2\sin^2\theta)}{2\sigma^2}P\right]\right\} \\
&= \frac{W}{2}\cdot\log_2\left[1+\frac{\rho_1+\rho_2}{2\sigma^2}P+\left(\rho_1\rho_2+\frac{\sin^2(2\theta)(\rho_1-\rho_2)^2}{4}\right)\frac{P^2}{4\sigma^4}\right].
\end{aligned}
\qquad ((16))
$$

Obviously, when $\theta=0$, the achievable rate is minimum as Equation 14; and when $\theta=\frac{\pi}{4}$, the achievable rate is maximum as Equation 15.

According to Lemma 1, we can come to the following theorem:

Theorem 1.

For any actual rank-2 MIMO fading channel with real-valued transmit signals, the ergodic achievable rate of JCMD-MIMO is greater than that of BICM-MIMO with the conventional uniform power allocation.

Proof.

For any actual rank-2 MIMO fading channel, the ergodic achievable rate is the mathematical average expectation of the above constant achievable rate over all possible channel eigenvalue realization. Generally speaking, the equal eigenvalue realization is only a small probability event. The case that the channel eigenvalues are different must be existed. According to Lemma 1, the ergodic achievable rate of JCMD-MIMO is greater than that of BICM-MIMO with the conventional uniform power allocation.

Generally speaking, assuming a rank-L MIMO with a descending-order eigenvalue vector $\bar{\rho}=\{\rho_1,\rho_2,\ldots,\rho_L\}$, Q-component interleaver only changes the order of Q components on L layers to another eigenvalue vector $\bar{\zeta}=\{\zeta_1,\zeta_2,\ldots,\zeta_L\}$, where $\bar{\zeta}$ is just another arrangement order of $\bar{\rho}$ corresponding to the output order of Q-component spatial interleaver.

Hence, due to the orthogonal modulation, JCMD-MIMO can be viewed as L parallel fading channels with an eigenvalue vector $\bar{\upsilon} = \frac{\bar{\rho}+\bar{\xi}}{2}$. So, the achievable rate of rank-L JCMD-MIMO is shown as the following:

$$\begin{aligned} C_L(\bar{\xi}) &= \frac{W}{2} \cdot \sum_{i=1}^{L} \log_2 \left[1 + \frac{(\rho_i + \xi_i)}{2\sigma^2 L} P\right] \\ &= \frac{W}{2} \cdot \log_2 \left[\prod_{i=1}^{L} \left(1 + \frac{(\rho_i + \xi_i)}{2\sigma^2 L} P\right)\right], \end{aligned} \tag{(17)}$$

where P is the total transmit power for L layers. Hence, the optimum problem of the achievable rate is to find the optimum $\bar{\xi}$ so as to maximize C_L. We can reach the following theorem:

Theorem 2.

To maximize the achievable rate of rank-L JCMD-MIMO with a descending-order eigenvalue vector $\bar{\rho} = \{\rho_1, \rho_2, \ldots, \rho_L\}$, the optimum Q-component interleaver vector $\bar{\omega} = \{\rho_L, \rho_{L-1}, \ldots, \rho_1\}$, that is to say, $\bar{\omega}$ should be in ascending order, which is just in reverse order of $\bar{\rho}$.

Proof.

Assuming someone claims to find a non-increasing-order vector $\bar{\rho}'$ other than $\bar{\omega}$ to have maximum achievable rate $C_L(\bar{\rho}')$, we must can find a pair of $\{\rho_i', \rho_{i+1}'\}$ from $\bar{\rho}'$ to satisfy $\rho_i' > \rho_{i+1}'$. And then, we can construct a new vector $\eta = \{\rho_1', \rho_2', \ldots, \rho_{i-1}', \rho_{i+1}', \rho_{i}', \rho_{i+2}', \ldots, \rho_L'\}$, which only changes the order of $\{\rho_i', \rho_{i+1}'\}$ from $\bar{\rho}'$. So, we can compute the difference of $\frac{C_L(\eta) - C_L(\bar{\rho}')}{W/2}$,

$$\begin{aligned} \frac{C_L(\eta) - C_L(\bar{\rho}')}{W/2} &= \sum_{k=1}^{L} \log_2\left[1 + \frac{(\rho_k + \eta_k)}{2\sigma^2 L} P\right] - \sum_{k=1}^{L} \log_2\left[1 + \frac{(\rho_k + \rho_k')}{2\sigma^2 L} P\right] \\ &= \log_2\left[1 + \frac{(\rho_i + \eta_i)}{2\sigma^2 L} P\right] + \log_2\left[1 + \frac{(\rho_{i+1} + \eta_{i+1})}{2\sigma^2 L} P\right] \\ &\quad - \log_2\left[1 + \frac{(\rho_i + \rho_i')}{2\sigma^2 L} P\right] - \log_2\left[1 + \frac{(\rho_{i+1} + \rho_{i+1}')}{2\sigma^2 L} P\right] \\ &= \log_2\left[1 + \frac{(\rho_i + \rho_{i+1}')}{2\sigma^2 L} P\right] + \log_2\left[1 + \frac{(\rho_{i+1} + \rho_i')}{2\sigma^2 L} P\right] \\ &\quad - \log_2\left[1 + \frac{(\rho_i + \rho_i')}{2\sigma^2 L} P\right] - \log_2\left[1 + \frac{(\rho_{i+1} + \rho_{i+1}')}{2\sigma^2 L} P\right]. \end{aligned} \tag{(18)}$$

Let a constant $c = \frac{P}{2\sigma^2 L}$, and then,

$$\frac{C_L(\eta) - C_L(\bar{\rho}')}{W/2} = \log_2 \frac{M_1}{M_2}, \qquad ((19))$$

where

$$M_1 = 1 + c\left(\rho_i + \rho_i' + \rho_{i+1} + \rho_{i+1}'\right) + c^2\left(\rho_i + \rho_{i+1}'\right)\left(\rho_{i+1} + \rho_i'\right) > 0, \qquad ((20))$$

$$M_2 = 1 + c\left(\rho_i + \rho_i' + \rho_{i+1} + \rho_{i+1}'\right) + c^2\left(\rho_i + \rho_i'\right)\left(\rho_{i+1} + \rho_{i+1}'\right) > 0. \qquad ((21))$$

So, the difference

$$M_1 - M_2 = c^2(\rho_i - \rho_{i+1})(\rho_i' - \rho_{i+1}'). \qquad ((22))$$

Because $\rho_i > \rho_{i+1}$ and $\rho_i' > \rho_{i+1}'$.

So, $M_1 - M_2 > 0$; Thus, $\frac{M_1}{M_2} > 1$; and then, $C_L(\eta) - C_L(\bar{\rho}') > 0$.

That is to say, we find a counter-example η to have more achievable rate than $C_L(\bar{\rho}')$, which contradicts the assumption of maximum achievable rate $C_L(\bar{\rho}')$. Therefore, $\overline{\omega}$ should be in ascending order, which is just in reverse order of $\bar{\rho}$.

So, in Section 2, the reverse Q-component spatial interleaver is applied as Equation 2 so as to have the maximum achievable rate. In fact, if L is an even number, the rank-L JCMD-MIMO with the reverse Q-component spatial interleaver can be viewed as $\frac{L}{2}$ parallel pairs of rank-2 JCMD-MIMO with eigenvalues $\{\rho_i, \rho_{L+1-i}\}$, where $i \in \left[1, \frac{L}{2}\right]$. Likewise, if L is an odd number, it can be viewed as the parallel combination of one SISO fading channel with eigenvalue $\left\{\rho_{\frac{L+1}{2}}\right\}$ and $\frac{L-1}{2}$ pairs of rank-2 JCMD-MIMO with eigenvalues $\{\rho_i, \rho_{L+1-i}\}$, where $i \in \left[1, \frac{L-1}{2}\right]$. Thus, the largest eigenvalue layer couples with the smallest eigenvalue layer, the second largest eigenvalue layer couples with the second smallest eigenvalue layer, and so on. Actually, BICM-MIMO is also one special case of JCMD-MIMO when $\bar{\zeta} = \bar{\rho}$.

Theorem 3.

As for the achievable rate of a rank-L JCMD-MIMO with a descending-order eigenvalue vector $\bar{\rho} = \{\rho_1, \rho_2, \ldots, \rho_L\}$, the upper bound is $C_L(\overline{\omega})$, where $\overline{\omega}$ is in reverse order of $\bar{\rho}$, and the lower bound is the BICM-MIMO achievable rate $C_L(\bar{\rho})$.

Proof.

According to Theorem 2, the maximum achievable rate is $C_L(\bar{\omega})$, we can get $C \leq C_L(\bar{\omega})$. In addition, $C_L(\bar{\rho})$ is the minimum achievable rate, which can be proved by the similar math skill as follows.

Assuming someone claims to find a non-descending-order vector $\bar{\rho}'$ other than $\bar{\rho}$ to have minimum achievable rate $C_L(\bar{\rho})$, we must can find a pair of $\{\rho'_i, \rho'_{i+1}\}$ from $\bar{\rho}'$ to satisfy $\rho'_i < \rho'_{i+1}$. And then, we can construct a new vector $\eta = \{\rho'_1, \rho'_2, \ldots, \rho'_{i-1}, \rho'_{i+1}, \rho'_{i,} \rho'_{i+2}, \ldots, \rho'_L\}$, which only changes the order of $\{\rho'_i, \rho'_{i+1}\}$ from $\bar{\rho}'$. So, we can compute the difference of $\frac{C_L(\eta) - C_L(\bar{\rho}')}{W/2}$,

$$\begin{aligned}
\frac{C_L(\eta) - C_L(\bar{\rho}')}{W/2} &= \sum_{k=1}^{L} \log_2\left[1 + \frac{(\rho_k + \eta_k)}{2\sigma^2 L} P\right] - \sum_{k=1}^{L} \log_2\left[1 + \frac{(\rho_k + \rho'_k)}{2\sigma^2 L} P\right] \\
&= \log_2\left[1 + \frac{(\rho_i + \eta_i)}{2\sigma^2 L} P\right] + \log_2\left[1 + \frac{(\rho_{i+1} + \eta_{i+1})}{2\sigma^2 L} P\right] \\
&\quad - \log_2\left[1 + \frac{(\rho_i + \rho'_i)}{2\sigma^2 L} P\right] - \log_2\left[1 + \frac{(\rho_{i+1} + \rho'_{i+1})}{2\sigma^2 L} P\right] \\
&= \log_2\left[1 + \frac{(\rho_i + \rho'_{i+1})}{2\sigma^2 L} P\right] + \log_2\left[1 + \frac{(\rho_{i+1} + \rho'_i)}{2\sigma^2 L} P\right] \\
&\quad - \log_2\left[1 + \frac{(\rho_i + \rho'_i)}{2\sigma^2 L} P\right] - \log_2\left[1 + \frac{(\rho_{i+1} + \rho'_{i+1})}{2\sigma^2 L} P\right].
\end{aligned} \tag{(23)}$$

Let a constant $c = \frac{P}{2\sigma^2 L}$, and then,

$$\frac{C_L(\eta) - C_L(\bar{\rho}')}{W/2} = \log_2 \frac{M_1}{M_2}, \tag{(24)}$$

where

$$M_1 = 1 + c\left(\rho_i + \rho_i' + \rho_{i+1} + \rho_{i+1}'\right) + c^2\left(\rho_i + \rho_{i+1}'\right)\left(\rho_{i+1} + \rho_i'\right) > 0, \quad ((25))$$

$$M_2 = 1 + c\left(\rho_i + \rho_i' + \rho_{i+1} + \rho_{i+1}'\right) + c^2\left(\rho_i + \rho_i'\right)\left(\rho_{i+1} + \rho_{i+1}'\right) > 0. \quad ((26))$$

So, the difference

$$M_1 - M_2 = c^2(\rho_i - \rho_{i+1})\left(\rho_i' - \rho_{i+1}'\right). \quad ((27))$$

Because $\rho_i > \rho_{i+1}$ and $\rho_i' < \rho_{i+1}'$.

So, $M_1 - M_2 < 0$; Thus, $\frac{M_1}{M_2} < 1$; and then, $C_L(\eta) - C_L(\bar{\rho}') < 0$.

That is to say, we find a counter-example η to have small achievable rate than $C_L(\bar{\rho}')$, which contradicts the assumption of minimum achievable rate $C_L(\bar{\rho}')$. Therefore, $C_L(\bar{\rho})$ is the minimum achievable rate.

AMI ANALYSIS FOR COMPLEX-VALUED QAM SIGNALS

For practical communication systems, the transmit signal **X** usually belongs to a finite alphabet (constellation signal set), such as the complex-valued QAM signal. AMI also varies with the rotation angle of QAM signals in the MIMO system. Assuming equiprobable QAM constellation inputs and the independent Rayleigh fading channel, the AMI of the coded modulation (CM) MIMO system before the demapper is called CM-AMI, which is irrelevant to the labeling and defined as [2]

$$I_{\mathrm{CM}} = I\left(\mathbf{X};\mathbf{Y}|\Lambda\right) = \sum_{l=1}^{N_L}\left\{m - E_{\mathbf{x}_l,\mathbf{y}_l,\lambda_l}\left[\log_2\frac{\sum_{\hat{\mathbf{x}}\in\chi} P\left(\mathbf{y}_l|\hat{\mathbf{x}},\lambda_l\right)}{P(\mathbf{y}_l|\mathbf{x}_l,\lambda_l)}\right]\right\}. \quad ((28))$$

The AMI after the demapper is called BICM-AMI, which is relevant to the labeling and defined as [2]

$$I_{\text{BICM}} = \sum_{k=1}^{m} I\left(C_k;\mathbf{Y}|\boldsymbol{\Lambda}\right)$$
$$= \sum_{l=1}^{N_L} \left\{ m - \sum_{k=1}^{m} E_{c_k,\mathbf{y}_l,\lambda_l} \left[\log_2 \frac{\sum_{\hat{\mathbf{x}}\in\chi} p(\mathbf{y}_l|\hat{\mathbf{x}},\lambda_l)}{\sum_{\hat{\mathbf{x}}\in\chi,\hat{c}_k=c_k} p(\mathbf{y}_l|\hat{\mathbf{x}},\lambda_l)} \right] \right\}. \qquad ((29))$$

AMI analysis is an effective method to calculate its achievable rate. From Equation 28, because $I(\mathbf{X};\mathbf{Y}|\boldsymbol{\Lambda})$ is independent from labeling, it implies that the AMI of CM system is not related to the labeling. However, for the BICM system, the BICM-AMI strongly depends on the labeling. Extensive literature has proven that Gray labeling is optimal for the non-ID BICM system in the AWGN channel [25]. Monte Carlo simulation techniques are useful tools to get the expectation of a complex function by ergodicity of the random variables [26]. The expectation operations in Equations 28 and 29 can be evaluated by using the Monte Carlo simulation techniques. Based on Equation 9, we can get the received symbol before the soft demapper by generating the random coded bits c_k (corresponding to the modulated symbol x_k^l on each layer), the random fading coefficient λ_k^l (SVD of i.i.d. Rayleigh-distributed random channel matrix $\mathbf{H}_k$), and the i.i.d. Gaussian noise random variable n_k^l. Thus, by using the Monte Carlo simulation techniques, we can estimate the expectation values of Equations 28 and 29 by ergodicity of (c_k, $\mathbf{H}_k$, nlknkl), where $P(\mathbf{y}_l|\mathbf{x}_l,\boldsymbol{\lambda}_l)$ can be calculated as Equation 12. As for complex-valued QAM signals, the optimum rotation angle usually is different from the value of the real-valued signals, and we can get it by maximizing AMI.

Figure 3 shows how 2×2 MIMO CM-AMI varies with the rotation angle θ for real-valued binary phase shift keying (BPSK) symbols. Since $I(\mathbf{X};\mathbf{Y}|\Lambda)=I(C_k;\mathbf{Y}|\Lambda)$, the I_{CM} and I_{BICM} are the same for BPSK. Both for low and high SNR, e.g., at SNR =−5 dB and SNR =−15 dB, the optimal rotation angles are always θ= 45°, which coincides with our analysis in Section 3.

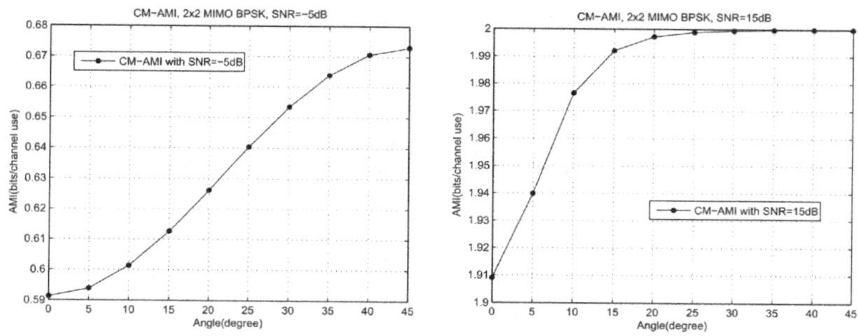

(a) SNR=-5dB (b) SNR=15dB

Figure 3: 2 × 2 MIMO CM-AMI I_{CM} vs. rotation angle θ for BPSK over Rayleigh fading channels. (**a**) SNR =−5 dB. (**b**) SNR = 15 dB.

Figures 4 and 5 show how 4×4 MIMO CM-AMI and BICM-AMI vary with the rotation angle θ for QPSK and 16QAM, respectively. The AMIs of proposed systems with two kinds of spatial Q-component interleavers (reverse and cyclic-shift) are plotted in the same figures. Examples of low SNR and high SNR are presented. In Figures 4 and 5, for both spatial Q interleaving algorithms, the optimal rotation angles just have slight difference in high SNR. For CM-AMI, the optimal angle of QPSK at SNR =−3 dB is 45°, while the optimal angle at SNR = 11 dB is about 30° and 29° for the cyclic-shift interleaver and the reverse interleaver, respectively. For BICM-AMI, the optimal angle of QPSK at SNR =−3 dB is 0°, while the optimal angle at SNR = 11 dB is about 26° and 27° for the cyclic-shift interleaver and the reverse interleaver, respectively.

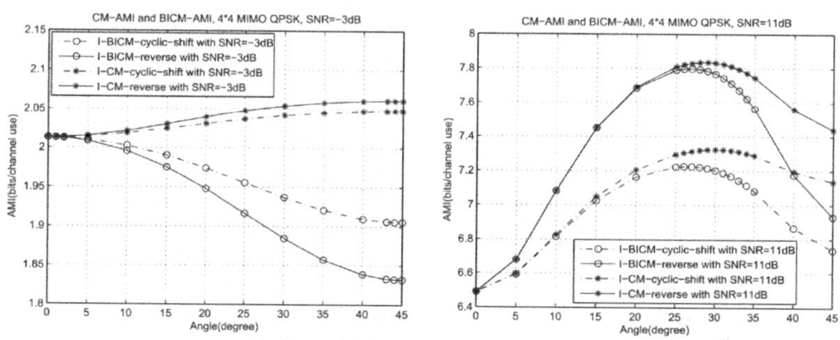

(a) SNR=-3dB (b) SNR=11dB

Figure 4: 4 × 4 MIMO CM-AMI I_{CM} and BICM-AMI I_{BICM} vs. rotation angle θ for QPSK over Rayleigh fading channels. (**a**) SNR =−3 dB. (**b**) SNR = 11 dB.

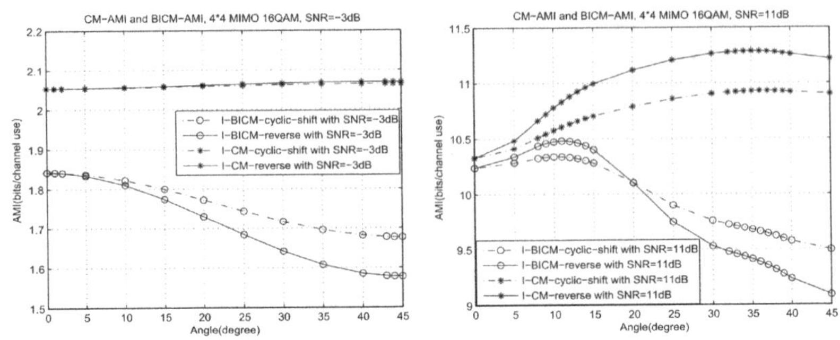

(a) SNR=-3dB (b) SNR=11dB

Figure 5: 4 × 4 MIMO CM-AMI I_{CM} and BICM-AMI I_{BICM} vs. rotation angle θ for 16QAM over Rayleigh fading channels. (**a**) SNR =−3 dB. (**b**) SNR = 11 dB.

Based on the AMI maximization criterion, the optimal angle corresponds to the maximum AMI value. Figure 6 shows the CM-AMI and BICM-AMI curves that correspond to the optimal angles in Figures 4 and 5 for 4×4 MIMO systems. In Figure 6, CM-AMI and BICM-AMI for the reverse interleaver described in Equation 2 are always not less than that of the cyclic-shift interleaver described in Equation 3 both for QPSK and 16QAM. These observations concide well with Theorem 2.

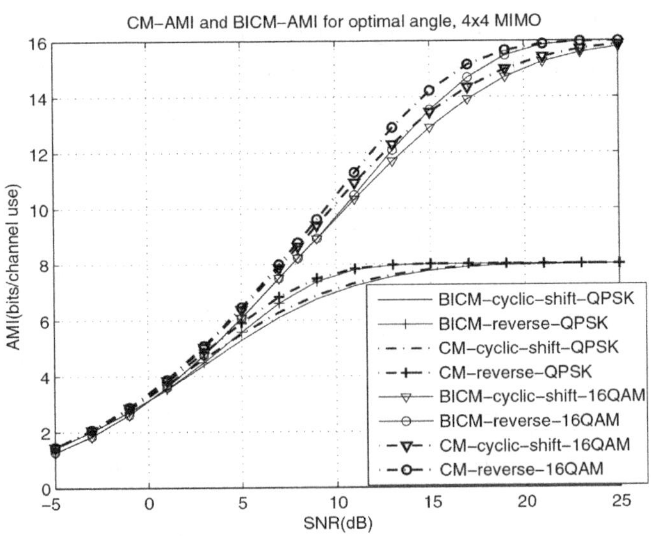

Figure 6: CM-AMI and BICM-AMI corresponding to the optimal angles for 4 × 4 MIMO.

Based on maximizing AMI, we can also obtain the relationship between optimal angle and SNR. The optimal angles for CM and BICM systems vary with SNR for QPSK and 16QAM, which is plotted in Figure 7. The optimal rotation angle of CM is always bigger than that of BICM. When SNR increases, the optimal angle of CM becomes smaller, but the optimal angle of BICM becomes bigger.

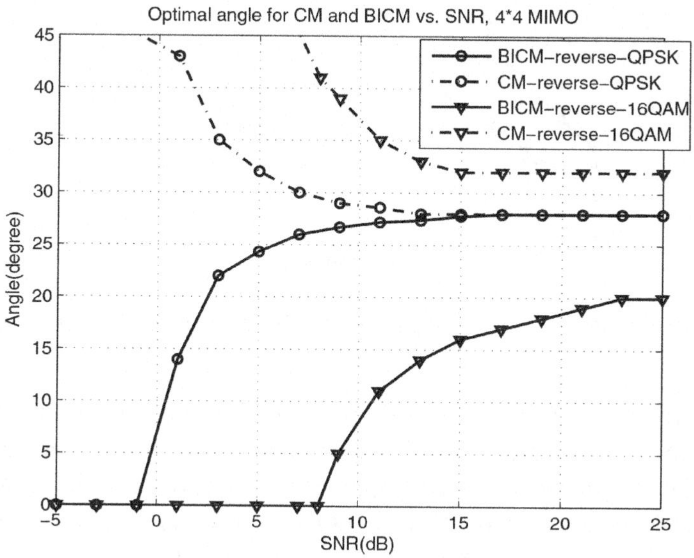

Figure 7: Optimal angle θ vs. SNR for QPSK and 16QAM over 4×4 Rayleigh fading channels.

For a given modulation order *m*, according to the AMI value *I* corresponding to optimal rotation angle in Figure 6, the optimal code rate for a given SNR can be calculated as $R = \frac{I}{m \cdot N_L}$. Therefore, we can get the relationship between the optimal code rate *R* and SNR, which is shown in Figure 8.

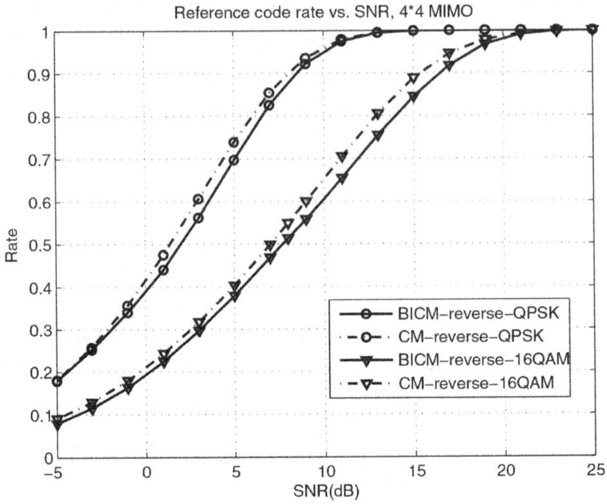

Figure 8: Code rate vs. SNR for QPSK and 16QAM over 4 × 4 Rayleigh fading channels.

Thus, for a given operating SNR, we can obtain the optimum rotation angle in Figure 7 and the optimal code rate in Figure 8. Furthermore, we can get the relationship between code rate R and optimal rotation angle, which is shown in Figure 9 for QPSK and 16QAM. It provides a good reference to select an optimal rotation angle for a given code rate. For instance, for code rate = 0.5, the optimal angles for 4×4 MIMO-BICM QPSK and 16QAM are 18° and 0°, respectively. According to Figures 7 and 9, at low SNR or low code rate, non-rotation is the best for BICM, which implies that the channel coding dominates the BICM performance, while 45° rotation is the best for CM, which indicates that the rotation symmetry affects the CM performance. However, at high SNR or high code rate, the signal space diversity dominates the performance both for BICM and CM, so BICM-AMI can approach CM-AMI, and the optimum angle of BICM with the optimum reverse spatial interleaver is also close to that of CM with the same interleaver. Note that the rotation angle is just related to the code rate and modulation. So, in practice, the rotation symbol mapper can be implemented through a look-up table that is calculated and stored in advance for a given modulation order and code rate, which is the same as the conventional symbol mapper. Hence, it does not introduce additional processing complexity and delay.

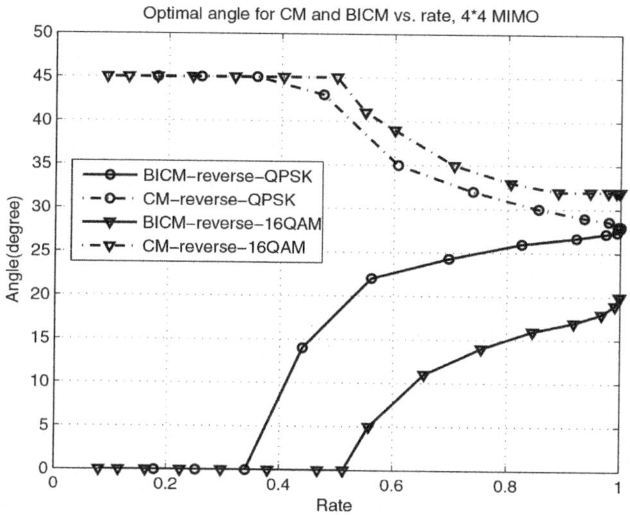

Figure 9: Optimal angle θ vs. code rate R for QPSK and 16QAM over 4×4 Rayleigh fading channels.

CODE DESIGN FOR JCMD-ID SYSTEM

For the non-ID BICM system, Gray mapping has been proved to be optimal. Based on the optimal rotation angle obtained by maximizing BICM-AMI and Gray mapping, the non-ID JCMD system only needs to consider the optimization of channel codes to achieve excellent performance. For the JCMD-ID system, besides the optimal rotation angle, it is also very crucial to choose a pair of a well-matched labeling and an outer channel code by some joint optimization.

The convergence behavior of the iterative demodulation and decoding can be analyzed by the EXIT chart method, which can describe the flow of extrinsic information between the demodulator and the decoder [27-29]. Several advantages of EXIT charts are summarized in [29]. The inputs to the demapper are the noise-corrupted channel observations and the *a priori* knowledge $A(c_k)$ on the unmapped bits. The demapper outputs channel and extrinsic information $E(c_k)$. According to Equations 12 and 19 in [27], the mutual information $I_{A1}=I(c_k;A(c_k))$ $(0 \leq I_{A1} \leq 1)$ between transmitted unmapped bits and the L-values $A(c_k)$ can be written as

$$I_{A1} = \frac{1}{2} \sum_{x=-1,1} \int_{-\infty}^{\infty} p_{A(c_k)} (\xi | c_k = x) \cdot \log_2 \times \frac{2p_{A(c_k)} (\xi | c_k = x)}{p_{A(c_k)} (\xi | c_k = -1) + p_{A(c_k)} (\xi | c_k = 1)} d\xi. \quad ((30))$$

The mutual information $I_{E1}=I(c_k;E(c_k))$ $(0 \leq I_{E1} \leq 1)$ can be written as

$$I_{E1} = \frac{1}{2} \sum_{x=-1,1} \int_{-\infty}^{\infty} p_{E(c_k)} (\xi | c_k = x) \cdot \log_2 \times \frac{2p_{E(c_k)} (\xi | c_k = x)}{p_{E(c_k)} (\xi | c_k = -1) + p_{E(c_k)} (\xi | c_k = 1)} d\xi. \quad ((31))$$

In [27], it turns out that the *a priori* input $A(c_k)$ is almost Gaussian distributed. Additionally, large interleavers keep the a priori L-values $A(c_k)$ fairly uncorrelated over many iterations. Hence, it seems appropriate to model the *a priori* input $A(c_k)$ by applying an independent Gaussian random variable with variance σ_{A1}^2 and mean zero (see Equations 1 to 10 in [27]). Thus, the conditional probability density function can be written as

$$p_{A(c_k)} (\xi | c_k = x) = \frac{1}{\sqrt{2\pi}\sigma_{A1}} \exp \left[- \frac{\left(\xi - \frac{\sigma_{A1}^2}{2} x \right)^2}{2\sigma_{A1}^2} \right] \quad ((32))$$

The mutual information $I_{E1}=I(c_k;E(c_k))$ can be viewed as a function of I_{A1}, SNR, and the rotation angle θ, i.e,

$$I_{E_1} = T_1(I_{A_1}, SNR, \theta). \quad ((33))$$

The function T_1 can not be expressed in a closed form. For a given value of the input mutual information $I_{A1}=I(c_k;A(c_k))$ $(0 \leq I_{A1} \leq 1)$ to the demodulator, to compute $T_1(I_{A_1}, SNR, \theta)\cdot$, the distributions $p_{E(c_k)} (\xi | c_k = x)$ are most conveniently determined by the Monte Carlo simulation (histogram measurements) proposed in [27,28]. The convenient method has been verified to allow an accurate prediction of the SNR-decoding threshold with low complexity [27]. On the other hand, the extrinsic transfer characteristics of the decoder can describe the input/output relationship between the input I_{A2} and the output I_{E2}, which is independent of SNR value. It can be computed by assuming the *a priori* input to be Gaussian distributed and applying the same method as used for the demapper characteristic T_1. Therefore, the transfer characteristic of the decoder is denoted by

$$I_{E_2} = T_2(I_{A_2}). \qquad ((34))$$

Qiuliang et al. [16] verified that the SSD technique can affect the demapper's EXIT curve for the SISO BICM-ID SSD system. For the MIMO system, we analyze the effect of the SSD technique to different demapper labelings. Different 16QAM labelings (natural, Gray, and reference [16]) are shown in Figure 10. For the 1/2-rate coded 16QAM 4×4 MIMO system, the corresponding 16QAM EXIT curves with 45° rotation and non-rotation are depicted in Figure 11. The doping technique is used for error-floor removal [30] and the doping rate P=100. For Natural and Gray labelings, the demappers' EXIT curves with rotation have a larger slope than that without rotation. For the referenced labeling, the demapper's EXIT curve with rotation is always above the curve without rotation.

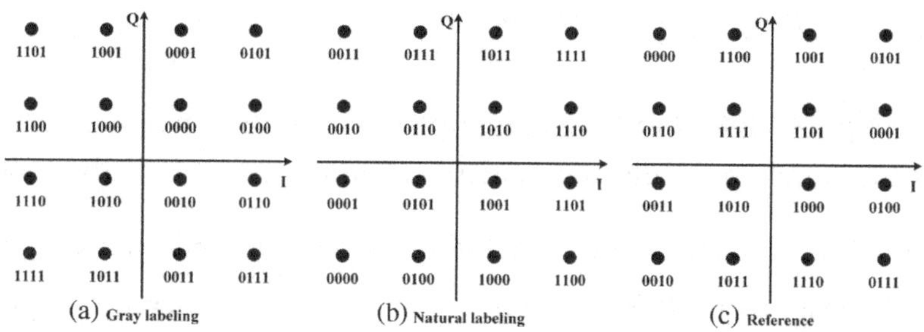

Figure 10: 16QAM labelings. (a) Gray labeling. (b) Natural labeling. (c) The reference labeling in [16].

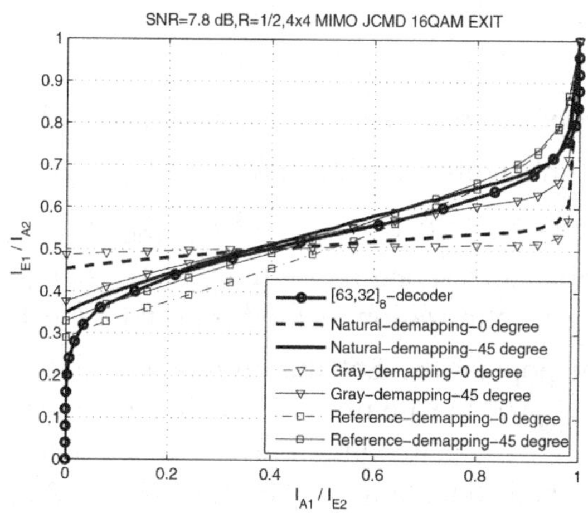

Figure 11: EXIT chart analysis for 16QAM at SNR = 7.8 dB.

Demapper-matched code design is very crucial for the JCMD-ID MIMO system with rotation. In order to approach the capacity, based on EXIT chart, we propose an optimization method of outer channel codes to match with a given demapper for the JCMD-ID system. We choose the BCC as the channel code. For a given SNR, if two EXIT curves of the demapper and the decoder do not intersect, the iterative decoding can converge, otherwise it cannot converge. Thus, the SNR which makes the two EXIT curves tangent is the SNR convergence threshold, which is also called pinch-off SNR. The objective of JCMD-ID optimization is to find the outer channel code that has the lowest SNR convergence threshold to match with the demapper.

We define the optimization function as follows.

$$\alpha(\mathbf{G}(N_{\text{Reg}}), SNR) = \min_{I_n \in [0,1], n=1,2,\ldots,\text{Num}} \left[T_1(I_n) - T_2^{-1}(I_n) \right], \quad ((35))$$

$$\overline{\text{SNR}} = \arg\min_{\text{SNR}} \left[\alpha \left(\mathbf{G} \left(N_{\text{Reg}} \right), \text{SNR} \right) > 0 \right], \quad ((36))$$

where $\mathbf{G}(N_{\text{Reg}})$ denotes the generator polynomial of BCC with N_{Reg} registers. Num is the number of selected statistical samples. $\overline{\text{SNR}}$ is the pinch-off SNR. For $\frac{1}{N_{\text{out}}}$ rate BCC, the objective is to find the optimum $\mathbf{G}_{\text{Opt.}}(N_{\text{Reg}})$ with the lowest pinch-off SNR from $(2^{N_{\text{Reg}}+1} - 1)^{N_{\text{out}}}$ generator polynomial candidates. However, the exhaustive searching of a channel code to well match the demapper's EXIT curve is trivial, especially for N_{Reg} is large.

GA is an efficient optimization algorithm, which is stochastic search techniques based on the mechanism of natural selection and natural genetics [31]. In GA, a genetic representation is required for the individuals in a population. Generator polynomials of BCC $\mathbf{G}(N_{\text{Reg}}) = [\mathbf{g}_1, \mathbf{g}_2, \ldots, \mathbf{g}_{N_{\text{out}}}]_2$ inherently provides a $(N_{\text{Reg}}+1)\times N_{\text{out}}$-length binary string $\mathbf{S}_g = < \mathbf{g}_1, \mathbf{g}_2, \ldots, \mathbf{g}_{N_{\text{out}}} >$. $\mathbf{g}_i$ is the $(N_{\text{Reg}}+1)$-length binary generator polynomial corresponding to ith $(1 \le i \le N_{\text{out}})$ output. Based on the genetic algorithm, an optimization method is proposed as follows.

Step 1: Initial population. Set the current iterative number (number of generations) $N_{\text{pop}}=0$, the number of candidate generator polynomials (population size) N_g, the maximum iterative number $N_{\max}$, crossover probability P_c, and mutation probability P_m. N_g binary strings

$\mathbf{S}_g^i =< \mathbf{g}_1^i, \mathbf{g}_2^i, \ldots, \mathbf{g}_{N_{\text{out}}}^i >.$ that correspond to the candidate polynomials are randomly initialized, which are denoted by a set $\mathbf{C}^{N_{\text{pop}}}$, where $\mathbf{g}_1^i, \mathbf{g}_2^i, \ldots, \mathbf{g}_{N_{\text{out}}}^i \in \left[1, 2^{N_{\text{Reg}}+1}\right), 1 \leq i \leq N_g$.

Step 2: Selection. Reduce the SNR by small steps Δ_{SNR} (e.g., 0.1 dB) and compute the pinch-off SNRs $\overline{\text{SNR}}_i$ of all the candidate generator polynomials $\mathbf{G}_i(N_{\text{Reg}}) = \left[\mathbf{g}_1^i, \mathbf{g}_2^i, \ldots, \mathbf{g}_{N_{\text{out}}}^i\right]_2$ in population, where $< \mathbf{g}_1^i, \mathbf{g}_2^i, \ldots, \mathbf{g}_{N_{\text{out}}}^i > \in \mathbf{C}^{N_{\text{pop}}}, 1 \leq i \leq N_g$. We use the pinch-off SNR to measure the fitness. The fitness function is associated with the maximum pinch-off SNR $\overline{\text{SNR}}_{\max} = \max_{1 \leq i \leq N_g} \left\{\overline{\text{SNR}}_i\right\}$, shown as

$$f\left(\mathbf{S}_g^i\right) = \overline{\text{SNR}}_{\max}(\text{dB}) + 0.1 - \overline{\text{SNR}}_i(\text{dB}).$$

((37))

N_g individuals are selected to breed a new generation with probability proportional to the fitness value. The probability that $\mathbf{S}_g^i$ is selected is $P(i) = \frac{f\left(\mathbf{S}_g^i\right)}{\sum_{k=1}^{N_g} f\left(\mathbf{S}_g^k\right)}$. Based on roulette wheel selection (RWS) algorithm, the tth ($1 \leq t \leq N_g$) individual selection follows the steps below.

A. Generate a uniform random number $\chi(t)$, $\chi(t) \in [0,1]$.

$\chi(t) \in [0, 1]$.

B. If $\sum_{i=0}^{k-1} P(i) \leq \chi(t) < \sum_{i=1}^{k} P(i)(1 \leq k \leq N_g)$,

B. If $P(0) = 0$, $\mathbf{S}_g^k$ is selected.

Step 3: Crossover. For two adjacent selected individuals, a random number κ_c from the range [0,1] is generated. Only when $\kappa_c < P_c$, the crossover operator is carried out. The crossover point is selected randomly. All bits beyond that point in either string are swapped between the two parent individuals, and then two children individuals are obtained.

Step 4: Mutation. For each individuals, a random number κ_m from the range [0,1] is generated. Only when $\kappa_m < P_m$, the mutation operator is carried out through one bit flip at random mutation position.

Step 5: Judgment. $N_{\text{pop}} = N_{\text{pop}} + 1$. After step 2~4, a new population $\mathbf{C}^{N_{\text{pop}}+1}$ is formed. If $N_{\text{pop}} < N_{\max}$, then go to step 2, otherwise stop. The generator polynomial with the lowest pinch-off SNR in $\mathbf{C}^{N_{\text{pop}}+1}$. is chosen as the optimum one.

Using the method above, a rate-half BCC code with $N_{\text{Reg}}=5$ is optimized for the natural labeling 16QAM JCMD-ID system. Based on the AMI analysis in Section 4, the optimal rotation angle for 0.5 rate 16QAM JCMD-ID system is 45°. The optimal generator polynomial is $[63,32]_8$. As shown in Figure 11, the natural labeling demapper's EXIT curve with 45° rotation keeps a narrow open tunnel with that of BCC decoder, while its non-rotation curve intersects with the decoder, which shows the effect of rotation to the ID system. The other two labeling demappers' EXIT curves with and without rotation always intersects with the decoder, so the Gray and reference labelings are not suitable for the BCC decoder. Therefore, the natural labeling with 45° rotation matches well with the $[63,32]_8$ BCC code and has the best performance.

SIMULATION RESULT

Results of Non-ID JCMD System on Fast Fading Channels

For the non-ID JCMD MIMO system, the optimal Gray labeling and powerful DVB-T2 LDPC coding are used to achieve excellent performance [32]. The optimal rotation angle is obtained by maximizing BICM-AMI in Figure 9. The size of coded block N is 64,800 bits. For the LDPC decoder, the log-belief propagation (BP) algorithm with 30 maximum iterations is utilized. In order to ensure the fairness of the comparison, the SVD precoding is implemented for both the conventional BICM MIMO system and proposed JCMD MIMO system.

Figure 12 shows the bit error rate (BER) performance comparisons of rate-1/2 LDPC-coded BPSK MIMO systems on the independent fast Rayleigh fading channel. The channel fading coefficients on each symbol are independent identical distributed Rayleigh random variables with the variance 1. BPSK is also a simple real-valued signal, so the optimal rotation angle is chosen as 45°. For 4×4 MIMO systems, JCMD with 45° rotation obtains significant SNR gains as compared with JCMD without rotation. For the target BER $=10^{-5}$, it can achieve 2.2 dB SNR gain. Note that the spatial interleaver, spatial Q-component interleaver, and time Q-component interleaver are all implemented in both JCMD scheme with and without rotation. That means the performance gain of JCMD does not exactly come from the interleavers compared with BICM scheme. We can still obtain significant gain through the optimal rotation and the matching of channel coding and labeling. Furthermore, the 4×4 JCMD MIMO system with the reverse spatial Q interleaver can obtain about 0.6 dB SNR gain compared with that with the cyclic-shift spatial Q interleaver. The results coincide well with the above analysis in Section 3.

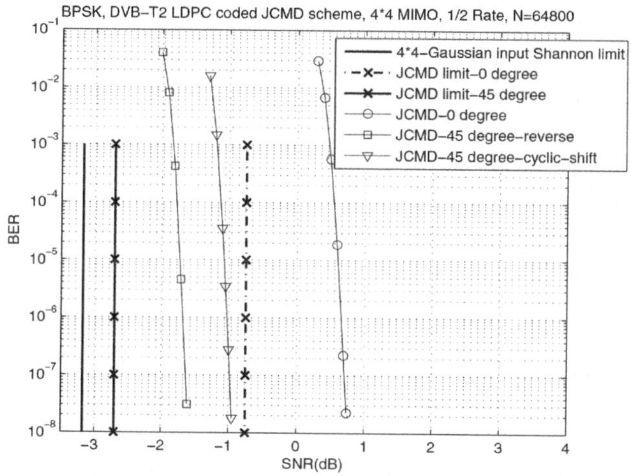

Figure 12: BER performance of rate-1/2 DVB-T2 LDPC coded, BPSK 4 × 4 JCMD MIMO systems on fast fading channels.

For the QPSK 4×4 MIMO system, the optimal angles for 1/2 and 3/4 rate are 18° and 25°, respectively. As shown in Figure 13, JCMD systems with the optimal rotation angle can obtain 0.6 and 3.9 dB SNR gains compared with the conventional BICM systems employing the ideal SVD precoding for low rate 1/2 and high rate 3/4, respectively. Given the optimal rotation angle, JCMD-MIMO with the reverse interleaver obtains significant 0.3 and 1.3 dB SNR gains at BER $=10^{-5}$ compared to that with the cyclic-shift interleaver for the code rate 1/2 and 3/4, respectively. Meanwhile, the proposed 1/2 rate JCMD system with the reverse interleaver is only about 1.1 dB SNR gap to the JCMD limit for 4×4 MIMO. For the 3/4 rate JCMD MIMO, the gap to the JCMD limit is reduced to 0.7 dB for 4×4 MIMO. SNR gains and gaps are summarized in Table 1. From the results, SNR gain becomes bigger for higher code rate.

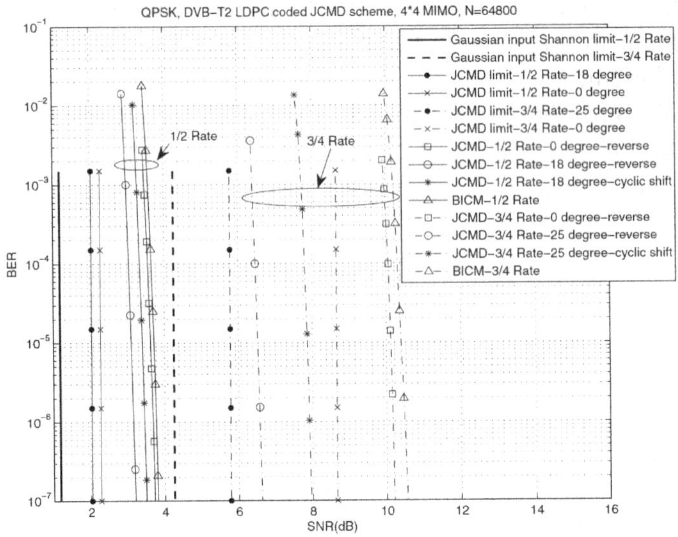

Figure 13: BER performance of rate-1/2 and 3/4 DVB-T2 LDPC coded, QPSK 4 × 4 JCMD MIMO systems on fast fading channels.

Table 1: SNR gains and gaps to the capacity for JCMD 4×4 MIMO systems

Parameters	Gains compared with BICM system (dB)	Gaps to the JCMD/Gaussian input Shannon limit (dB)
BPSK, 1/2-rate	2.2	1.0/1.5
QPSK, 1/2-rate	0.6	1.1/1.9
QPSK, 3/4-rate	3.9	0.7/2.3

Results of JCMD-ID System on Fast Fading Channels

For the JCMD-ID 16QAM MIMO system, the optimal rotation angle is obtained by maximizing CM-AMI in Figure 9, and it is 45° at 1/2 rate for 4×4 MIMO systems. In order to confirm our optimization method for JCMD-ID MIMO system, simulations are carried out with proposed $[63,32]_8$ BCC-coded JCMD-ID scheme on fast fading channels for 4×4 MIMO systems. The size of coded block N=64,800 bits and the maximal global iterative number is 30. The powerful 1/2 rate DVB-T2 LDPC-coded Gray-labeling BICM and BICM-ID schemes with the same block size are also simulated as the reference, and the same ideal SVD precoding method is used for the conventional BICM and BICM-ID schemes and the proposed JCMD-ID scheme.

The BER performance comparisons are shown in Figure 14.

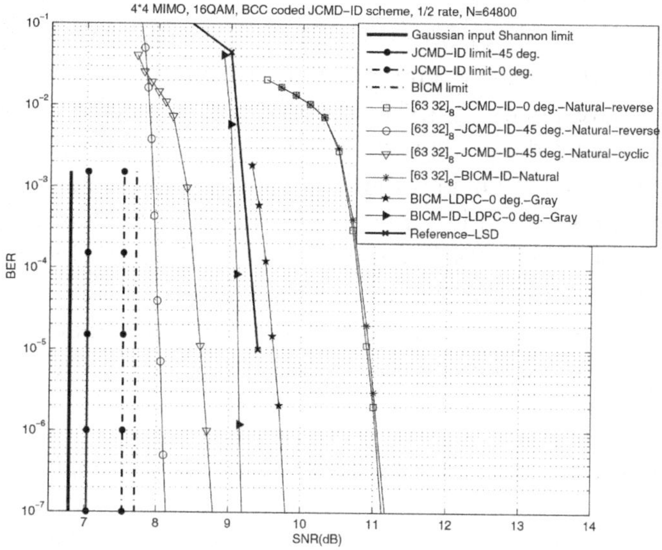

Figure 14: BER performance of rate-1/2, 16QAM 4 × 4 JCMD-ID MIMO systems on fast fading channels.

For the 4×4 MIMO system with natural labeling, the optimized BCC $[63,32]_8$ coded JCMD-ID scheme with 45° rotation and reverse Q interleaver exhibits excellent performance, which is only 1.3 and 1.1 dB away from the Gaussian-input capacity and JCMD-ID limit, respectively. It can obtain about 2.9 dB SNR gain compared with the BCC-coded BICM-ID scheme and JCMD-ID without rotation, which coincides with the above EXIT analysis. In addition, the scheme with the reverse interleaver obtains 0.4 dB SNR gain at BER $=10^{-5}$ compared with that with cyclic-shift Q interleaver, which also proves the above analysis. Furthermore, the optimized JCMD-ID scheme also outperforms the DVB-T2 LDPC-coded BICM-ID scheme and the turbo-coded BICM-ID scheme in [24] by 0.9 and 1.4 dB gains, respectively.

For the JCMD-ID and conventional BICM-ID schemes, the main complexity lies in the channel decoding. The Max-Log-MAP algorithm is a simplified algorithm for the decoding of turbo code and BCC. For each decoding iteration, it requires $10 \times 2^{N_{\mathrm{Reg}}} + 11$ additions and $5 \times 2^{N_{\mathrm{Reg}}} - 2$

maximum operations [33]. The BCC decoding only needs one iteration (one Max-Log-MAP operation), while turbo decoding needs eight iterations (16 Max-Log-MAP operations). For the Log-BP algorithm of LDPC decoding, each iteration requires $Md_c(d_c-1)+Nd_v(d_v-1)+Nd_c$ additions and Md_c^2 look-up table operations, where M is the number of the parity bits, and d_c and d_v are the average degree distributions of check nodes and variable nodes, respectively [34]. LDPC decoding requires 30 iterations. Thus, we can obtain the average decoding complexity comparison for each information bit, as shown in Table 2. Assuming the equal complexity of the three operations (addition, max, look-up), the decoding complexity of BCC is just is 44.3% and 28.2% of turbo and LDPC decoding, respectively. Therefore, the optimized BCC has much lower complexity compared with the DVB-T2 LDPC and turbo codes. This indicates that the optimized scheme with BCC coding obtains much better performance with much lower complexity.

Table 2: Average decoding complexity for each information bit

Operation	Optimized BCC	Turbo used in [24]	DVB-T2 LDPC
Additions	331	816	997.5
Max ops.	158	288	0
Look-ups	0	0	735

CONCLUSIONS

A high-spectral-efficient JCMD scheme over MIMO fading channels is proposed. By jointly optimizing the component interleaver, the rotation modulation, and the BCC code, this scheme exhibits excellent performance. An optimum spatial component interleaver is proposed to maximize the achievable rate. For real-valued signals, we prove that the achievable rate of JCMD MIMO is greater than that of the conventional BICM MIMO scheme and π4π4 is the optimal rotation angle. For the rotated QAM, the optimal rotation angles are investigated for the MIMO system according to the maximizing AMI criterion. For the JCMD-ID MIMO system, a simple GA-based search algorithm of BCC generator polynomials is also proposed to match the rotation modulation. Simulation results prove that this new scheme can significantly outperform the conventional turbo-coded BICM-

ID scheme over MIMO fading channels by 1.4 dB SNR gain, while it has much lower complexity. In a word, the proposed JCMD scheme is simple, efficient, and robust for the future wireless communication systems.

ACKNOWLEDGEMENTS

This work is sponsored by the National Natural Science Fund (61171101), National Great Science Specific Project (2009ZX03003-011-03) of People's Republic of China and 2014 Doctorial Innovation Fund of BUPT (CX201426) and the Fundamental Research Funds for the Central Universities.

REFERENCES

1. C Felita, M Suryanegara, in *Proceedings of International Conference on QiR (Quality in Research): 25-28 June 2013; Yogyakarta*. 5 g key technologies: identifying innovation opportunity (IEEE,Piscataway, 2013), pp. 235–238.
2. G Caire, G Taricco, E Biglieri, Bit-interleaved coded modulation. IEEE Trans. Inf. Theory. 44(3), 927–946 (1998).
3. X Li, JA Ritcey, Bit-interleaved coded modulation with iterative decoding. IEEE Commun. Lett. 1(6), 169–171 (1997).
4. B Vucetic, J Yuan, *Space-Time Coding* (John Wiley & Sons, Inc., England, 2003).
5. G Foschini, Layered space-time architecture for wireless communication in a fading environment when using multi-element antennas. Bell Labs Tech. J. 1(2), 41–59 (1996).
6. HE Gamal, AR Hammons, The layered space-time architecture: a new perspective. IEEE Trans. Inf. Theory. 1, 2321–2334 (2001).
7. J Boutros, E Viterbo, Signal space diversity: a power and bandwidth efficient diversity technique for the Rayleigh fading channel. IEEE Trans. Inf. Theory. 44(4), 1453–1467 (1998).
8. CA Nour, C Douillard, in *Proceedings of 5th International Symposium on Turbo Codes and Related Topics: 1-5 Sept. 2008; Lausanne*. Improving BICM performance of QAM constellations for broadcasting applications (IEEE,Piscataway, 2008), pp. 55–60.
9. NF Kiyani, UH Rizvi, JH Weber, GJM Janssen, in *Proceedings of IEEE Wireless Communications and Networking Conference: 11-15 March 2007; Kowloon*. Optimized rotations for LDPC-coded MPSK constellations with signal space diversity (IEEE,Piscataway, 2007), pp. 677–681.
10. NF Kiyani, JH Weber, in *Proceedings of IEEE Symposium on Communications and Vehicular Technology in the Benelux: 15-15 Nov. 2007; Delft*. OFDM with BICM-ID and rotated MPSK constellations and signal space diversity (IEEE,Piscataway, NJ, USA, 2007), pp. 1–4.
11. NF Kiyani, JH Weber, Exit chart analysis of iterative demodulation and decoding of MPSK constellations with signal space diversity. J. Commun. 3(3), 43–50 (2008).

12. NH Tran, HH Nguyen, T Le-Ngoc, Performance of BICM-ID with signal space diversity. IEEE Trans. Wireless Commun. 6(5), 1732–1742 (2007).
13. M Zhenzhou, S Zhiping, Z Chong, Z Zhongpei, in *Proceedings of International Conference on Communications, Circuits and Systems: 25-27 May 2008; Fujian*. Design of signal space diversity based on non-binary LDPC code (IEEE,Piscataway, 2008), pp. 31–34.
14. W Zhanji, W Wenbo, Improved coding-rotated-modulation orthogonal frequency division multiplexing system. IET Commun. 6(3), 272–280 (2012).
15. W Zhanji, W Wenbo, in *Proceedings of 2010 Global Mobile Congress: 18-19 Oct. 2010; Shanghai*. A novel joint-coding-modulation-diversity OFDM system (IEEE,Piscataway, 2010), pp. 1–6.
16. X Qiuliang, S Jian, P Kewu, Y Fang, W Zhaocheng, Coded modulation with signal space diversity. IEEE Trans. Wireless Commun. 10(2), 660–668 (2011).
17. N Sharma, CB Papadias, Improved quasi-orthogonal codes through constellation rotation. IEEE Trans. Wireless Commun. 51(3), 332–335 (2003).
18. W An Zhong, Z Jian Kang, in *Proceedings of 11th Canadian Workshop on Information Theory (CWIT): 13-15 May 2009; Ottawa*. Novel rotation angle for quasi-orthogonal space-time block codes (IEEE,Piscataway, 2009), pp. 213–216.
19. W Su, X-G Xia, Signal constellations for quasi-orthogonal space time block codes with full diversity. IEEE Trans. Inf. Theory. 50(10), 2331–2347 (2004).
20. D Dung Ngoc, C Tellambura, in *Proceedings of IEEE Global Telecommunications Conference (GLOBECOM): Dec. 2005; St. Louis*. Optimal rotations for quasi-orthogonal STBC with two-dimensional constellations (IEEE,Piscataway, 2005), pp. 2317–2321.
21. L Yueqian, M Salehi, in *Proceedings of 46th Annual Conference on Information Sciences and Systems (CISS): 21-23 March 2012; Princeton*. Coded MIMO systems with modulation diversity for block-fading channels (IEEE,Piscataway, 2012), pp. 1–5.
22. H Lee, A Paulraj, MIMO systems based on modulation diversity. IEEE Trans. Wireless Commun. 58(12), 3045–3049 (2010).
23. H Soon Up, C Jinyong, J Sungho, R Ho Jin, S Jongsoo, in *Proceedings*

of IEEE International Symposium on Broadband Multimedia Systems and Broadcasting: 13-15 May 2009; Bilbao. Performance evaluation of MIMO-OFDM with signal space diversity over frequency selective channels (IEEE,Piscataway, NJ, USA, 2009), pp. 1–5.

24. BM Hochwald, S ten Brink, Achieving near-capacity on a multiple-antenna channel. IEEE Trans. Commun. 51(3), 389–399 (2003).
25. SY Goff, Signal constellation for bit-interleaved coded modulation. IEEE Trans. Inf. Theory. 49(1), 307–313 (2003).
26. FM Gardner, JD Baker, *Simulation Techniques: Models of Communication Signal and Processes* (Wiley, New York, 1995).
27. S ten Brink, Convergence behavior of iteratively decoded parallel concatenated codes. IEEE Transa. Commun. 49(10), 1727–1737 (2001).
28. S ten Brink, Designing iterative decoding schemes with the extrinsic information transfer chart. AEU Int. J. Electron. Commun. 54(6), 389–398 (2000).
29. A Ashikhmin, G Kramer, S ten Brink, Extrinsic information transfer functions: model and erasure channel properties. IEEE Trans. Inf. Theory. 50(11), 2657–2673 (2004).
30. S Pfletschinger, F Sanzi, Error floor removal for bit-interleaved coded modulation with iterative detection. IEEE Trans. Wireless Commun. 5(11), 3174–3181 (2006).
31. P Guo, W Xuezhi, H Yingshi, in *Proceedings of 3rd International Conference on Biomedical Engineering and Informatics: 16-18 Oct. 2010; Yantai*. The enhanced genetic algorithms for the optimization design (IEEE,Piscataway, 2010), pp. 2990–2994.
32. Frame structure channel coding and modulation for a second generation digital terrestrial television broadcasting system (DVB-T2). ETSI Std. DVB Document A122 [S] (June 2008).
33. P Robertson, E Villebrun, P Hoeher, in *Proceedings of IEEE International Conference on Communications: 18-22 Jun 1995; Seattle*. A comparison of optimal and sub-optimal map decoding algorithms operating in the log domain, (1995), pp. 1009–1013.
34. C Jinghu, A Dholakia, E Eleftheriou, MPC Fossorier, Reduced-complexity decoding of LDPC codes. IEEE Trans. Commun. 53(8), 1288–1299 (2005).

18

Orthogonal Frequency Division Multiplexing Techniques Comparison for Underwater Optical Wireless Communication Systems

Jie Lian [1] , Yan Gao [2] , Peng Wu [3] and Dianbin Lian [4]

[1] Department of Electrical and Computer Engineering, University of Virginia, Charlottesville, VA 22903, USA
[2] Information Engineering School, Xi'an University, Xi'an 710065, China
[3] Xi'an Modern Control Technology Research Institute, Xi'an 710065, China
[4] Electronics and Information School, Northwestern Polytechnical University, Xi'an 710072, China

ABSTRACT

Optical wireless communication is an energy-efficient and cost-effective solution for high-speed and highly-secure wireless connections. In this paper, we compare, discuss, and analyze three popular optical orthogonal frequency division multiplexing (OFDM) techniques, such as DC-biased optical OFDM (DCO-OFDM), asymmetrically-clipped optical OFDM

Citation: Lian, J.; Gao, Y.; Wu, P.; Lian, D. "Orthogonal Frequency Division Multiplexing Techniques Comparison for Underwater Optical Wireless Communication Systems". Sensors 2019, 19, 160.

(ACO-OFDM), and unipolar OFDM (U-OFDM), for underwater optical wireless communication systems. The peak power constraint, bandwidth limit of the light source, turbulence fading underwater channel, and the channel estimation error are taken into account. To maximize the achievable data propagation distance, we propose to optimize the modulation index that controls the signal magnitude, and a bitloading algorithm is applied. This optimization process trades off the clipping distortion caused by the peak power constraint and the signal to noise ratio (SNR). The SNR and clipping effects of the three compared OFDM techniques are modeled in this paper. From the numerical results, DCO-OFDM outperforms ACO- and U-OFDM when the transmitted bit rate is high compared to the channel bandwidth. Otherwise, U-OFDM can provide a longer propagation distance or requires less transmitted power.

Keywords: optical wireless communications, underwater communications, OFDM, peak power constraint, propagation distance, BER, band-limited channel, turbulence fading

INTRODUCTION

In recent years, underwater wireless communications (UWC) has attracted more and more research attention due to the increasing civilian and military demands [1,2,3]. For example, underwater vehicles need high-speed and long-distance wireless connections between each other. The underwater sensor network needs a reliable link to transmit the collected data back to the ground. Radio-frequency (RF) waves, acoustic waves, and optical waves are three possible transmission media for UWC. Due to the high security and transmission speed, underwater optical wireless communication (UOWC) becomes a more and more promising technique, supporting highly-reliable wireless underwater connections [4].

Compared with optical wireless communications, underwater acoustic communication has a longer application history due to its low attenuation level and long propagation distance [1]. However, the slow data rate and severe delay are the drawbacks for underwater acoustic systems. Although RF communication can provide a much higher data rate than acoustic systems, the extremely limited propagation range due to the high attenuation of radio waves in the water limits its application. UOWC is an emerging technique for underwater systems with a high-speed transmission for moderate distance communication. Although UOWC can provide a longer propagation distance than underwater RF systems, an extremely long-range underwater communication should currently rely on acoustic systems. Therefore, in UOWC, most of the research focuses on improving the transmission data rate and propagation distance [5,6,7].

A light-emitting diode (LED) working as a high power-efficient light source has attracted more attention recently and shows a big potential in underwater optical wireless communication systems [8,9,10,11]. Single-carrier frequency division multiple access (SC-FDMA) can achieve a lower peak to average power ratio than OFDM, which has the potential to be beneficial for underwater optical wireless communications as well [12,13,14,15]. Since LED is a band-limited device, multi-carrier modulation is more suitable for the LED-based UOWC systems. Orthogonal frequency division multiplexing (OFDM) is the key technique for 4G wireless communications, which is also widely used in many applications. Due to its resistance to inter-symbol interference (ISI) and high spectral efficiency, OFDM is a candidate technique for optical wireless communications. In OFDM techniques, *M*-ary quadrature amplitude modulation (M-QAM) is commonly used for a high efficiency of bandwidth. Unlike the conventional RF OFDM, the OFDM for the optical wireless communication systems can only transmit real and non-negative signals. Nowadays, optical OFDM has attacked much attention, both in academia and industry [16,17,18]. Recently, there have been many popular optical OFDM techniques such as DC-biased optical OFDM (DCO-OFDM), asymmetrically-clipped optical OFDM (ACO-OFDM), and unipolar OFDM (U-OFDM) [19,20,21]. To avoid the negative signals, DCO-OFDM adds a DC bias to the original bipolar real-valued OFDM signals. However, since the DC-biased signals cannot totally get rid of negative-valued information, the signals under zero must be hard clipped. ACO-OFDM and U-OFDM are recently-proposed optical OFDM techniques for intensity modulation and direct detection systems [20]. For ACO-OFDM, only odd frequency subcarriers are used for modulating data, and even frequency subcarriers are set to zero. By that, the zero clipping effects can only introduce interference on the even subcarriers [22]. In U-OFDM, the values of positive and negative information are transmitted successively; therefore, there is no zero clipping distortion.

Underwater optical OFDM has attracted much research attention recently since OFDM is a candidate technique to provide high-speed transmission. Some related work based on underwater optical OFDM has been proposed

[11,23,24]. The researchers presented an experimental demonstration to show that the underwater OFDM can support up to a 15.36-Mbps data rate for transmitting 4 m when the bandwidth of the LED is 10 MHz [11]. Using a a two-stage injection-locked 405-nm blue light laser diode transmitter, a data rate can be achieved of several gigabits per second [24].

Since the light sources for optical wireless communications are nonlinear and power limited, both the zero and peak power clipping should be considered. In this paper, we compare, analyze, and discuss both the zero and peak power limit distortions for DCO-, ACO-, and U-OFDM in the applications of underwater optical wireless communication systems. The propagation distance, bit error rate (BER), and data throughput are used as evaluation criteria in this paper.

The contributions of this paper are listed below:

- A comparison of the state-of-the-art optical OFDM techniques for underwater optical wireless communication systems is discussed based on the propagation distance, BER, and data throughput.
- Considering the turbulence fading underwater channel, the performances of ACO-, DCO-, and U-OFDM using the channels with different bandwidths are tested.
- For different transmitted bit rates, the achievable propagation distances of DCO-, ACO-, and U-OFDM with the band-limited channel are discussed.
- The underwater channel uncertainty caused by the pointing error is modeled and tested.

The notations in this paper are summarized and shown in Table 1, and the rest of the paper is organized as follows. The research problem is described in Section 2. System models and performance analysis are described in Section 3. Section 4 considers the band-limited channel and discusses the system performance. In Section 5, numerical results are shown. Finally, conclusion and future work are discussed in Section 6.

Table 1: Notation list. DCO, DC-biased optical; ACO, asymmetrically-clipped optical

$a(\lambda)$:	absorption attenuation
$b(\lambda)$:	scattering attenuation

α:	clipping coefficient
β:	modulation index
h:	channel loss
Ar:	size of the photodetector
d:	propagation distance
N:	Number of subcarriers for OFDM
$x_D[m]$	mth sample for the DCO-OFDM signal
$x_U[m]$	mth sample for the UCO-OFDM signal
$x_A[m]$	mth sample for the ACO-OFDM signal
b_{DC}	DC bias for DCO-OFDM
f_D	probability density function of the DCO-OFDM signal
f_U	probability density function of the UCO-OFDM signal
f_A	probability density function of the ACO-OFDM signal
$\bar{P}$	average optical power
P_m	peak transmitted power

RESEARCH PROBLEM

This section describes the differences between RF and optical OFDM, the underwater optical wireless communications channel model, and the impacts of the power and bandwidth limitations of LEDs.

RF OFDM vs. Optical OFDM

OFDM is a popular technique that is designed for RF systems due to its robustness to intersymbol interference and frequency selective channels

[25]. The optical OFDM works as a candidate for high-speed transmission and has attracted more and more attention [26]. However, the conventional RF OFDM cannot be directly adopted in optical systems. For RF systems, both negative and positive complex-valued signals can be transmitted. Unlike the RF systems, optical wireless communication systems can only generate non-negative and real-valued signals. Therefore, Hermitian symmetry is usually used to generate real-valued signals [22]. Compared with RF OFDM, the bandwidth efficiency of optical OFDM is reduced by half due to the Hermitian symmetry.

Underwater Optical Channel Model

Due to the unique physical characteristics of underwater light propagation, absorption and scattering are the two major factors that impact the channel loss. The absorption and scattering coefficients highly depend on the water types. The work in [27] investigated the attenuation coefficients for different water types, such as pure sea, clear ocean, coastal, and harbor. In Table 2, some typical attenuation coefficients are illustrated. From the results in this table, the pure sea water has the smallest absorption and scattering coefficients, which result in the least signal attenuation.

Table 2: Typical attenuation coefficients for different water types [27]

Water Types	Absorption (m^{-1})	Scattering (m^{-1})
Pure Sea Water	0.053	0.003
Clear Ocean	0.114	0.037
Coastal Ocean	0.179	0.219
Turbid Harbor	0.295	1.875

Light with different wavelengths experience different signal attenuation. Figure 1 shows the attenuation coefficient in pure sea water [28].

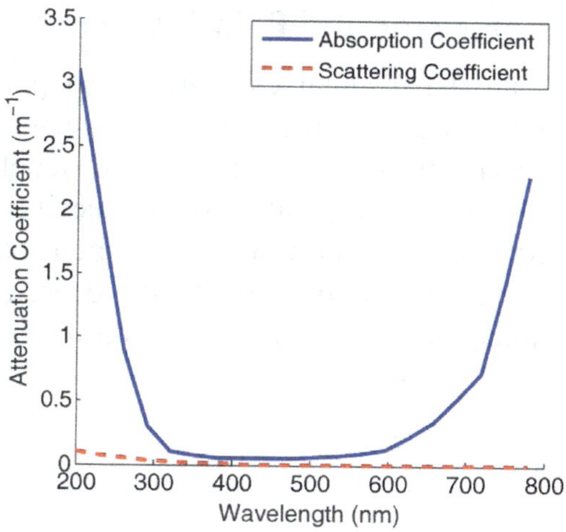

Figure 1: Absorbing and scattering coefficients for different wavelengths in pure sea water [28].

For different light wavelengths and water conditions (deep water or harbor area), the absorption and scattering may vary. In this paper, the effects of water conditions are beyond the scope; therefore, the effect of light wavelength is considered only. In UOWC, the overall attenuation of absorption and scattering can be described as [4]:

$$c(\lambda) = a(\lambda) + b(\lambda), \tag{1}$$

where a(λ) and b(λ) are absorption and scattering attenuation coefficients, respectively. The unit of the overall attenuation coefficient, c(λ), is m^{-1}. The channel loss can be represented as:

$$h_o(d) = L(\lambda, d)\frac{A_r \cos\theta \cos\varphi}{2\pi d^2}, \tag{2}$$

where A_r is the size of the photodetector (PD) in the receiver. θ and φ are the irradiation and incident angles between the LED and the user. *d* is the propagation distance between the LED and the user. L(λ,d) represents the link attenuation, which is calculated as:

$$L(\lambda, d) = \exp(-c(\lambda) \cdot d). \tag{3}$$

To simplify the problem, we assume that the propagation distance and attenuation of absorption and scattering mainly impact the channel loss.

Channel Estimation Error and Pointing Error

In this paper, we take the channel estimation and pointing errors into account. Since the transmitters and receivers are fluctuant in the underwater environment, the pointing errors can result in the channel estimation error. As shown in Figure 2, the actual position of the receiver varies over time around the estimated location; therefore, due to the central limit theorem (CLT), we can model the channel uncertainty, Δh, caused by pointing error as a Gaussian distributed variable. Thus, the overall channel loss can be modeled as $h = \tilde{h}h_o(d) + \Delta h$.

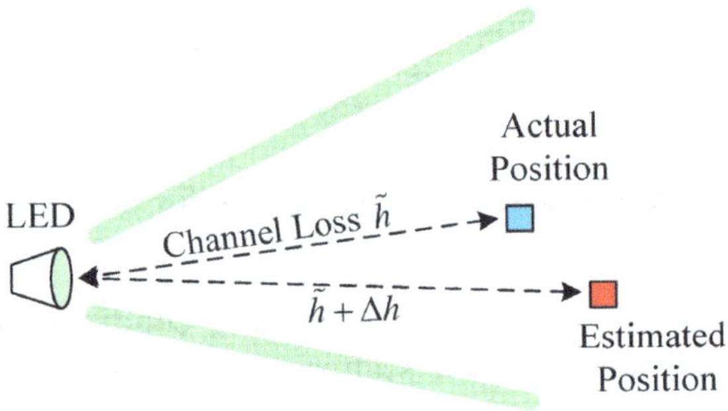

Figure 2: Pointing error caused by the fluctuation of transmitters and receivers.

Considering a turbulence scenario, $\tilde{h}$ represents the fading coefficient of the channel loss, which can be modeled with the log-normal distribution as [29]:

$$f_{\tilde{h}}(\tilde{h}) = \frac{1}{\tilde{h}\sqrt{2\pi\sigma_x^2}} \exp\left(\frac{-(\ln(\tilde{h}) - \mu_x)^2}{8\sigma_x^2}\right), \quad (4)$$

where μ_x and σ_x^2 are the mean and variance of the log-normal distribution factor $x = \frac{1}{2}\ln(h)$ [30]. In this paper, we assume the pointing error is independent between LEDs and users, and this uncertainty can be modeled as a Gaussian random process with zero mean and variance σ2h.

Power and Bandwidth Constraints of LEDs

LEDs working as lighting sources and transmitters are nonlinear and power-limited devices with a bandwidth constraint.

Since the optical OFDM signals can be modeled as Gaussian random variables, the magnitude can be very large. Due to the peak transmitted power constraint of LEDs, the high-valued signals greater than this power limit must be clipped. As shown in Figure 3, the signals beyond the peak power and below zero must be hard clipped. This clipping distortion is unrecoverable and can severely degrade the OFDM system performance.

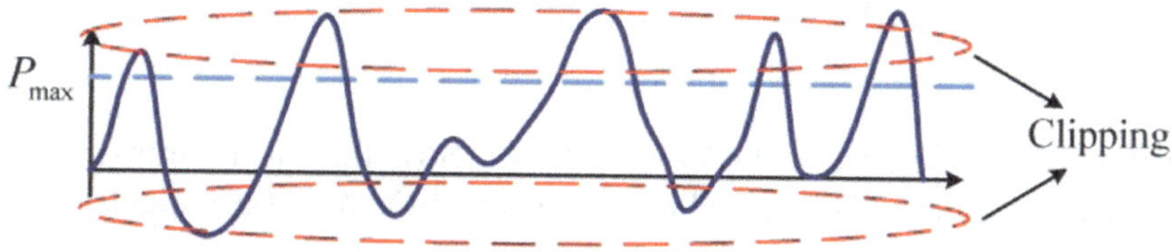

Figure 3: Illustration of zero and peak power limits for bipolar signals.

Due to the slow rise-time of the LED, its bandwidth is limited, which impacts the data throughput of UOWC systems. In this paper, we compare DCO-, ACO-, and U-OFDM in UOWC systems with considerations of zero and peak power constraints, as well as the bandwidth limit.

ANALYSIS OF OPTICAL OFDM FOR UNDERWATER COMMUNICATIONS

This section describes and analyzes the principles and performances of DCO-, ACO-, and U-OFDM in underwater communication systems.

DCO-OFDM

DCO-OFDM is a basic method of optical OFDM and is widely used due to its low complexity [31,32]. Since only the non-negative signals can be transmitted, DCO-OFDM signals are derived by adding a DC bias to the bipolar OFDM signals. Figure 4 shows the block diagram of DCO-OFDM. In this figure, Parts (a) and (b) illustrate the transmitter and receiver structures, respectively. Hermitian symmetry is used to generate real signals for optical wireless communication systems at the transmitter. However, considering the peak power limit, the added DC bias cannot totally avoid zero clipping. In the meantime, adding the DC can introduce peak power clipping distortion. At the receiver, the DC bias needs to be removed first, then the M-QAM data can be recovered and demodulated.

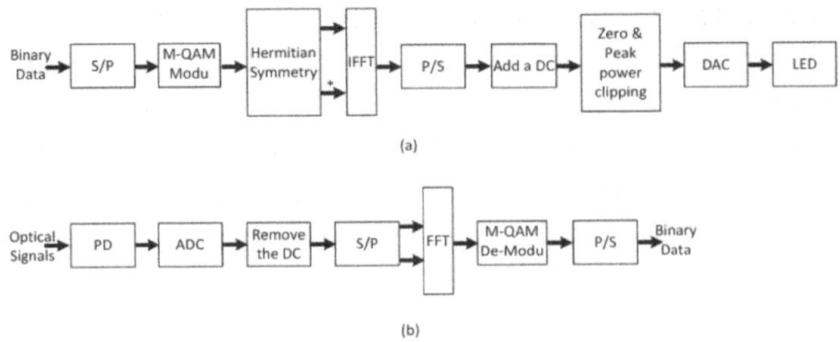

Figure 4: Block diagram of DCO-OFDM. (a) Structure of the transmitter; (b) structure of receiver [33]. P, parallel; S, serial; M-QAM, *M*-ary quadrature amplitude modulation.

In Figure 4a, after the inverse fast Fourier transform (IFFT), the discrete data for the kth subcarrier component, x[k], can be represented as:

$$x[k] = \sum_{i=0}^{N-1} X_i \exp\left(\frac{j2\pi ki}{N}\right), \forall k = 0, 1, \cdots, N-1, \tag{5}$$

where N represents the number of subcarriers used for modulation. Xi is the data after M-QAM mapping for the ith subcarrier. Due to the Hermitian symmetry, we set $X_{N-1-i} = X_i^*, \forall i = 0, 1, \cdots, N-1$. After the parallel to serial (P/S) converter, the mth sample of one DCO-OFDM can be represented as:

$$x_D[m] = \beta \sum_{k=0}^{N-1} x[k]\delta[m-k] + b_{DC}, \tag{6}$$

where b_{DC} is the additive DC bias and β is a coefficient that controls the scale of the DCO-OFDM signals. Since we assume that the number of subcarriers, N, is usually very large (at least 64), $x_D[m]$ can be modeled as a Gaussian random variable with b_{DC} mean and σ_x^2 variance. Considering the zero and peak power clipping, the probability density function (pdf) of the DCO-OFDM signal can be represented as:

$$f_D(x, b_{DC}) = \frac{1}{\sqrt{2\pi\sigma_x^2}} \exp\left(\frac{-(x - b_{DC})^2}{2\sigma_x^2}\right) \Pi(x, P_m) + \frac{1}{2}\text{erfc}\left(\frac{b_{DC}}{\sqrt{2\sigma_x^2}}\right)\delta(x) + \frac{1}{2}\text{erfc}\left(\frac{P_m - b_{DC}}{\sqrt{2\sigma_x^2}}\right)\delta(x - P_m), \quad (7)$$

where $\Pi(x,P_m)=u(x)-u(x-P_m)$ is defined as a rectangular function and u(x) is the unit step function. erfc(x) is the complementary error function, which is defined as:

$$\text{erfc}(x) = \frac{2}{\sqrt{\pi}} \int_x^{\infty} \exp(-y^2)dy. \quad (8)$$

Figure 5 shows the pdf of the DCO-OFDM signal numerically. In this paper, simulation and analytical results are perfectly fit. From the results, zero clipping distortion is smaller than the peak clipping distortion when the DC bias is greater than half of the peak transmitted power. Otherwise, zero clipping distortion is greater than the peak clipping distortion.

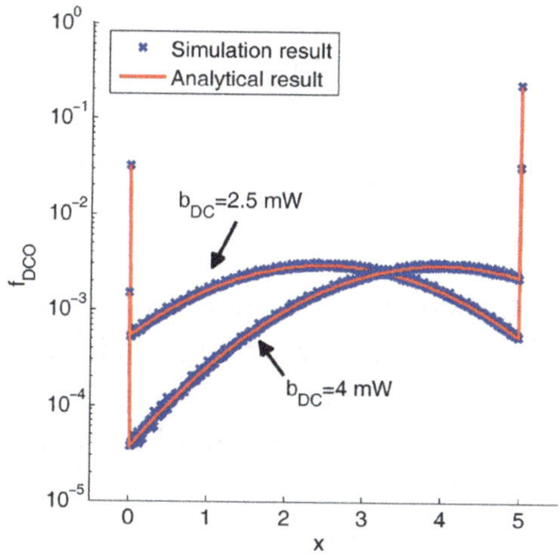

Figure 5: Simulation and analytical results of the pdf of the DCO-OFDM signal, β=20 and Pm=5 mW.

Since the DCO-OFDM signal is modeled as a Gaussian random variable and the LED works as a nonlinear device, the sampled received signal then can be modeled based on the Bussgang theorem as [26,34]:

$$y_{\mathrm{DCO}}[m] = rh\left(\alpha_{\mathrm{D}}(b_{\mathrm{DC}}) \cdot x_{\mathrm{D}}[m] + c_{\mathrm{D}}[m]\right) + n[m],$$
$$= r(\tilde{h}h_o + \Delta h)\left(\alpha_{\mathrm{D}}(b_{\mathrm{DC}}) \cdot x_{\mathrm{D}}[m] + c_{\mathrm{D}}[m]\right) + n[m] \quad (9)$$

where r is the optical power to current converting rate. $\alpha_D(b_{DC})$ represents the clipping coefficient, which is a function of the DC bias, b_{DC}. The calculation of $\alpha_D(b_{DC})$ is derived in the Appendix A, the result of which is shown as:

$$\alpha_{\mathrm{D}}(b_{\mathrm{DC}}) = 1 - \frac{1}{2}\left(\operatorname{erfc}\left(\frac{-P_{\mathrm{m}} + b_{\mathrm{DC}}}{\sqrt{2\sigma_x^2}}\right) + \operatorname{erfc}\left(\frac{b_{\mathrm{DC}}}{\sqrt{2\sigma_x^2}}\right)\right). \quad (10)$$

In Equation (9), $c_D[m]$ is clipping noise, and n[m] is the sum of shot noise and thermal noise. $c_D[m]$ and n[m] are both modeled as Gaussian random variables with zero mean. The variance of the clipping noise can be calculated by:

$$\begin{aligned}
\sigma_{\mathrm{D}}^2 &= \int_{-\infty}^{0} x^2 \cdot f_{\mathrm{D}}(x)dx + \int_{P_{\mathrm{m}}}^{\infty} (x - P_{\mathrm{m}})^2 \cdot f_{\mathrm{D}}(x)dx \\
&= \sqrt{\frac{\sigma_x^2}{8\pi}}(b_{\mathrm{DC}} - P_{\mathrm{m}})\exp\left(\frac{-(b_{\mathrm{DC}} - P_{\mathrm{m}})^2}{2\sigma_x^2}\right) \\
&- \sqrt{\frac{\sigma_x^2}{8\pi}} b_{\mathrm{DC}} \exp\left(\frac{-b_{\mathrm{DC}}^2}{2\sigma_x^2}\right) + \frac{1}{2}\left(\sigma_x^2 + (b_{\mathrm{DC}} - P_{\mathrm{m}})^2\right) \\
&+ \frac{1}{4}\operatorname{erfc}\left(\frac{b_{\mathrm{DC}}}{\sqrt{2\sigma_x^2}}\right) - \frac{1}{4}\operatorname{erfc}\left(\frac{b_{\mathrm{DC}} - P_{\mathrm{m}}}{\sqrt{2\sigma_x^2}}\right)
\end{aligned} \quad (11)$$

and the variance of n[m] is:

$$\sigma_n^2 = N_o R_s + 2qr\bar{P}R_s, \quad (12)$$

where N_o and R_s are thermal noise spectral density and symbol rate, respectively. q and $\bar{P}$ are the electronic charge and average received optical power, respectively. $\bar{P}$ is derived in the Appendix A. Therefore, the signal to noise ratio (SNR) using DCO-OFDM after FFT in the receiver, as shown in Figure 4b, can be calculated as:

$$\gamma_{\mathrm{D}}(d, \beta, b_{\mathrm{DC}}) = \frac{(\tilde{h}h_o(d)\beta\alpha_{\mathrm{D}}(b_{\mathrm{DC}}))^2\sigma_x^2}{N((\tilde{h}^2 h_o^2(d) + \sigma_h^2)\sigma_{\mathrm{D}}^2 + \sigma_n^2)}, \quad (13)$$

which is a function of β, d, and the DC bias. Considering the fading underwater channel, the average SNR can be calculated as:

$$\bar{\gamma}_{\mathrm{D}}(d, \beta, b_{\mathrm{DC}}) = \int_0^{\infty} \gamma_{\mathrm{D}}(d, \beta, b_{\mathrm{DC}}) f_{\tilde{h}}(\tilde{h}) d\tilde{h} \quad (14)$$

Since the propagation distance is a concern in underwater optical wireless communication systems, we can obtain the optimal transmission distance, d∗, by optimizing the coefficient, β. The optimal transmission distance can be found by:

$$d^* = \max_{\beta, b_{DC}} d$$

$$s.t.\ \mathrm{BER}(d, \beta, b_{DC}) < B^{\max}, \qquad (15)$$

where B^{max} is the allowed maximum bit error rate (BER) and BER(d,β) represents the BER using M-QAM given by a transmission distance and β, which approximates as:

$$\mathrm{BER}(d,\beta,b_{DC}) \approx \frac{\sqrt{M}-1}{\sqrt{M}\log_2\left(\sqrt{M}\right)} \mathrm{erfc}\left(\sqrt{\frac{3\bar{\gamma}_D(d,\beta,b_{DC})}{2(M-1)}}\right), \qquad (16)$$

where *M* is the modulation constellation size for M-QAM.

ACO-OFDM

ACO-OFDM is proposed for optical wireless communications, which can totally avoid the zero clipping distortion [22]. By using only the odd frequency subcarriers modulating data and leaving the even subcarriers blank, a bipolar signal is generated, as shown in Figure 6. Although zero clipping affects the ACO-OFDM signal, the interference caused by the zero clipping only happens on the even subcarriers. Therefore, no extra DC bias is needed. However, due to the zero clipping, the power of the data modulated using ACO-OFDM is reduced by half [22].

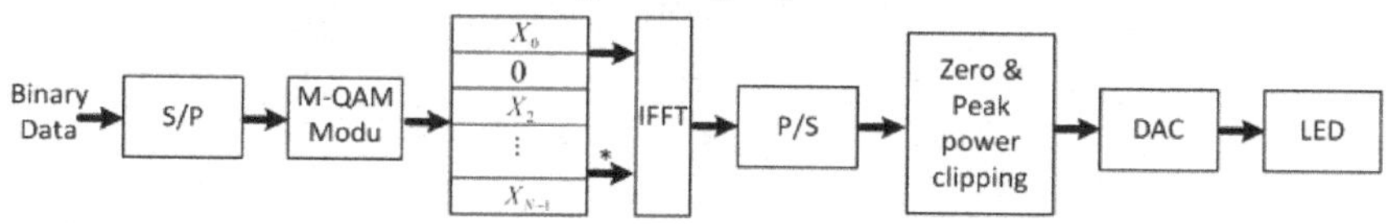

Figure 6: Block diagram of the ACO-OFDM transmitter [35].

Similar to the DCO-OFDM, we can derive the pdf of the ACO-OFDM signal, which is represented as:

$$f_{\mathrm{A}}(x) = \frac{1}{\sqrt{\pi\sigma_x^2}}\exp\left(\frac{-x^2}{\sigma_x^2}\right)\Pi(x, P_{\mathrm{m}}) + \frac{1}{2}\delta(x) + \frac{1}{2}\mathrm{erfc}\left(\frac{P_{\mathrm{m}}}{\sqrt{\sigma_x^2}}\right)\delta(x - P_{\mathrm{m}}). \quad (17)$$

Figure 7 shows the simulation and analytical results of the ACO-OFDM signal's pdf. From the results, a larger β can introduce more peak power clipped distortions, as well as a higher SNR. Thus, there is a compromise of β. Based on the Bussgang theorem, the clipping coefficient of ACO-OFDM can be obtained by:

$$\alpha_{\mathrm{A}} = \frac{1}{2} - \frac{1}{2}\mathrm{erfc}\left(\frac{P_{\mathrm{m}}}{\sqrt{2\sigma_x^2}}\right). \quad (18)$$

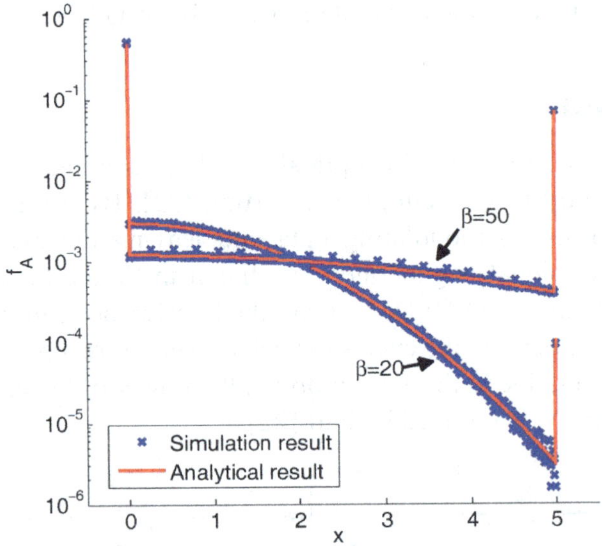

Figure 7: Simulation and analytical results of the pdf of the ACO-OFDM signal.

Due to the peak transmitted power constraint, the variance of clipping noise of ACO-OFDM can be calculated by:

$$\sigma_{\mathrm{A}}^2 = \int_{P_{\mathrm{m}}}^{\infty}(x - P_{\mathrm{m}})^2 \cdot f_{\mathrm{A}}(x)dx = \frac{1}{2}\left(P_{\mathrm{m}}^2 + \sigma_x^2\right)\mathrm{erfc}\left(\frac{P_m^2}{2\sigma_x^2}\right) - \frac{1}{2}\sigma_x P_{\mathrm{m}}\exp\left(\frac{-P_{\mathrm{m}}^2}{2\sigma_x^2}\right). \quad (19)$$

The received SNR using ACO-OFDM can be calculated by:

$$\gamma_{\mathrm{A}}(d,\beta)=\frac{(\tilde{h}h_o(d)\beta\alpha_{\mathrm{A}})^2\sigma_x^2}{2N((\tilde{h}^2h_o^2(d)+\sigma_{\tilde{h}}^2)\sigma_{\mathrm{A}}^2+\sigma_{\tilde{n}}^2)}. \tag{20}$$

Then, the average SNR can be calculated by:

$$\bar{\gamma}_{\mathrm{A}}(d,\beta)=\int_0^{\infty}\gamma_{\mathrm{A}}(d,\beta)f_{\tilde{h}}(\tilde{h})d\tilde{h} \tag{21}$$

U-OFDM

The unipolar OFDM works as a novel optical OFDM technique, which was proposed in [21]. It has advantages over DCO- and ACO-OFDM. Unlike ACO-OFDM, all the subcarriers can be used for modulating. Being superior to DCO-OFDM, there is no zero clipping for U-OFDM.

U-OFDM successively transmits the positive and negative information using two time slots to avoid the zero clipping distortions. A block diagram of U-OFDM is shown in Figure 8. When transmitting the positive values of signals, zeros are set at the negative values' positions. Then, the absolute value of the negative information is sent. To recover the data easily, the two time slots should have the same length. For example, if the bipolar information is [1,−2,−4,2,−5,3,0,−1], the U-OFDM transmits [1,0,0,2,0,3,0,0] using the first slot. Then, [0,2,4,0,5,0,0,1] are transmitted in the following slot. Thus, only the peak power constraint can affect the signal. At the receiver, the signal received from the first and second slots can be used to reconstruct the original bipolar signal. Obviously, U-OFDM requires a twice larger bandwidth than DCO-OFDM when transmitting the same bit rate using the equivalent modulation scheme.

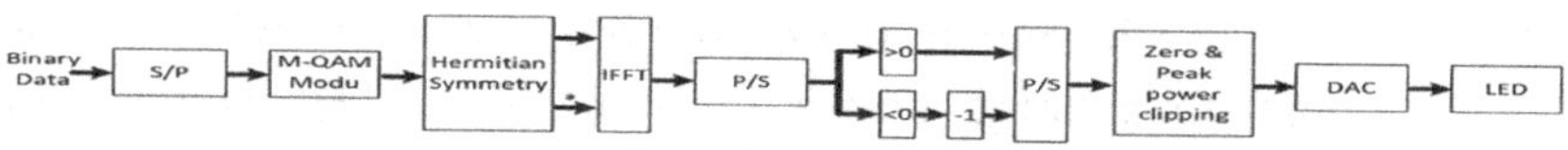

Figure 8: Block diagram of the unipolar (U)-OFDM transmitter [21].

Similar to DCO- and ACO-OFDM, we can calculate the clipping coefficient, clipping noise, and SNR for U-OFDM. After computing, the clipping coefficient for U-OFDM can be represented as:

$$\alpha_{\mathrm{U}}=1-\operatorname{erfc}\left(\frac{P_{\mathrm{m}}}{\sqrt{2\sigma_x^2}}\right). \tag{22}$$

Since U-OFDM signal is unipolar, the pdf can be represented as:

$$f_{\mathrm{U}}(x) = \frac{1}{\sqrt{2\pi\sigma_x^2}} \exp\left(\frac{-x^2}{2\sigma_x^2}\right) \Pi(x, P_{\mathrm{m}}) + \frac{1}{2}\delta(x) + \frac{1}{2}\mathrm{erfc}\left(\frac{P_{\mathrm{m}}}{\sqrt{2\sigma_x^2}}\right) \delta(x - P_{\mathrm{m}}). \tag{23}$$

Thus, the variance of clipping noise can be calculated and obtained as:

$$\sigma_{\mathrm{U}}^2 = \int_{P_{\mathrm{m}}}^{\infty} (x - P_{\mathrm{m}})^2 \cdot f_{\mathrm{U}}(x) dx = \left(P_{\mathrm{m}}^2 + \sigma_x^2\right) \mathrm{erfc}\left(\frac{P_m^2}{2\sigma_x^2}\right) - \sigma_x P_{\mathrm{m}} \exp\left(\frac{-P_{\mathrm{m}}^2}{2\sigma_x^2}\right). \tag{24}$$

The received SNR of U-OFDM can be represented as:

$$\gamma_{\mathrm{U}}(d, \beta) = \frac{(\tilde{h} h_o^2 \beta \alpha_{\mathrm{U}})^2 \sigma_x^2}{2N((\tilde{h}^2 h_o^2 + \sigma_{\tilde{h}}^2)\sigma_{\mathrm{U}}^2 + \sigma_n^2)}, \tag{25}$$

and the average SNR for U-OFDM is calculated by:

$$\bar{\gamma}_{\mathrm{U}}(d, \beta) = \int_0^{\infty} \gamma_{\mathrm{U}}(d, \beta) f_{\tilde{h}}(\tilde{h}) d\tilde{h}. \tag{26}$$

Since U-OFDM transmits positive and negative signals successfully, after reconstructing the signal at the receiver, the additive Gaussian noise variance is twice that of ACO- or DCO-OFDM.

Modulation Index Optimization

The modulation index, β, in DCO-, ACO-, and U-OFDM can be optimized to improve the SINR. Since β controls the scale of the OFDM signals, a larger β can provide a higher signal power. However, the larger scale of the OFDM signal can introduce more clipping distortion, which is caused by the high peak to average power ratio of the OFDM signal and the peak transmitted power constraint of the LED. Therefore, the modulation index needs to be optimized to trade off the signal power and the clipping distortion. To achieve a high data rate, the modulation constellation size can be adaptively adjusted based on the SINR for a given β.

BAND-LIMITED CHANNEL

In this section, we discuss and analyze the effects of the band-limited underwater optical channel. Due to the scattering and multi-reflection effects of the underwater environment, the channel is fading and frequency selective. Since LEDs are used as transmitters in the underwater optical wireless communication systems, the band-limited LED dominates the overall bandwidth of the systems. Usually, the 3-dB bandwidth of a white-color LED is limited only to a few tens of MHz. Thus, we assume that the overall bandwidth of the underwater optical wireless communication system is round 15 MHz [36].

As shown in Figure 9, the LED can be modeled as a first order low-pass filter. Due to the low-pass characteristics, the SNR for each subcarrier is different. In addition, the phase distortion introduced by the band-limited channel can affect the modulated M-QAM data. Thus, a single-tap equalizer and a bitloading algorithm in the frequency domain can be applied to correct the phase distortion and maximize the total throughput. We assume that the channel response can be perfectly estimated; thus, the phase distortion can be totally eliminated. As shown in Figure 10, the single-tap equalizer is set after the FFT. Using a bitloading algorithm, the channel spectrum can be efficiently utilized.

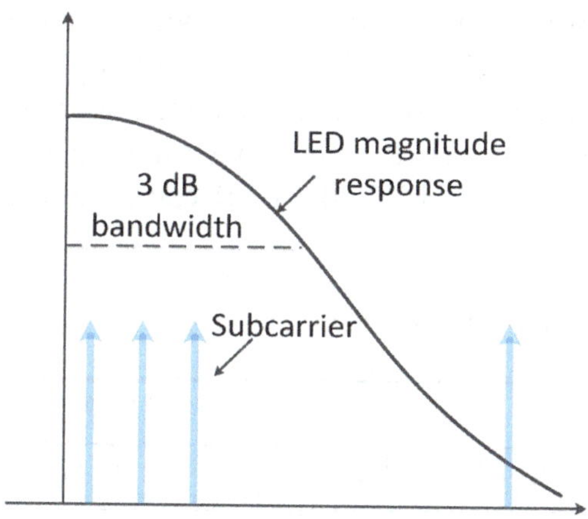

Figure 9: An illustration of the LED magnitude response.

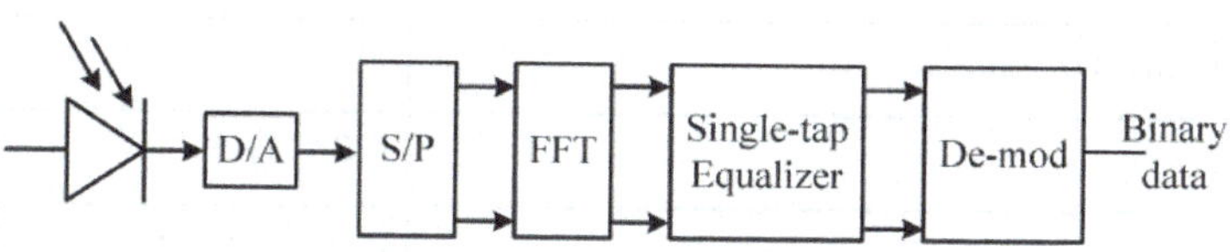

Figure 10: A block diagram of the optical OFDM receiver with a single-tap equalizer.

NUMERICAL RESULTS

In this section, numerical results of the comparison of the DCO-, ACO-, and U-OFDM techniques for UOWC systems are shown. The parameters used to obtain the numerical results are shown in Table 3. This is a baseline for all the numerical results in this paper. Since the attenuation coefficient varies over temperature, concentration of salt, and spectrum of transmitted light, to simplify the problem, we assume the pure sea water condition is used [4]. The light with a wavelength from 480 nm–520 nm is used, which has the lowest absorption rate in sea water. We also assume that the thermal noise is dominated. For the photodetector (PD), a lens can be integrated to increases the effective area and collect more power. For the numerical results, the bandwidth limitation is taken into account. We mainly consider the bandwidth limitation from LEDs since the PD's bandwidth can be much broader compared with the symbol rate tested in this paper [37,38]. The numerical results illustrated in this paper are based on the theory discussed above. We have tested and verified the theoretical results by using simulation. The forward error control (FEC) codes need to be applied in a practical situation for a better BER performance, which does not change the relative results among DCO-, ACO-, and U-OFDM obtained in this paper. Therefore, there is no need to consider the advantages of FEC in this paper. In addition, we set a 10−3 BER as the constraint when maximizing the bit rate, since, with the help of FEC, the BER can be easily and significantly improved.

Table 3: Parameters used in this paper

Responsivity	0.45 A/W
Equivalent area of the PD using a lens	10−4m2
Thermal noise density, N0	1.3×10−13 mW/Hz
Radiation optical power of each lamp, Pm	10 W
LED semi-angle	30°
a(λ)	0.053 (m−1)
b(λ)	0.003 (m−1)
c(λ)	0.056 (m−1)
Number of subcarriers, *N*	128

The numerical results of BER using Equation (16) are shown in Figure 11. In this figure, the normalized modulation index is used, which is β/N/Pm. Figure 11a–c illustrates the average BER of the optical OFDM tested under different channel conditions, respectively. In general, an increasing modulation index first leads to a better BER performance for all the tested optical OFDM. Then, continued increasing of the modulation index can introduce more clipping distortion caused by the peak power constraint; therefore, the BER worsens. Since an increasing modulation index improves the SNR, yet introduces more clipping distortion, an optimal modulation index needs to be found.

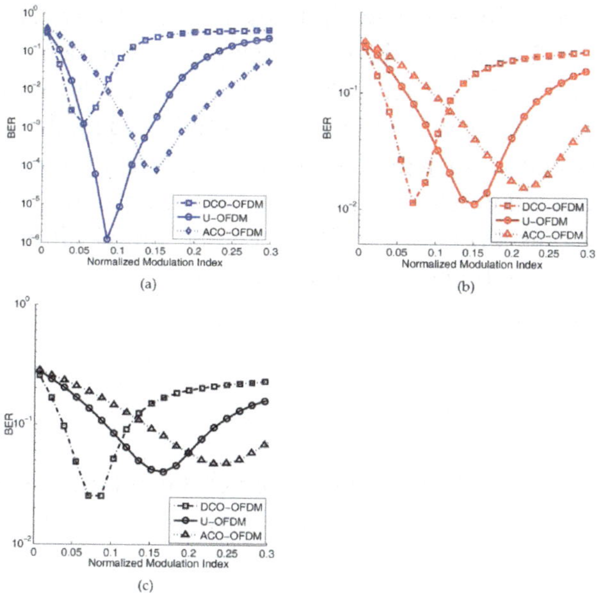

Figure 11: Average BER comparison of DCO-, ACO-, and U-OFDM with Rs=30 Msps and 64-QAM. Propagation distance is 1 m. (a) Ideal channel with no bandwidth limit; (b) the 3-dB bandwidth of the channel is 25 MHz; (c) the 3-dB bandwidth of the channel is 10 MHz.

The ideal channel (no bandwidth constraint) case is shown in Figure 11a. Due to the less nonlinear distortion and higher power efficiency of U-OFDM compared with DCO- and ACO-OFDM, the BER performance of U-OFDM is better than DCO- and ACO-OFDM. Considering the channel bandwidth constraint, the BER performances of U- and ACO-OFDM become worse than DCO-OFDM. When the 3-dB bandwidth of the channel is 25 MHz, as shown in Figure 11b, the minimum BER of U-OFDM is similar to DCO-

OFDM. For a more narrow bandwidth case (3 dB bandwidth is 10 MHz), as shown in Figure 11c, the minimum BER of U-OFDM is higher than that of DCO-OFDM since U-OFDM requires a wider bandwidth than DCO-OFDM when transmitting the same data rate. ACO-OFDM is even worse than U-OFDM due to its lower bandwidth utilization efficiency. In summary, due to different channel bandwidths, the best optical OFDM scheme can be selected to obtain a better BER performance.

How to choose the optimal modulation index for DCO-, ACO-, and U-OFDM is a problem that needs to be discussed. Figure 12 shows the simulation results of the optimal normalized modulation index for different noise spectral density. In this result, the transmitted peak power and the data rate are fixed. Figure 12a,b shows the relationship between the modulation index and the noise level for two different band-limited channels, respectively. From the simulation results, we use cubic equations to fit and model the relationship between the optimal normalized modulation index and noise spectral density based on the simulation results. Comparing (a) and (b) in Figure 12, the two different bandwidths illustrate a similar phenomenon. The fitting equations for DCO-, U-, and ACO-OFDM are represented as:

$$\begin{aligned}
\text{DCO-OFDM}:\frac{\beta}{NP_{\max}} &= 6.7\times10^{21}\times N_0^3 - 3.6\times10^{14}\times N_0^2 + 7.8\times10^{6}\times N_0 + 0.061\\
\text{U-OFDM}:\frac{\beta}{NP_{\max}} &= 1.3\times10^{22}\times N_0^3 - 6.9\times10^{14}\times N_0^2 + 1.2\times10^{7}\times N_0 + 0.12\\
\text{ACO-OFDM}:\frac{\beta}{NP_{\max}} &= 2.5\times10^{22}\times N_0^3 - 1.1\times10^{15}\times N_0^2 + 1.8\times10^{7}\times N_0 + 0.17,
\end{aligned} \tag{27}$$

where these equations can be used to predict the optimal modulation index for DCO-, ACO-, and U-OFDM in noisy environments with different noise levels.

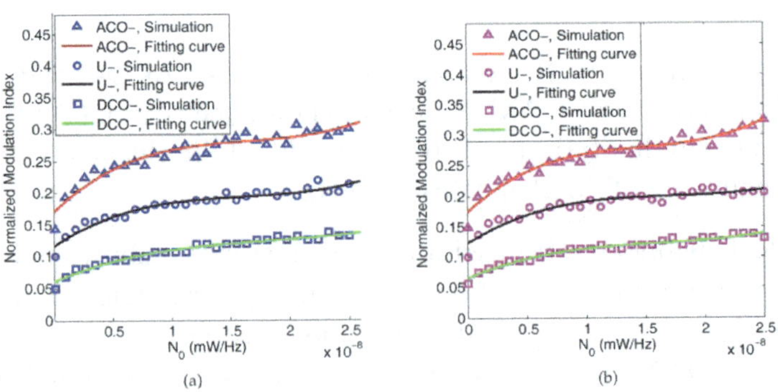

Figure 12: Normalized modulation index of the minimum BER for different noise levels. (a) The 3-dB bandwidth of the channel is 25 MHz; (b) the 3-dB bandwidth of the channel is 10 MHz.

Figure 13 shows analytical results of the required power using DCO-, ACO-, and U-OFDM for different propagation distances with a 10-MHz, 3-dB bandwidth. In this figure, the cases with and without the bitloading algorithm are compared. Since the bandwidth tested is narrow, DCO-OFDM has a better performance than ACO- and U-OFDM, which is consistent, as shown in Figure 11. Without using the bitloading algorithm, the required power of the three optical OFDM schemes tested is about 30% higher than the case with bitloading. Since a single LED can only provide about tens of Watts [39,40], a LED-array with multiple LEDs needs to be used when a high power is required.

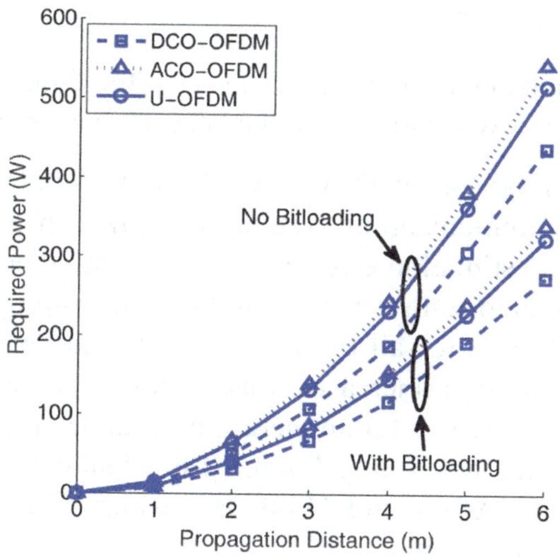

Figure 13: Required transmitted power for different propagation distances for a 10−3 BER. The 3-dB bandwidth is 10 MHz. The bit rate is 100 Mbps.

Figure 14 shows the maximum propagation distance achieved using different transmission bit rates through a band-limited channel. In this result, a bitloading algorithm is used [41]. When the symbol rate is low compared to the channel bandwidth, U-OFDM can support a longer propagation distance than ACO- and DCO-OFDM due to its higher power efficiency and less nonlinear distortion than ACO- and DCO-OFDM. For the higher bit rates case, DCO-OFDM outperforms ACO- and U-OFDM due to its higher spectral efficiency. Depending on different required bit rates and channel bandwidth, we can choose the proper optical OFDM technique to obtain a good propagation and BER performance.

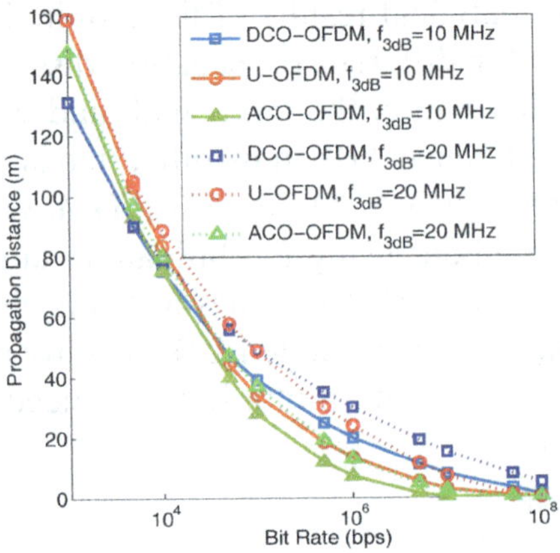

Figure 14: The maximum propagation distance for different bit rates. Peak transmitted power is fixed, and a bitloading algorithm is used.

A bit rate comparison of DCO-, ACO-, and U-OFDM in underwater optical wireless communication systems is shown in Figure 15. In this figure, the channel estimation error caused by pointing error is considered, and the bitloading algorithm is used to improve the throughput. In general, the maximum data rate for all the techniques tested drops as the channel uncertainty variance, σ2h, increases. Based on the theoretical analysis, the SNR dramatically decreases for an increasing channel uncertainty variance. For this result, since a severely-limited LED bandwidth is used, DCO-OFDM outperforms ACO- and U-OFDM, which is consistent with the previous analysis and results.

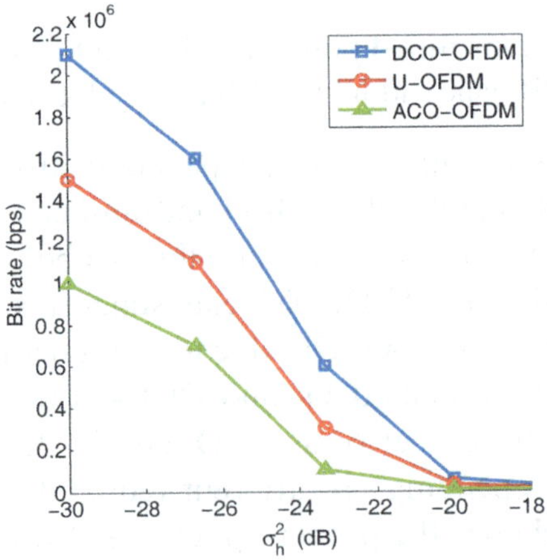

Figure 15: The maximum bit rate for DCO-, ACO-, and U-OFDM with an increasing σ2h. The propagation distance is 20 m, and the 3-dB bandwidth of the LED is 10 MHz.

CONCLUSIONS

In this paper, we compare, discuss, and analyze DCO-, ACO-, and U-OFDM in underwater optical communication systems. Since the peak transmitted power is limited, distortions caused by the peak power are taken into account. We propose to optimize the modulation index and compromise the clipping effects and SNR. By optimizing the SNR, the achievable transmission distance is obtained by the given transmission bit rate and BER performance. Considering the bandwidth limitation, we find the relationship between the optimal modulation index and the noise spectral density, which can guide us to achieve the best BER performance in different noisy environments. For the channels with different bandwidths, we compare the performances of DCO-, ACO-, and U-OFDM in underwater OWC systems. When the required bit rate is high (compared to the bandwidth), DCO-OFDM can provide a longer propagation distance than ACO- and U-OFDM. Otherwise, U-OFDM outperforms ACO- and DCO-OFDM.

Appendix A

This Appendix gives the derivation of the clipping coefficient and the average optical power of the DCO-, ACO-, and U-OFDM signals.

$$\begin{aligned}\alpha_{\mathrm{D}}(b_{\mathrm{DC}}) &= \frac{1}{\sigma_x^3\sqrt{2\pi}}\int_{-\infty}^{\infty} x\psi_{\mathrm{D}}(x)\exp\left(\frac{-(x-b_{\mathrm{DC}})^2}{2\sigma_x^2}\right)dx \\ &= \frac{1}{\sigma_x^3\sqrt{2\pi}}\left(\int_0^{P_m} x^2\exp\left(\frac{-(x-b_{\mathrm{DC}})^2}{2\sigma_x^2}\right)dx\right. \\ &\left.+P_{\mathrm{m}}\int_{P_m}^{\infty} x\exp\left(\frac{-(x-b_{\mathrm{DC}})^2}{2\sigma_x^2}\right)dx\right) \\ &= 1-\frac{1}{2}\left(\mathrm{erfc}\left(\frac{-P_{\mathrm{m}}+b_{\mathrm{DC}}}{\sqrt{2\sigma_x^2}}\right)+\mathrm{erfc}\left(\frac{b_{\mathrm{DC}}}{\sqrt{2\sigma_x^2}}\right)\right).\end{aligned} \quad \text{(A1)}$$

where:

$$\psi_{\mathrm{D}}(x)=\begin{cases}0, & x<0\\ x, & 0<x<P_{\mathrm{m}}\\ P_{\mathrm{m}}, & x>P_{\mathrm{m}},\end{cases} \quad \text{(A2)}$$

$$\begin{aligned}\alpha_{\mathrm{U}} &= \frac{1}{\sigma_x^3\sqrt{2\pi}}\int_{-\infty}^{\infty} x\psi_{\mathrm{U}}(x)\exp\left(\frac{-x^2}{2\sigma_x^2}\right)dx \\ &= \frac{1}{\sigma_x^3\sqrt{2\pi}}\left(-P_{\mathrm{m}}\int_{-\infty}^{-P_{\mathrm{m}}} x\exp\left(\frac{-x^2}{2\sigma_x^2}\right)dx\right. \\ &+ \int_{-P_{\mathrm{m}}}^{P_{\mathrm{m}}} x^2\exp\left(\frac{-x^2}{2\sigma_x^2}\right)dx \\ &\left.+P_{\mathrm{m}}\int_{P_{\mathrm{m}}}^{\infty} x\exp\left(\frac{-x^2}{2\sigma_x^2}\right)dx\right) \\ &= 1-\mathrm{erfc}\left(\frac{P_{\mathrm{m}}}{\sqrt{2\sigma_x^2}}\right),\end{aligned} \tag{A3}$$

where:

$$\psi_{\mathrm{U}}(x)=\begin{cases}-P_{\mathrm{m}}, & x<-P_{\mathrm{m}}\\ x, & -P_{\mathrm{m}}<x<P_{\mathrm{m}}\\ P_{\mathrm{m}}, & x>P_{\mathrm{m}}.\end{cases} \tag{A4}$$

For DCO-OFDM, the average optical power can be calculated by:

$$\begin{aligned}\bar{P}(b_{\mathrm{DC}}) &= \int_0^{P_{\mathrm{m}}} x f_{\mathrm{D}}(x, b_{\mathrm{DC}})dx \\ &= \int_0^{P_{\mathrm{m}}} x\frac{1}{\sqrt{2\pi\sigma_x^2}}\exp\left(\frac{-(x-b_{\mathrm{DC}})^2}{2\sigma_x^2}\right)dx \\ &+ P_{\mathrm{m}}\frac{1}{2}\mathrm{erfc}\left(\frac{P_{\mathrm{m}}-b_{\mathrm{DC}}}{\sqrt{2\sigma_x^2}}\right) \\ &= \frac{b_{\mathrm{DC}}}{2}\left(\mathrm{erfc}\left(\frac{b_{\mathrm{DC}}}{\sqrt{2\sigma_x^2}}\right)-\mathrm{erfc}\left(\frac{b_{\mathrm{DC}}-P_{\mathrm{m}}}{\sqrt{2\sigma_x^2}}\right)\right) \\ &+ \frac{\sigma_x}{\sqrt{2\pi}}\left(\exp\left(\frac{-b_{\mathrm{DC}}^2}{2\sigma_x^2}\right)-\exp\left(\frac{-(b_{\mathrm{DC}}-P_{\mathrm{m}})^2}{2\sigma_x^2}\right)\right) \\ &+ P_{\mathrm{m}}\frac{1}{2}\mathrm{erfc}\left(\frac{P_{\mathrm{m}}-b_{\mathrm{DC}}}{\sqrt{2\sigma_x^2}}\right)\end{aligned} \tag{A5}$$

AUTHOR CONTRIBUTIONS

J.L. and Y.G. contribute this work equally. The individual contributions of all the authors are represented as follows. "conceptualization, J.L. and Y.G.; methodology, J.L.; software, J.L.; validation, J.L. and Y.G.; investigation, J.L. and Y.G.; writing–original draft preparation, J.L., Y.G., P.W. and D.L.; visualization, Y.G., P.W. and D.L.; resources, P.W. and D.L.

REFERENCES

1. Zeng Z., Fu S., Zhang H., Dong Y., Cheng J. A Survey of Underwater Optical Wireless Communications. IEEE Commun. Surv. Tutor. 2017;19:204–238. doi: 10.1109/COMST.2016.2618841.
2. Kaushal H., Kaddoum G. Underwater Optical Wireless Communication. IEEE Access. 2016;4:1518–1547. doi: 10.1109/ACCESS.2016.2552538.
3. Cochenour B.M., Mullen L.J., Laux A.E. Characterization of the Beam-Spread Function for Underwater Wireless Optical Communications Links. IEEE J. Ocean. Eng. 2008;33:513–521. doi: 10.1109/JOE.2008.2005341.
4. Zhang H., Dong Y. General Stochastic Channel Model and Performance Evaluation for Underwater Wireless Optical Links. IEEE Trans. Wirel. Commun. 2016;15:1162–1173. doi: 10.1109/TWC.2015.2485990.
5. Shen C., Guo Y., Sun X., Liu G., Ho K., Ng T.K., Alouini M., Ooi B.S. Going beyond 10-meter, Gbit/s underwater optical wireless communication links based on visible lasers; Proceedings of the 2017 Opto-Electronics and Communications Conference (OECC) and Photonics Global Conference (PGC); Singapore. 31 July–4 August 2017; pp. 1–3.
6. Wang C., Yu H.Y., Zhu Y.J. A Long Distance Underwater Visible Light Communication System With Single Photon Avalanche Diode. IEEE Photonics J. 2016;8:7906311. doi: 10.1109/JPHOT.2016.2602330.
7. Alkhasraji J., Tsimenidis C. Coded OFDM over short range underwater optical wireless channels using LED; Proceedings of the OCEANS 2017; Aberdeen, UK. 19–22 June 2017; pp. 1–7.
8. Tsonev D., Chun H., Rajbhandari S., McKendry J., Videv S., Gu E., Haji M., Watson S., Kelly A., Faulkner G., et al. A 3 Gb/s Single LED OFDM-Based Wireless VLC Link Using a Gallium Nitride μLED. IEEE Photonics Technol. Lett. 2014;26:637–640. doi: 10.1109/LPT.2013.2297621.
9. Akhoundi F., Salehi J.A., Tashakori A. Cellular Underwater Wireless Optical CDMA Network: Performance Analysis and Implementation Concepts. IEEE Trans. Commun. 2015;63:882–891. doi: 10.1109/TCOMM.2015.2400441.
10. Tang S., Dong Y., Zhang X. Impulse Response Modeling for Underwater Wireless Optical Communication Links. IEEE

Trans. Commun. 2014;62:226–234. doi: 10.1109/TCOMM.2013.120713.130199.

11. Hessien S., Tokgoz S.C., Anous N., Boyacu A., Abdallah M., Qaraqe K.A. Experimental Evaluation of OFDM-Based Underwater Visible Light Communication System. IEEE Photonics J. 2018;10:7907713. doi: 10.1109/JPHOT.2018.2871958.
12. Cicalo S., Tralli V. Adaptive Resource Allocation With Proportional Rate Constraints for Uplink SC-FDMA Systems. IEEE Commun. Lett. 2014;18:1419–1422. doi: 10.1109/LCOMM.2014.2332997.
13. Tsiropoulou E.E., Kapoukakis A., Papavassiliou S. Energy-efficient subcarrier allocation in SC-FDMA wireless networks based on multilateral model of bargaining; Proceedings of the 2013 IFIP Networking Conference; Brooklyn, NY, USA. 22–24 May 2013; pp. 1–9.
14. Ahani G., Yuan D., Ding W. On SC-FDMA resource allocation with power control; Proceedings of the 2016 IEEE 21st International Workshop on Computer Aided Modelling and Design of Communication Links and Networks (CAMAD); Toronto, ON, Canada. 23–25 October 2016; pp. 112–116.
15. Eleni Tsiropoulou E., Aggelos Kapoukakis S.P. Uplink resource allocation in SC-FDMA wireless networks: A survey and taxonomy, Computer Networks. Comput. Netw. 2016;96:1–28. doi: 10.1016/j.comnet.2015.12.006.
16. Azhar A., Tran T., O'Brien D. A Gigabit/s Indoor Wireless Transmission Using MIMO-OFDM Visible-Light Communications. IEEE Photonics Technol. Lett. 2013;25:171–174. doi: 10.1109/LPT.2012.2231857.
17. Wu L., Zhang Z., Dang J., Liu H. Adaptive Modulation Schemes for Visible Light Communications. J. Lightwave Technol. 2015;33:117–125. doi: 10.1109/JLT.2014.2374171.
18. Elgala H., Mesleh R., Haas H. Indoor optical wireless communication: Potential and state-of-the-art. IEEE Commun. Mag. 2011;49:56–62. doi: 10.1109/MCOM.2011.6011734.
19. Dissanayake S.D., Armstrong J. Comparison of ACO-OFDM, DCO-OFDM and ADO-OFDM in IM/DD Systems. J. Lightwave Technol. 2013;31:1063–1072. doi: 10.1109/JLT.2013.2241731.

20. Armstrong J., Schmidt B.J.C. Comparison of Asymmetrically Clipped Optical OFDM and DC-Biased Optical OFDM in AWGN. IEEE Commun. Lett. 2008;12:343–345. doi: 10.1109/LCOMM.2008.080193.
21. Tsonev D., Sinanovic S., Haas H. Novel Unipolar Orthogonal Frequency Division Multiplexing (U-OFDM) for Optical Wireless; Proceedings of the 2012 IEEE 75th Vehicular Technology Conference (VTC Spring); Yokohama, Japan. 6–9 May 2012; pp. 1–5.
22. Armstrong J., Lowery A.J. Power efficient optical OFDM. Electr. Lett. 2006;42:370–372. doi: 10.1049/el:20063636.
23. Bai J., Li Y., Cheng W., Yang Y., Duan Z., Wang Y. PAPR reduction for IM/DD-OFDM signals in underwater wireless optical communication system; Proceedings of the 2018 13th IEEE Conference on Industrial Electronics and Applications (ICIEA); Wuhan, China. 31 May–2 June 2018; pp. 1837–1840.
24. Lu H., Li C., Lin H., Tsai W., Chu C., Chen B., Wu C. An 8 m/9.6 Gbps Underwater Wireless Optical Communication System. IEEE Photonics J. 2016;8:7906107. doi: 10.1109/JPHOT.2016.2601778.
25. Schmidl T.M., Cox D.C. Robust frequency and timing synchronization for OFDM. IEEE Trans. Commun. 1997;45:1613–1621. doi: 10.1109/26.650240.
26. Armstrong J. OFDM for Optical Communications. J. Lightwave Technol. 2009;27:189–204. doi: 10.1109/JLT.2008.2010061.
27. Gabriel C., Khalighi M., Bourennane S., Leon P., Rigaud V. Channel modeling for underwater optical communication; Proceedings of the 2011 IEEE GLOBECOM Workshops (GC Wkshps); Houston, TX, USA. 5–9 December 2011; pp. 833–837.
28. Al-Kinani A., Wang C., Zhou L., Zhang W. Optical Wireless Communication Channel Measurements and Models. IEEE Commun. Surv. Tutor. 2018;20:1939–1962. doi: 10.1109/COMST.2018.2838096.
29. Jamali M.V., Akhoundi F., Salehi J.A. Performance Characterization of Relay-Assisted Wireless Optical CDMA Networks in Turbulent Underwater Channel. IEEE Trans. Wirel. Commun. 2016;15:4104–4116. doi: 10.1109/TWC.2016.2533616.
30. Akhoundi F., Jamali M.V., Hassan N.B., Beyranvand H., Minoofar A., Salehi J.A. Cellular Underwater Wireless Optical CDMA Network: Potentials and Challenges. IEEE Access. 2016;4:4254–4268. doi: 10.1109/ACCESS.2016.2593398.

31. Zhang M., Zhang Z. An Optimum DC-Biasing for DCO-OFDM System. IEEE Commun. Lett. 2014;18:1351–1354. doi: 10.1109/LCOMM.2014.2331068.
32. Carruthers J.B., Kahn J.M. Multiple-subcarrier modulation for nondirected wireless infrared communication. IEEE J. Sel. Areas Commun. 1996;14:538–546. doi: 10.1109/49.490239.
33. Dimitrov S., Sinanovic S., Haas H. A comparison of OFDM-based modulation schemes for OWC with clipping distortion; Proceedings of the 2011 IEEE GLOBECOM Workshops (GC Wkshps); Houston, TX, USA. 5–9 December 2011; pp. 787–791.
34. Price R. A useful theorem for nonlinear devices having Gaussian inputs. IRE Trans. Inf. Theory. 1958;4:69–72. doi: 10.1109/TIT.1958.1057444.
35. Dimitrov S., Sinanovic S., Haas H. Double-Sided Signal Clipping in ACO-OFDM Wireless Communication Systems; Proceedings of the 2011 IEEE International Conference on Communications (ICC); Kyoto, Japan. 5–9 June 2011; pp. 1–5.
36. Zeng L., Minh H.L., O'Brien D., Faulkner G., Lee K., Jung D., Oh Y. Equalisation for high-speed Visible Light Communications using white-LEDs; Proceedings of the 2008 6th International Symposium on Communication Systems, Networks and Digital Signal Processing; Graz, Austria. 25 July 2008; pp. 170–173.
37. Hu S., Gao Q., Gong C., Xu Z. Efficient Visible Light Sensing in Eigenspace. IEEE Commun. Lett. 2018;22:994–997. doi: 10.1109/LCOMM.2018.2808498.
38. Fahs B., Chowdhury A.J., Hella M.M. A 12-m 2.5-Gb/s Lighting Compatible Integrated Receiver for OOK Visible Light Communication Links. J. Lightwave Technol. 2016;34:3768–3775. doi: 10.1109/JLT.2016.2587598.
39. Popoola W.O. Impact of VLC on Light Emission Quality of White LEDs. J. Lightwave Technol. 2016;34:2526–2532. doi: 10.1109/JLT.2016.2542110.
40. Zhang L., Zhang W., Sun J. VLC system implementation with white LEDs; Proceedings of the 2017 IEEE/CIC International Conference on Communications in China (ICCC); Qingdao, China. 22–24 October 2017; pp. 1–6.

41. Zhang Y.J., Letaief K.B. Multiuser adaptive subcarrier-and-bit allocation with adaptive cell selection for OFDM systems. IEEE Trans. Wirel. Commun. 2004;3:1566–1575. doi: 10.1109/TWC.2004.833501.

INDEX

D

E

F

G

H

I

L

M

Printed in the United States
By Bookmasters